LABORATORY MANUAL FOR INTRODUCTORY GEOLOGY

THIRD EDITION

ALLAN LUDMAN
Queens College, New York

STEPHEN MARSHAK
University of Illinois

W. W. NORTON & COMPANY
NEW YORK • LONDON

Copyright © 2015, 2012, 2010 by W. W. Norton & Company, Inc.
All rights reserved.
Printed in the United States of America.
Third Edition

Editor: Eric Svendsen
Senior Project Editor: Thomas Foley
Associate Production Director: Benjamin Reynolds
Developmental Editor: Sunny Hwang
Copy Editor: Chris Curioli
Managing Editor, College: Marian Johnson
Managing Editor, College Digital Media: Kim Yi
Media Editors: Robin Kimball and Rob Bellinger
Associate Media Editor: Cailin Barrett Bressack
Media Editorial Assistants: Victoria Reuter and Ruth Bolster
Marketing Manager, Geology: Meredith Leo
Design Director: Hope Miller Goodell
Designer: Lissi Sigillo
Photography Editor: Evan Luberger
Photo Researcher: Lynn Gadson
Permissions Manager: Megan Jackson
Editorial Assistants: Rachel Goodman and Lindsey Thomas
Composition and page layout by codeMantra
Illustrations by Precision Graphics/Lachina
Project manager at Precision Graphics/Lachina: Terri Hamer
Manufacturing by LSC–Kendallville, IN
Topographic maps created by Mapping Specialists—Madison, WI

Permission to use copyrighted material is included in the backmatter of this book.

ISBN: 978–0–393–93791–6

W. W. Norton & Company, Inc., 500 Fifth Avenue, New York, NY 10110
wwnorton.com

W. W. Norton & Company Ltd., 15 Carlisle Street, London W1D 3BS

5 6 7 8 9 0

CONTENTS

PREFACE

This laboratory manual is based on our collective 70+ years of teaching and coordinating introductory geology courses—experience that has helped us to understand both how students best learn geologic principles and which strategies help instructors arouse student interest in order to enhance the learning process. This manual provides: (1) an up-to-date, comprehensive background that focuses on the hands-on tasks at the core of any introductory geology course; (2) clearer and more patient step-by-step explanations than can be accommodated in textbooks; (3) a text and exercises that engage students in thinking like a geologist, in order to solve real-life problems important to society; and (4) the passion and excitement that we still feel after decades as geologists and teachers.

Students often ask us how we maintain this enthusiasm. Our answer is to share with them both the joys and the frustrations of facing and solving real-world geologic problems. You will find many of those types of problems in the following pages—modified for the introductory nature of the course, but still reflecting their challenges and the rewards of solving them. This manual brings that experience directly to the students, *engaging* them in the learning process by *explaining* concepts clearly and providing many avenues for further *exploration*.

Unique Elements

As you read through this manual, you will find elements and a pedagogical approach that distinguish it from others, including:

Hands-on, inquiry-based pedagogy

We believe that students learn science best by doing science, not by just memorizing facts. Beginning in Chapter 1 and continuing in each subsequent chapter, students are guided through real-world geologic puzzles in order to understand concepts more deeply and to start learning to think like a geologist. In Chapter 9, for example, students work out the rules of contour lines for themselves by comparing a topographic map with a digital elevation model of the same area. In Chapter 5, they reason out the cooling rates for plutons of different sizes and shapes, and so on.

Innovative exercises that engage students and provide instructors with choice

Tiered exercises are carefully integrated into the text, leading students to understand concepts for themselves by first reading about the concept and then immediately using what they learned in the accompanying exercise. These unique exercises engage students because they show how important geologic principles are to our every-day activities.

There are more exercises per chapter than can probably be completed in a single lab session. This is done intentionally, to provide options for instructors during class or as potential out-of-class assignments. The complexity and rigor of the exercises increases within each chapter, enabling instructors to use the manual for both non-majors and potential majors alike. Just assign the exercises that are most appropriate for your student population.

Superb illustration program

Readers have come to expect a superior illustration program in any Norton geology text, and this manual does not disappoint. The extensive and highly illustrative photos, line drawings, maps, and DEMs continue the tradition of Stephen Marshak's *Earth: Portrait of a Planet* and *Essentials of Geology*.

Reader-friendly language and layout

Our decades of teaching introductory geology help us to identify the concepts that are most difficult for students to understand. The conversational style of this manual and the use of many real-world analogies help to make these difficult concepts clear and enhance student understanding. The crisp, open layout makes the book more attractive and reader-friendly than other laboratory manuals, which are crammed with pages of multiple-column text.

Unique mineral and rock labs

Students learn the difference between minerals and rocks by classifying Earth materials in a simple exercise (Ch. 3). That exercise then leads to the importance of physical properties and a logical system for identifying minerals. In Chapter 4, students make intrusive and extrusive igneous "rocks," clastic and chemical sedimentary "rocks," and a foliated metamorphic "rock" to understand how the rock-forming processes are indelibly recorded in a rock's texture. The goal in studying rocks in Chapters 5–7 is to interpret the processes and conditions by which they formed, not just to be able to find their correct names.

Digital elevation models (DEMs)

Digital elevation models are used to enhance the understanding of contour lines and to build map-reading skills.

Improvements to the Third Edition

We are grateful to the many adopters of the First and Second Editions for their detailed feedback and helpful suggestions, which we have incorporated into this improved Third Edition. This edition represents a significant revision, based on the comments of reviewers and users who asked for an even more visual, hands-on, thought-provoking, and easier-to-use experience. We assembled a team that included the authors, two reviewers, and three professional editors to both update the content and make certain that all of the improvements were integrated seamlessly into the Third Edition. Throughout this revision, we have retained the core approach of the earlier editions—interspersing exercises throughout the text, so that the manual follows the logical sequence of laboratory activities. Adopters have praised this organization and following their suggestions, we have strengthened this approach. Other improvements to the Third Edition include:

- *A completely revised map program:* All of the topographic maps in this edition are brand new and were created specifically for this lab manual by professional mapmakers. They reproduce clearly, are easy to use, and each is tailored to the specific lab exercise that it accompanies. We continue to produce maps in a full page format so that they are easy to use with their corresponding exercises.

■ *New and significantly revised figures that explain important concepts:* Figures throughout the text have been revised and enlarged, and new figures added to better illustrate the geologic concepts for today's more visual learners.

■ *"What Do You Think" exercises:* Each chapter now contains a special "What Do You Think" scenario where students are asked to make a decision, offer a recommendation, or express their opinion—often as if they were a geologic consultant. These exercises are designed to help students see the connection of geology to the "real" world, to spark discussion in the lab, and to connect geology with other disciplines such as economics, ethics, and history.

■ *Improved pedagogy:* Chapters now begin with clear learning objectives that facilitate student assessment and at the request of reviewers, we have also simplified the figure numbering to improve ease of use for both students and instructors.

■ *Simplified and improved mineral charts and flowcharts:* The mineral charts and flowcharts in Chapter 3 have greatly benefited from the recommendations of several users, making them easier for students to understand and use.

■ *New and revised content based on reviewer feedback:* Many chapters have been reorganized to simplify and improve their approach. In particular, Chapter 2 (Plate Tectonics), Chapter 6 (Metamorphic Rocks), and Chapter 15 (Structural Geology) have all been strengthened by new sequencing and additional basic exercises. Other chapter-by-chapter changes include:

Chapter 1: The introductory material has been reorganized to start with a student-focused mini-case.

Chapter 2: The simplistic cut-and-paste exercises have been removed, the exercises were reordered to group examples of each type of plate boundary together, and the figures used in several exercises have been improved.

Chapter 3: New photographs were added to illustrate cleavage and hardness. The determinative tables and flow charts have been improved based on user suggestions.

Chapter 4: An improved diagram and explanation of the rock cycle is included and better photos of rock textures have been added.

Chapter 5: We have added a section on volcanoes and volcanic hazards and improved the Bowen's Reaction Series and igneous rock classification diagrams.

Chapter 6: This chapter has a reorganized structure that introduces sedimentary processes and rock types more clearly, and easier-to-use classification diagrams for sedimentary rocks have been added.

Chapter 7: This chapter was significantly restructured in order to present metamorphism in a more logical and student-friendly way.

Chapter 8: An improved illustration program clarifies basic map elements.

Chapter 9: A new exercise explaining isolines using the contours of forest growth has been included.

Chapter 10: New photos showing the range of stream sinuosity and figures showing stream features and stream evolution have been added, as well as improved exercises for the principles of stream behavior.

Chapter 11: A new section and exercise on climate change and glacial retreat is now included.

Chapter 12: The maps and exercises have been revised for greater clarity.

Chapter 13: Diagrams showing the evolution of arid landforms have been improved.

Chapter 14: An expanded explanation of factors affecting shoreline landforms is included, as well as a more comprehensive and up-to-date treatment of coastal hazards and strategies to protect shorelines. This chapter has also been updated to include the effects of Hurricane Sandy on the east coast of the United States.

Chapter 15: This chapter has been shortened and simplified in response to reviewer comments, with revised exercises that better explain the basics of rock deformation and the interpretation of geologic maps.

Chapter 16: Improved figures showing the propagation of seismic body and surface waves is now included and other figures have been improved to aid student completion of the exercises.

Chapter 17: A new exercise that applies paleoenvironmental indicators to reconstruct the Cretaceous paleogeography of North America has been added.

- *Revisions to more closely match terminology with other Marshak texts:* While this book is designed to work with any introductory textbook, terminology in particular was reviewed for consistency with Stephen Marshak's *Earth: Portrait of a Planet* and *Essentials of Geology.*

- *New electronic lab versions, pre-lab worksheets, and supplements:* Because some schools now need to offer the lab in a distance-learning environment, we have created a new coursepack compatible with all of the major learning management systems. First, coursepacks include electronic review sheets that you can assign to students before the lab, thus guaranteeing they have reviewed the basic concepts. Then, for help with your distance-learning courses, we have prepared electronic versions of many of the lab exercises. While these labs require students to have the text in print or electronic format to access figures and maps, questions have been tailored to work better in an on-line environment and to work with your learning management system. Please consult with your W. W. Norton sales rep, or review the Instructor's Manual for a complete list of these exercises. A revised instructor's solution manual and new figures in electronic formats are available for download at the Norton Instructor Resource site: wwwnorton.com/instructors.

Supplements

(available for download at wwnorton.com/instructors)

Coursepacks. Available at no cost to professors or students, Norton Coursepacks bring high-quality Norton digital media into your course. This new supplement includes:

- *Prelab quizzes,* available as autograded assignments or printable worksheets, are designed to assess if students have prepared their pre-lab material.

- *On-line versions of selected lab exercises* (a complete list is in the Instructor's Manual). For professors that need to offer this course in a blended- or distance-learning environment, we have adapted the best exercises for these formats into our coursepacks. Responses are either autograded or written to require brief responses. (*Note:* students still need either a print or an electronic version of the lab manual to access figures and background reading.) We have also constructed the labs to work with typical rock kits that can be purchased from many suppliers.

Instructor's Manual, available in electronic format. The revised Instructor's Manual contains word files of the solutions to each exercise, teaching tips for each lab, and a detailed conversion guide showing changes between the Second and Third Editions

Electronic figures. All figures, photographs, charts, and maps in this text are available for you to download and incorporate in your presentations, handouts, or online courses.

Animations of core concepts in geology, which are available to download or stream from our site.

Acknowledgments

We are indebted to the talented team at W. W. Norton & Company whose zealous quest for excellence is matched only by their ingenuity in solving layout problems, finding that special photograph, and keeping the project on schedule. We also appreciate the professors who provided accuracy reviews and feedback for earlier editions: Pete Wehner of Austin Community College–Northridge; Daniel Imrecke, Jinny Sisson, and Julia Smith Wellner of the University of Houston; Kurt Wilkie and Amanda Stahl of Washington State University; Michael Rygel of SUNY–Potsdam; and Karen Koy of Missouri Western State University. We also appreciate the detailed chapter-by-chapter review of the Second Edition by the core teaching faculty and teaching assistants at the University of Houston. We would also like to thank Nathalie Brandes of Lone Star College and Geoffrey W. Cook of the University of California, San Diego, who worked closely with the authors and a team of editors at W. W. Norton to proofread and check the accuracy of the Third Edition.

We are very grateful to and would like to thank all of the following expert reviewers for their input and expertise in making this Lab Manual the best it can be:

Stephen T. Allard, *Winona State University*
Richard Aurisano, *Wharton County Junior College*
Miriam Barquero-Molina, *University of Missouri*
Theodore Bornhorst, *Michigan Tech University*
Nathalie Brandes, *Lone Star College*
Lee Anne Burrough, *Prairie State College*
Geoffrey W. Cook, *University of California, San Diego*
Winton Cornell, *University of Tulsa*
Juliet Crider, *University of Washington*
John Dassinger, *Chandler-Gilbert Community College*
Meredith Denton-Hedrick, *Austin Community College*
Mark Evans, *Central Connecticut State University*
Todd Feeley, *Montana State University*
Jeanne Fromm, *University of South Dakota*
Lisa Hammersley, *Sacramento State University*
Bernie Housen, *Western Washington University*
Daniel Imrecke, *University of Houston*
Jacalyn Gorczynski, Texas *A&M University–Corpus Christi*
Michael Harrison, *Tennessee Technological University*
Daniel Hembree, *Ohio University*
Ryan Kerrigan, *University of Maryland*
Karen Koy, *Missouri Western State University*
Heather Lehto, *Angelo State University*
Jamie Macdonald, *Florida Gulf Coast University*
John A. Madsen, *University of Delaware*

Lisa Mayo, *Motlow State Community College*
Amy Moe Hoffman, *Mississippi State University*
Kristen Myshrall, *University of Connecticut*
David Peate, *University of Iowa*
Alfred Pckarck, *St. Cloud State University*
Elizabeth Rhodes, *College of Charleston*
Anne Marie Ryan, *Dalhousie University*
Ray Russo, *University of Florida*
Mike Rygel, *SUNY Potsdam*
Jinny Sisson, *University of Houston*
Roger Shew, *University of North Carolina - Wilmington*
Amanda Stahl, *Washington State University*
Alexander Stewart, *St. Lawrence University*
Christiane Stidham, *SUNY Stony Brook*
Lori Tapanila, *Idaho State University*
JoAnn Thissen, *Nassau Community College*
Peter Wallace, *Dalhousie University*
Pete Wehner, *Austin Community College–Northridge*
Julia Smith Wellner, *University of Houston*
Kurt Wilkie, *Washington State University*
Andrew H. Wulff, *Western Kentucky University*
Victor Zabielski, *Northern Virginia Community College*

1

Setting the Stage for Learning about the Earth

This sunset over the red cliffs of Horseshoe Bend on the Colorado River in Page, Arizona shows the Earth System at a glance—air, water, and rock all interacting together to produce this stunning landscape.

LEARNING OBJECTIVES

- Understand challenges geologists face when studying a body as large and complex as the Earth
- Practice basic geologic reasoning and strategies
- Introduce the concept of the Earth System, and begin to learn how energy and matter are connected through major cycles
- Use concepts of dimension, scale, and order of magnitude to describe the Earth
- Review the materials and forces you will encounter while studying the Earth
- Learn how geologists discuss the ages of geologic materials and events and how we measure the rates of geologic processes
- Become familiar with the types of diagrams and images used by geologists

MATERIALS NEEDED

- Triple-beam or electronic balance
- 500-ml graduated cylinder
- Metric ruler
- Calculator
- Compass

1.1 Thinking Like a Geologist

Learning about the Earth is like training to become a detective. Both geologists and detectives need keen powers of observation, curiosity about slight differences, broad scientific understanding, and instruments to analyze samples. And both ask the same questions: What happened? How? When? Although much of the logical thinking is the same, there are big differences between the work of a detective and that of a geologist. A detective's "cold" case may be 30 years old, but "old" to a geologist means hundreds of millions or billions of years. To a detective, a "body" is a human body, but to a geologist, a body may be a mountain range or a continent. Eyewitnesses can help detectives, but for most of Earth's history there weren't any humans to act as eyewitnesses for geologists. To study the Earth, geologists must therefore develop different strategies from those of other kinds of investigators. The overall goal of this manual is to help you look at the Earth and think about its mysteries like a geologist.

To illustrate geologic thinking, let's start with a typical geologic mystery. Almost 300 years ago, settlers along the coast of Maine built piers (like the modern pier shown in **FIGURE. 1.1**) to load and unload ships. Some of these piers are now submerged to a depth of 1 meter (39 inches) below sea level.

FIGURE 1.1 Subsidence along the coast of Maine.

Tourists might not think twice about this before heading for a lobster dinner at the local restaurant, but a geologist would want to know how rapidly the pier was submerged and what caused the submergence. How would a geologist go about tackling this problem? To get started, let's try a sample exercise designed to present some basic geologic reasoning and to show the types of real world problems that it can solve.

Name: _____   Section: _____
Course: _____   Date: _____

2015   1715

The figure on the left illustrates a pier whose walkway sits 1 meter below the ocean's surface today. Since we weren't there when it was built 300 years ago, we have to make some assumptions—geologists often do this to make estimates. So let's assume that the pier's walkway was originally 1 m *above* sea level at high tide, as many are built today (illustrated in the figure on the right), and that submergence occurred at a constant rate. With these assumptions, calculating the rate of submergence for the past 300 years becomes simple arithmetic.

(a) The rate of submergence is the total change in the elevation of the pier (_____ m) divided by the total amount of time involved (_____ years) and is therefore _____ cm/yr. (Remember, 1 m = 100 cm.)

Now consider a problem this might solve:

(b) A local restaurant owner is considering the purchase of a pier, whose walkway is 50 cm above the high-water mark, for use in outdoor events. The owner has been advised that piers with walkways less than 30 cm above the high-water mark should be avoided because they can be flooded by storms and very high tides. If submergence continues at the rate you calculated, how many years will pass before the high-water mark is less than 30 cm from the base of the deck? _____ yrs.

? What Do You Think   Now it's time to really try thinking like a geologist. Given your answers to (a) and (b), would you recommend the restaurant owner purchase this pier? In a sentence or two, explain why. Then describe another issue that you think the owner should investigate before they make their decision.

You've just tried your first geology problem, Congratulations! A geologist, however, would also want to explain *why* the piers were submerged. When faced with such a problem, geologists typically try to come up with as many explanations as possible. For example, which of the following explanations could account for the submergence?

☐ The sea level has risen.

☐ The land has sunk.

☐ Both the sea level and land have risen, but sea level has gone up more.

☐ Both the sea level and land have sunk, but the land has sunk more.

☐ Over time, the piers gradually sink in the mud of the seafloor.

If you thought all five choices might be right (correctly!), you realize that explaining submergence along the Maine coast may be more complicated than it seemed at first. To find the answer, you need more **data**—more observations and/or measurements. One way to obtain more data would be to see if submergence is restricted to Maine or the east coast of North America or if it is perhaps worldwide. As it turns out, submergence is observed worldwide, suggesting that the first choice above (sea-level rise) is the most probable explanation, but not necessarily the only one.

With even more data, we could answer questions like, When did submergence happen? Was the sea-level rise constant? Maybe all submergence occurred in the first 100 years and then stopped. Or perhaps it began slowly and then accelerated. Unfortunately, we may not be able to answer all of these questions because, unlike television detectives who always get the bad guys, geologists don't always have enough data and must often live with uncertainty. We still do not have the answers to many questions about the Earth.

1.1.1 The Scientific Method

Like all scientists (and most people trying to find answers to problems they have identified), geologists follow a logical process that you are probably familiar with—the scientific method. You did so instinctively in Exercise 1.1 and will do so many times throughout this course. The scientific method begins with making observations of Earth features or processes—in Exercise 1.1, for example, we observed that a colonial pier is now below sea level.

STEP 1 Observing an Earth feature or process (e.g., the submerged pier).

STEP 2 Recognizing that a problem exists and asking questions about that problem. Usually the problem is that we don't understand how what we've observed came to be. In Exercise 1.1, by how much has the pier been submerged? How fast did submergence take place? Why was the pier submerged? We respond with the steps that follow.

STEP 3 Collecting more data to see if the observation is valid and to help us understand what is going on. In Exercise 1.1, for instance, it isn't just one pier being submerged, but many along the Maine coast.

STEP 4 Proposing tentative answers to those questions called **hypotheses** (singular, **hypothesis**). Some versions of the scientific method suggest that we propose a single hypothesis. When we first look at problems, we usually come up with more than one hypothesis, a practice called *multiple working hypotheses*.

STEP 5 Testing the hypotheses by getting more data, which may support some, rule out others, or possibly lead to new hypotheses. Some of this testing can be in the form of a formal laboratory experiment, but there are also other forms, such as field trips to gain additional information, detailed measurements where there had been only eyeball estimates, and so forth.

STEP 6 If your test supports the hypothesis, continue to make additional tests to further verify your result. If your test does not support the hypothesis, then revise the hypothesis and try again, or propose an entirely new hypothesis.

The cycle of observations and experiments illustrated schematically in **FIGURE 1.2** enables us to eliminate many of the original hypotheses and leads us to propose new ones until, sometimes, only a single viable explanation remains. Even then we are not satisfied and continue to test it, using new technology as it becomes available. If after years of rigorous testing a hypothesis has withstood all tests, it is elevated to the level of a **theory**, like the germ theory of disease, the theory of evolution, or, as you will learn in Chapter 2, the plate tectonic theory for the origin and evolution of mountains, oceans, and continents.

FIGURE 1.2 The scientific method.

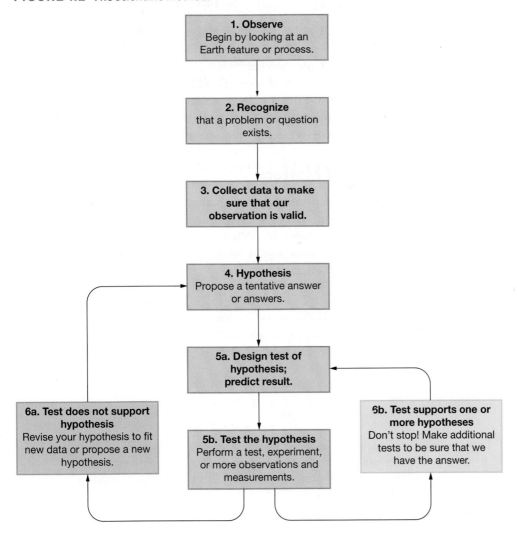

1.1.2 The Challenge of Studying a Planet

Problems like submergence along the Maine coast pose challenges to geologists and geology students who are trying to learn about the Earth. These challenges require us to:

- understand the many kinds of materials that make up the Earth and how they behave.
- be aware of how energy causes changes at Earth's surface and beneath it.
- consider features at a wide range of sizes and scales—from the atoms that make up rocks and minerals to the planet as a whole.
- think in *four* dimensions, because geology involves not just the three dimensions of space but also an enormous span of *time*.
- realize that some geologic processes occur in seconds but others take millions or billions of years and are so slow that we can detect them only with very sensitive instruments.

The rest of this chapter examines these challenges and how geologists cope with them. You will learn basic geologic terminology and how to use tools of observation and measurement that will be useful throughout your geologic studies. Some terms will probably be familiar to you from previous science classes.

1.2 Studying Matter and Energy

Earth is a dynamic planet. Unlike the airless Moon, which has remained virtually unchanged for billions of years, Earth's gases, liquids, and solids constantly move from one place to another. They also change from one state to another through the effects of heat, gravity, other kinds of energy, and living organisms. We refer to all of Earth's varied materials and the processes that affect them as the **Earth System**, and the first step in understanding the Earth System is to understand the nature of matter and energy and how they interact with one another.

1.2.1 The Nature of Matter

Matter is the "stuff" of which the Universe is made; we use it to refer to any material making up the Universe. Geologists, chemists, and physicists have shown that matter consists of ninety-two naturally occurring elements and that some of these elements are much more abundant than others. Keep the following definitions in mind as you read further (**TABLE 1.1**).

TABLE 1.1 Basic definitions

- An **element** is a substance that cannot be broken down chemically into other substances.
- The smallest piece of an element that still has all the properties of that element is an **atom**.
- Atoms combine with one another chemically to form **compounds**; the smallest possible piece of a compound is called a **molecule**.
- Atoms in compounds are held together by **chemical bonds**.
- A simple **chemical formula** describes the combination of atoms in a compound. For example, the formula H_2O shows that a molecule of water contains two atoms of hydrogen and one of oxygen.

Matter occurs on Earth in three states—solid, liquid, or gas. Atoms in *solids*, like minerals and rocks, are held in place by strong chemical bonds. As a result, solids retain their shape over long periods. Bonds in *liquids* are so weak that atoms or molecules move easily, and as a result, liquids adopt the shape of their containers. Atoms or molecules in *gases* are barely held together at all, so a gas expands to fill whatever container it is placed in. Matter changes from one state to another in many geologic processes, as when the Sun evaporates water to produce water vapor, or when water freezes to form ice, or when lava freezes to become solid rock.

We describe the amount of matter in an object by indicating its **mass** and the amount of space it occupies by specifying its **volume**. The more mass packed into a given volume of matter, the greater the **density** of the matter. You notice density differences every day: it's easier to lift a large box of popcorn than a piece of rock of the same size because the rock is much denser; it has much more mass packed into the same volume and therefore weighs much more.

1.2.2 Distribution of Matter in the Earth System

Matter is stored in the Earth System in five major **reservoirs** (**FIG. 1.3a**). Most gas is in the **atmosphere**, a semi-transparent blanket composed of about 78% nitrogen (N_2) and 21% oxygen (O_2), with minor amounts of carbon dioxide (CO_2), water vapor (H_2O), ozone (O_3), and methane (CH_4). Nearly all liquid occurs as water in

FIGURE 1.3 The Earth System.

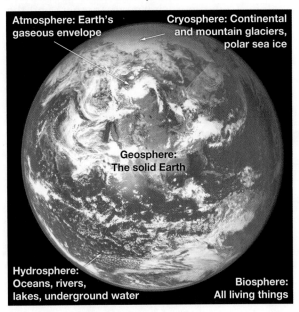

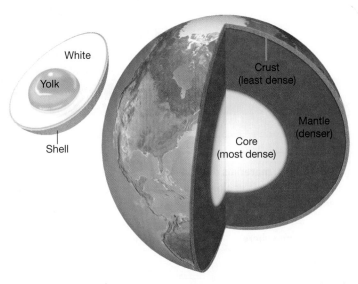

(a) Earth's major reservoirs of matter.

(b) An early image of Earth's internal layers. The hard-boiled egg analogy for the Earth's interior. Earth's interior is denser than the mantle and crust.

the **hydrosphere**—Earth's oceans, rivers, lakes, and groundwater, which is found in cracks and pores beneath the surface. Frozen water makes up the **cryosphere**, including snow, thin layers of ice on the surface of lakes or oceans, and huge masses of ice in glaciers and the polar ice caps.

The solid Earth is called the **geosphere**, which geologists divide into concentric layers like those in a hard-boiled egg (**FIG. 1.3b**). The outer layer, the **crust**, is relatively thin, like an eggshell, and consists *mostly* of rock. Below the crust is the **mantle**, which also consists *mostly* of different kinds of rock and, like the white of an egg, contains most of Earth's volume. We say *mostly* because about 2% of the crust and mantle has melted to produce liquid material called **magma** (known as **lava** when it erupts on the surface). The central part of the Earth, comparable to the egg yolk, is the **core**. The outer core consists mostly of a liquid alloy of iron and nickel, and the inner core is a solid iron-nickel alloy.

Continents make up about 30% of the crust and are composed of relatively low-density rocks. The remaining 70% of the crust is covered by the oceans. Oceanic crust is both thinner and denser than the crust under the continents. Three types of solid materials are found at the surface: **bedrock**, a solid aggregate of minerals and rocks attached to the Earth's crust; **sediment**, unattached mineral grains such as boulders, sand, and clay; and **soil**, sediment and rock modified by interactions with the atmosphere, hydrosphere, and organisms so that it can support plant life.

The **biosphere** is the realm of living organisms, extending from a few kilometers below Earth's surface to a few kilometers above. Geologists have learned that organisms—from bacteria to mammals—are important parts of the Earth System. They exchange gases with the atmosphere, absorb and release water, break rock into sediment, and help convert sediment and rock to soil.

The movement of materials from one reservoir to another is called a **flux** and happens in many geologic processes. For example, rain is a flux in which water moves from the atmosphere to the hydrosphere. Rates of flux depend on the materials, the reservoirs, and the processes involved. In some cases, a material moves among several reservoirs but eventually returns to the first. We call such a path a

geologic cycle. In your geology class you will learn about several cycles, such as the rock cycle (the movement of atoms from one rock type to another) and the hydrologic cycle (the movement of water from the hydrosphere to and from the other reservoirs). Exercises 1.2 and 1.3 will help you understand the distribution of matter.

| EXERCISE 1.2 | Reservoirs in the Earth System |

Name: _____ Section: _____
Course: _____ Date: _____

What Earth materials did you encounter in the past 24 hours? List at least ten in the following table without worrying about the correct geologic terms (for example, "dirt" is okay *for now*). Place each Earth material in its appropriate reservoir and indicate whether it is a solid (S), liquid (L), or gas (G).

Atmosphere	Hydrosphere	Geosphere	Cryosphere	Biosphere

1.2.3 Energy in the Earth System

Natural disasters in the headlines remind us of how dynamic the Earth is: rivers flood cities and fields, lava and volcanic ash bury villages, earthquakes topple buildings, and hurricanes ravage coastal regions. However, many geologic processes are much slower and less dangerous, like the movement of ocean currents and the almost undetectable creep of soil downhill. All are caused by energy, which acts on matter to change its character, move it, or split it apart.

Energy for the Earth System comes from (1) Earth's *internal* heat, which melts rock, causes earthquakes, and builds mountains (some of this heat is left over from the formation of the Earth, but some is being produced today by radioactive decay); (2) *external* energy from the Sun, which warms air, rocks, and water on the Earth's surface; and (3) the pull of Earth's gravity. Heat and gravity, working independently or in combination, drive most geologic processes. Exercise 1.4 explores this idea.

Heat energy is a measure of the degree to which atoms or molecules move about (vibrate) in matter—including in solids. When you heat something in an oven, for example, the atoms in the material vibrate faster and move farther apart. Heat energy drives the flux of material from one state of matter to another or from one reservoir of the Earth System to another. For example, heating ice causes **melting** (solid → liquid; cryosphere → hydrosphere) and heating water causes **evaporation** (liquid → gas; hydrosphere → atmosphere). Cooling slows the motion, causing **condensation** (gas → liquid, atmosphere → hydrosphere) or **freezing** (liquid → solid, hydrosphere → cryosphere).

Name: _____ Section: _____

Course: _____ Date: _____

Even without a geology course, you already have a sense of how water moves from one reservoir to another in the Earth System. Based on your experience with natural phenomena on Earth, complete the following table that describes fluxes associated with the *hydrologic cycle*. First describe what happens in the process using plain language, then describe how matter may have moved between Earth's reservoirs. The first example is given.

Process	What happens?	Did matter change reservoir? Movement from _____ to _____
Sublimation	*Solid ice becomes water vapor.*	*Yes. cryosphere to atmosphere*
Ice melting		
Evaporation of a puddle		
Water freezing		
Plants absorbing water		
Raindrop formation		
Cloud formation		
Steam erupting from a volcano		

Name: _____   Section: _____
Course: _____   Date: _____

Some of the heat that affects geologic processes comes from the Sun and some comes from inside the Earth. What role does each of these heat sources play in Earth processes?

(a) If you take off your shoes on a beach or any sandy environment and walk on it on a hot, sunny day, is the sand hot or cold? Why?

_____

- Now, dig down in the sand just a few inches. What do you feel now, and why?

_____

- What does this suggest about the depth to which heat from the Sun can penetrate the Earth?

_____

- Based on this conclusion, is the Sun's energy or Earth's internal heat the cause of melting rock within the Earth? Explain.

_____

_____

(b) The deeper down one goes into mines or drill holes, the hotter it gets. This temperature increase is called the **geothermal gradient**. Does this phenomenon support or contradict your conclusion in (a)? Explain.

_____

_____

- In the upper 10 km of the crust, the geothermal gradient is typically about 25°C per km, but it can range from 15°C/km to 50°C/km. In the chart below, plot these three geothermal gradients (15°C/km, 25°C/km, and 50°C/km) for the upper 10 km of the Earth. Use one line for each gradient.

Depth (km) vs Temperature (°C)

- The deepest mine on Earth penetrates to a depth of about 2 km. Using the geothermal gradients you just drew, what range of temperatures would you expect in the mine? Explain your answer. How hot is it in the bottom of this mine? What assumptions did you make to come up with this answer?

_____

_____

_____

Gravity, as Isaac Newton showed more than three centuries ago, is the force of attraction that every object exerts on other objects. The strength of this force depends on the amount of mass in each object and how close the objects are to one another. The greater the mass and the closer the objects are, the stronger the gravitational attraction. The smaller the mass and the farther apart the objects are, the weaker the attraction. The Sun's enormous mass produces a force of gravity sufficient to hold Earth and the other planets in their orbits. Earth's gravitational force is far less than the Sun's, but it is strong enough to hold the Moon in orbit, hold you on its surface, cause rain or volcanic ash to fall, and enable rivers and glaciers to flow.

1.3 Units for Scientific Measurement

Before we begin to examine components of the Earth System scientifically, we must first consider its dimensions and the units used to measure them. We can then examine the challenges of scale that geologists face when studying Earth and the atoms of which it is made.

1.3.1 Units of Length and Distance

If you described this book to a friend as being "big," would your friend have a clear picture of its size? Is it big compared to a quarter or to a car? Without providing a frame of reference for the word *big*, your friend wouldn't have enough information to visualize the book. A *scientific* description would much more accurately give the book's dimensions of length, width, and thickness using units of distance.

People have struggled for thousands of years to describe size in a precise way with standard units of measurement. Scientists everywhere and people in nearly all countries except the United States use the **metric system** to measure length and distance. The largest metric unit of length is the kilometer (km), which is divided into smaller units: 1 km = 1,000 meters (m); 1 m = 100 centimeters (cm); 1 cm = 10 millimeters (mm). Metric units differ from each other by a factor of 10, making it very easy to convert one unit into another. For example, 5 km = 5,000 m = 500,000 cm = 5,000,000 mm. Similarly, 5 mm = 0.5 cm = 0.005 m = 0.000005 km.

The United States uses the U.S. customary system (sometimes known as the English Unit System) to describe distance. Distances are given in miles (mi), yards (yd), feet (ft), and inches (in), where 1 mi = 5,280 ft; 1 yd = 3 ft; and 1 ft = 12 in. As scientists, we use metric units in this book, but when appropriate, equivalents are also given (in parentheses).

Appendix 1.1, at the end of this chapter, provides basic conversions between U.S. Customary and metric units.

1.3.2 Other Dimensions, Other Units

Distance is just one of the dimensions of the Earth that you will examine during this course. We still need other units to describe other aspects of the Earth, its processes, and its history: units of time, velocity, temperature, mass, and density.

Time is usually measured in seconds (s), minutes (min), hours (h), days (d), years (yr), centuries (hundreds of years), and millennia (thousands of years). A year is the amount of time it takes for the Earth to complete one orbit around the Sun. Because the Earth is very old, geologists also have to use much larger units of time: thousand years ago (abbreviated **ka**, for "kilo-annum"), million years ago (**Ma**, for "mega-annum"), and billion years ago (**Ga**, for "giga-annum"). The 4,570,000,000-year age of the Earth can thus be expressed as 4.57 Ga, or 4,570 Ma.

Velocity, or the rate of change of the position of an object, is described by units of distance divided by units of time, such as meters per second (m/s), feet per second (ft/s), kilometers per hour (km/h), or miles per hour (mph). You will learn later that the velocity at which geologic materials move ranges from extremely slow (mm/yr) to extremely fast (km/s).

Temperature is a measure of how hot an object is relative to a standard. It is measured in degrees Celsius (°C) in the metric system and degrees Fahrenheit (°F) in the U.S. customary system. The reference standards are the freezing and boiling points of water: 0°C and 100°C or 32°F and 212°F, respectively. Note that there are 180 Fahrenheit degrees between freezing and boiling but only 100 Celsius degrees. A change of 1°C is thus 1.8 times larger than a change of 1°F (180°/100°). To convert Fahrenheit to Celsius or vice versa, see Appendix 1.1.

Mass refers to the amount of matter in an object, and **weight** refers to the force with which one object is attracted to another. The weight of an object on Earth therefore depends not only on its mass but also on the strength of Earth's gravitational force. Objects that have more mass than others also weigh more on Earth because of the force of Earth's gravity. While the mass of an object remains the same whether it is on the Earth or on the Moon, the object *weighs* less on the Moon because of the Moon's weaker gravity.

Grams and kilograms (1 kg = 1,000 g) are the units of mass in the metric system; the U.S. Customary System uses pounds and ounces (1 lb = 16 oz). For those who don't read the metric equivalents on food packages, 1 kg = 2.2046 lb and 1 g = 0.0353 oz.

We saw earlier that the **density (δ)** of a material is a measure of how much mass is packed into each unit of volume, generally expressed as units of g/cm^3. We instinctively distinguish unusually low-density materials like Styrofoam and feathers from low-density materials like water ($\delta = 1 \ g/cm^3$) and high-density materials like steel ($\delta = \sim 7 \ g/cm^3$) because the former feel very light *for their sizes* and the latter feel unusually heavy *for their sizes* (**FIG. 1.4**).

FIGURE 1.4 Weight of materials with different densities.

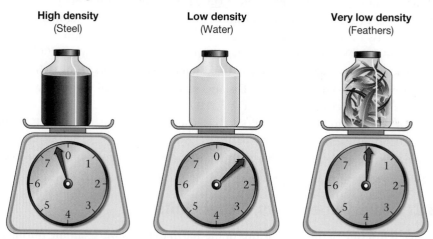

To measure the density of a material, we need to know its mass and volume. We measure mass on a balance or scale, and we can use simple mathematical formulas to determine the volumes of cubes, bricks, spheres, or cylinders. For example, to calculate the volume of a bar of gold, you would multiply its length times its width, times its height (**FIG. 1.5a**). But for an irregular chunk of rock, it is easiest to submerge the rock in a graduated cylinder partially filled with water. Measure the volume of water before the rock is added and then with the rock in the cylinder. The rock displaces a volume of water equivalent to its own volume, so simply subtract the volume of the water before the rock was added from that of the water plus rock to obtain the volume of the rock (**FIG. 1.5b**). The density of a rock can be calculated simply from the definition of density: density = mass ÷ volume.

FIGURE 1.5 Measuring the volume of materials.

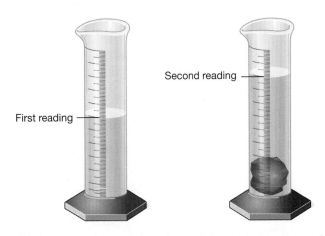

(a) For a rectangular solid, volume = length × width × height.

(b) For an irregular solid, the volume of the rock is the volume of the displaced liquid, which equals the difference between the first and second readings.

Thousands of years ago, the Greek scientist Archimedes recognized that density determines whether an object floats or sinks when placed in a liquid. For example, icebergs float in the ocean because ice is less dense than water. Nearly all rocks sink when placed in water because the density of most rocks is at least 2.5 times greater than the density of water. Exercise 1.5 helps you practice your understanding of density.

EXERCISE 1.5   Measuring the Density of Earth Materials

Name: _____   Section: _____
Course: _____   Date: _____

(a) You are given a graduated cylinder, a balance, and a container of water. Determine the density of water. _____ g/cm^3

(b) Your instructor will provide you with two samples of rock: *granite*, a light-colored rock that makes up a large amount of the continental crust, and *basalt*, a dark-colored rock that makes up most of the oceanic crust and the lower part of the continental crust.
- What is the density of granite?   _____ g/cm^3
- What is the density of basalt?   _____ g/cm^3
- What is the ratio of the density of granite to the density of basalt?   _____

(c) Most modern ships are made of steel, which has a density of about 7.85 g/cm^3. This is much greater than the density of water, so how can ships float?

_____

_____

1.4 Expressing Earth's "Vital Statistics" with Appropriate Units

Now that you are familiar with some of the units used to measure the Earth, we can look at some of the planet's "vital statistics."

■ *The Solar System:* Earth is the third planet from the Sun in a Solar System of eight planets (Pluto has been reclassified as a "dwarf planet"). It is one of four terrestrial planets, along with Mercury, Venus, and Mars, whose surfaces are made of rock rather than of frozen gases. Earth is, on average, 150,000,000 km (93,000,000 mi) from the Sun.

- *Earth's orbit:* It takes Earth 1 year (365.25 days, or 3.15×10^7 seconds) to complete one slightly elliptical orbit around the Sun.

- *Earth's rotation:* It takes Earth 1 day (24 hours, or 86,400 seconds) to rotate once on its axis. The two points where the axis intersects the surface of the planet are the north and south **geographic poles**. At present, Earth's axis is tilted at about 23.5° to the plane in which it orbits the Sun (the **plane of the ecliptic**), but Earth wobbles a bit so this tilt changes over time.

- *Shape:* Our planet is almost, but not quite, a sphere. Earth's rotation causes it to bulge slightly at the equator. The equatorial radius of 6,400 km (~4,000 mi) is 21 km (~15 mi) longer than its polar radius.

- *Temperature:* Earth's surface temperature ranges from 58°C (136°F) in deserts near the equator to −89°C (−129°F) near the South Pole. *Average* daily temperature in New York City ranges from 0°C (32°F) in the winter to 24°C (76°F) in the summer. The geothermal gradient is about 15°C to 30°C per km in the upper crust, and temperatures may reach 5000°C at the center of the Earth.

- *Some additional dimensions:* The highest mountain on Earth is Mt. Everest, at 8,850 m (29,035 ft) above sea level. The average depth of the world's oceans is about 4,500 m (14,700 ft), and the deepest point on the ocean floor—the bottom of the Mariana Trench in the Pacific Ocean—is 11,033 m (35,198 ft) below the surface. The *relief*, the vertical distance between the highest and lowest points, is thus just a little less than 20 km (12 mi). It would take only about 10 minutes to drive this distance at highway speed! This relief is extremely small compared to the overall size of the planet.

1.5 The Challenge of Scale when Studying a Planet

Geologists deal routinely with objects as incredibly small as atoms and others as incredibly large as the Appalachian Mountains or the Pacific Ocean. Sometimes we have to look at a feature at different scales, as in **FIGURE 1.6a, b**, to understand it fully.

FIGURE 1.6 The white cliffs of Dover, seen at two different scales.

(a) The chalk cliffs of Dover, England; the person in the foreground gives an idea of the size of the cliffs.

(b) Microscopic view of the chalk (plankton shells) that the cliffs are made of. The eye of a needle gives an idea of the minuscule sizes of the shells that make up the cliffs.

The challenge of scale is often a matter of perspective: to a flea, the dog on which it lives is its entire world; but to a parasite inside the flea, the flea is *its* entire world. For most of our history, humans have had a flea's-eye view of the world, unable to recognize its dimensions or shape from our vantage point on its surface. As a result, we once thought Earth was flat and at the center of the Solar System. It's easy to laugh at such misconceptions now, but Exercise 1.6 gives a feeling for the challenges of scale that still exist.

1.5.1 Scientific Notation and Orders of Magnitude

Geologists must cope with the enormous ranges in the scale of distance (atoms to sand grains to planets), temperature (below 0°C in the cryosphere and upper atmosphere to more than 1000°C in some lavas, to millions of degrees Celsius in the Sun), and velocity (continents moving at 2 cm/yr to light moving at 299,792 km/s). We sometimes describe scale in approximate terms, and sometimes more precisely. For example, the terms *mega-scale, meso-scale,* and *micro-scale* denote enormous, moderate, and tiny features, respectively, but don't tell exactly how large or small they are, because they depend on a scientist's frame of reference. For example, *mega-scale* to an astronomer might mean intergalactic distances, but to a geologist it may mean the size of a continent or a mountain range. *Micro-scale* could refer to a sand grain, a bacterium, or an atom. Geologists commonly approximate scale in terms that clearly specify the frame of reference. For example, your instructor may use "outcrop-scale" for a feature in a single exposure of rock or "hand-specimen-scale" for a rock the size of your fist. Exercise 1.6 helps you practice with scale.

EXERCISE 1.6 **The Challenge of Perspective and Visualizing Scale**

Name: _____ Section: _____
Course: _____ Date: _____

The enormous difference in size between ourselves and our planet gives us a limited perspective on large-scale features and makes understanding major Earth processes challenging. To appreciate this challenge, consider the relative sizes of familiar objects (use Appendix 1.1 for conversions):

1 mm

1 m

Relative sizes of a dog and a flea.

12,800 km

2 m

Relative sizes of the Earth and a tall human being.

(a) Assume that the flea is 1 mm long and that a dog is 1 m long.
- To relate this to our English system of measurement, how long is this flea in inches? _____ in
- How many times larger is the dog than the flea? _____
- How long is the bar scale representing the flea? _____
- How long would the bar representing the dog have to be, if the dog and the flea were shown at the same scale? _____

(continued)

Name: _____   Section: _____

Course: _____   Date: _____

(b) Now, think about the relative dimensions of a geologist and the Earth.
- How many times larger than the geologist is the Earth? _____
- How large would the drawing of the Earth have to be, if it were drawn at the same scale as the geologist? Give your answer in kilometers and in miles. _____ km _____ mi
- Based on the relative sizes of flea and dog versus human and Earth, does a flea have a better understanding of a dog than a human has of the Earth? Or vice versa? Explain.

_____

_____

_____

Geologists use a system based on powers of 10 to describe things spanning the entire range of scales that we study, sometimes using the phrase **orders of magnitude** to indicate differences in scale. A feature that is an order of magnitude larger than another is 10 times larger; a feature one-tenth the size of another is an order of magnitude *smaller*. Something 100 times the size of another is two orders of magnitude larger, and so on. **TABLE 1.2** shows that the range of dimensions in our Universe spans an almost incomprehensible 44 orders of magnitude, from the diameter of the particles that make up an atom (about 10^{-18} m across) to the radius of the observable Universe (about 10^{26} m).

TABLE 1.2   Orders of magnitude defining lengths in the Universe (in meters).

~2×10^{26}	Radius of the observable Universe
2.1×10^{22}	Distance to the nearest galaxy (Andromeda)
9.0×10^{20}	Diameter of the Milky Way
1.5×10^{11}	Diameter of Earth's orbit
6.4×10^{6}	Radius of the Earth
5.1×10^{6}	East–west length of the United States
8.8×10^{3}	Height of Mt. Everest above sea level (Earth's tallest mountain)
1.7×10^{0}	Average height of an adult human
1.0×10^{-3}	Diameter of a pinhead
6.0×10^{-4}	Diameter of a living cell
2.0×10^{-6}	Diameter of a virus
1.0×10^{-10}	Diameter of a hydrogen atom (the smallest atom)
1.1×10^{-14}	Diameter of a uranium atom's nucleus
1.6×10^{-15}	Diameter of a proton (a building block of an atomic nucleus)
~1×10^{-18}	Diameter of an electron (a smaller building block of an atom)

We simplify the numbers that describe such a wide range with a method called **scientific notation** based on powers of 10. In scientific notation, 1 is written as 10^0, 10 as 10^1, 100 as 10^2, and so on. Numbers less than 1 are shown by negative exponents: for example, $\frac{1}{10} = 0.1 = 10^{-1}$; $\frac{1}{100} = 0.01 = 10^{-2}$; and $\frac{1}{1,000} = 0.001 = 10^{-3}$. A positive exponent tells how many places the decimal point has to be moved to the right of the number, and a negative exponent how many places to the left.

In scientific notation, the 150,000,000-km (93,000,000-mi) distance from the Earth to the Sun is written as 1.5×10^8 km (9.3×10^7 mi)—start with 1.5 and move the decimal eight places to the *right*, adding zeroes as needed. Similarly, a small number such as 0.0000034 would be written as 3.4×10^{-6} (start with 3.4 and move the decimal six places to the *left*, adding zeroes as needed).

Exercise 1.7 practices scientific notation and gives a sense of our perspective relative to the Earth's basic processes.

EXERCISE 1.7   Moving Along

Name: _____   Section: _____
Course: _____   Date: _____

Complete the following calculations to get a sense of the distances that the Earth travels and the speed at which it moves.

Earth's orbit. Earth orbits the Sun in an elliptical path. To simplify calculation, here we assume a circular orbit with a radius of 150,000,000 km (93,000,000 mi).

Earth's rotation. Earth rotates on its axis every 24 hours.

(a) For simplicity, in this exercise we picture the Earth's orbit around the Sun as a circle, with a radius of 150,000,000 km (93 million mi). It takes 1 year for the Earth to orbit the Sun.
   • What distance does the Earth travel during a single complete orbit? Remember that the circumference of a circle is $2\pi r$, and that $\pi \approx 3.14$. Orbital distance = _____ km (_____ mi).
   • What is the velocity of the Earth as it orbits the Sun? Give your answer in both kilometers per hour and miles per hour. _____ km/h _____ mph
   • Express the above results using scientific notation. _____ km/h _____ mph

(b) Earth rotates on its axis in 1 day. At what velocity is a person standing at the equator in the figure above (Person 1) moving due to our planet's rotation? _____ km/h _____ mph

(c) Now consider a person standing on Earth's surface at a point halfway between the equator and the geographic pole (Person 2). Is this person moving faster or slower than the person standing at the equator? Explain your answer.

_____

_____

(continued)

Name: _____   Section: _____

Course: _____   Date: _____

(d) Many components of the Earth System are too *small* for us to comprehend. For example, atoms are too small to be seen even with the most powerful optical microscope. A sodium atom has an approximate diameter of 2×10^{-8} cm. The diameter of the period at the end of this sentence is approximately 4×10^{-4} cm.
- If you lined up sodium atoms one next to the other, how many sodium atoms would it take to span the diameter of a period? _____ Express this number in scientific notation: _____
- How many orders of magnitude larger is the period than the atom? _____

1.5.2 Coping with Scale Problems: Maps, Diagrams, and Models

Figure 1.6 showed how we use microscopes to help see very *small* features. One way to cope with the challenge of very *large* geologic features is to construct scale models like that in **FIGURE 1.7a**. Using these models, we can re-create the conditions of floods or the ground shaking during earthquakes to understand these processes better. Scaled-down diagrams like the one of the geothermal gradient in Exercise 1.4 help show relationships within the Earth, and we use maps, aerial photographs, and satellite images to visualize the Earth's surface (**FIG. 1.7b**). Depending on the degree to which we scale down the area, these images can show the entire Earth on a single sheet of paper or just the part of it we wish to study. If you have a good high-speed Internet connection, you can download *Google Earth*™ or *NASA World Wind* to see spectacular images of any part of the world at almost any scale. Exercise 1.8 gives simple examples of how scaled-down maps and diagrams are made.

FIGURE 1.7 Coping with scale problems associated with very large features.

(a) A scale model of St. Paul Harbor, Alaska, used to study water and sediment movement.

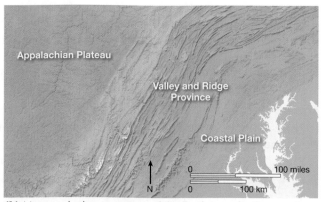

(b) Maps scale down portions of the Earth to make them easier to visualize and understand.

Name: _____   Section: _____
Course: _____   Date: _____

(a) The United Kingdom is about 900 km long from north to south. By how many orders of magnitude would you have to scale down a map of the United Kingdom so it would fit on a sheet of paper 20 cm long? _____

(b) Earth's radius is about 6,500 km. Using a compass and ruler, draw a cross section of the Earth in the left-hand box below, using Point E to locate the center of the Earth. *Indicate the scale that you used in the lower left corner of the box.*

(c) The Moon has a radius of 1,750 km. At the same scale that you used in (a), draw a cross section of the Moon in the right-hand box below, using Point M to locate the center of the Moon.

(d) Compare the size of the Moon with that of Earth's core, which has a radius of 3,400 km. Which is larger and by approximately how much?

• E   • M

■■■■ _____ km

1.6 The Challenge of Working with Geologic Time

Geologists have amassed a large body of evidence showing that Earth is about 4.57 billion years old (4,570,000,000, or 4.57×10^9 years). This enormous span is referred to as **geologic time (FIG. 1.8)**, and understanding its vast scope is nearly impossible for humans, who live less than 100 years. Our usual frame of reference for time is based on human lifetimes: a war fought two centuries ago happened two lifetimes ago, and 3,000-year-old monuments were built about 30 lifetimes ago. To help visualize geologic time, we can compare its scale to more familiar dimensions. For example, if geologic time (4.57 billion years) was represented by a 1-km-long field, each millimeter would represent 4,570 years, each centimeter 45,700 years, and each meter 4,570,000 years. If all of geologic time was condensed to a single year, each second would represent 145 years. Exercise 1.9 helps demonstrate this scale.

Name: _____   **Section:** _____
Course: _____   **Date:** _____

Complete the following calculations to appreciate the duration of geologic time and how slow processes can lead to major changes.

(a) If you want to represent the 4.57 billion years of geologic time with a roll of masking tape so that 1 inch represents 1 million years, how long would the tape have to be? *Answer using scientific notation.* _____ in _____ mi _____ km

(b) If you want to represent geologic time with a 24-hour clock, how many years would be represented by a single second? _____ a minute? _____ an hour? _____

(c) If you saved a penny a day, how long would it take you to save $1 million? _____ days
   • If you collected a penny a year for a hundred years, how much money would you have? $ _____
   • How much would you have in a million years? $ _____ a billion years? $ _____

FIGURE 1.8 Geologic time line. Milestones in the history of the Earth.

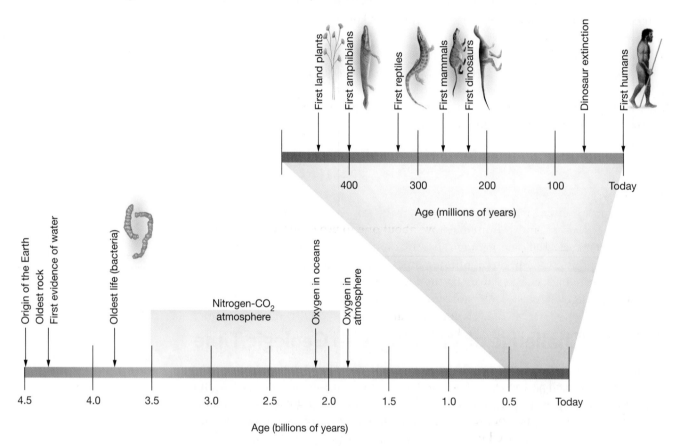

1.6.1 Uniformitarianism: The Key to Understanding the Geologic Past

The challenge in studying Earth history is that many of the processes that made the rocks, mountains, and oceans we see today occurred millions or billions of years ago. How can we figure out what those processes were if they happened so long ago? The answer came nearly 300 years ago when a Scottish geologist, James Hutton,

FIGURE 1.9 The principle of uniformitarianism.

(a) Ripple marks in 145-million-year-old sandstone at Dinosaur Ridge, Colorado.

(b) Ripple marks in modern sand on the shore of Cape Cod, Massachusetts.

noted that some features observed in rocks resembled those forming in modern environments (**FIG. 1.9a, b**). For example, the sand on a beach or a stream's bed commonly forms a series of ridges called ripple marks. You can see identical ripple marks preserved in ancient layers of sandstone, a rock made of sand grains.

Based on his observations, Hutton proposed the **principle of uniformitarianism** as a basis for interpreting Earth history. According to uniformitarianism, most ancient geologic features formed by the same processes as modern ones. The more we know about the physical and chemical processes by which modern features form, the more we can say about the ancient processes. This principle is often stated more succinctly as "the present is the key to the past."

1.6.2 Rates of Geologic Processes

Many geologic processes occur so slowly that it's difficult to know they are happening. For example, mountains rise and are worn away (**eroded**) at rates of about 1 to 1.5 *mm*/yr, and continents move about one to two orders of magnitude faster—1 to 15 *cm*/yr. Some geologic processes are much faster. A landslide or volcanic eruption can happen in minutes; an earthquake may be over in seconds; and a meteorite impact takes just a fraction of a second.

Before the significance of uniformitarianism became clear, scholars thought that Earth was about 6,000 years old. Uniformitarianism made scientists rethink this estimate and consider the possibility that the planet is much older. When geologists observed that it takes a very long time for even a thin layer of sand to be deposited or for erosion to deepen a modern river valley, they realized that it must have taken much more than all of human history to deposit the layers of rock in the Grand Canyon. Indeed, the submergence of coastal Maine discussed at the opening of this chapter turns out to be one of the *faster* geologic processes. Understanding rates of geologic processes thus helped us recognize Earth's vast age hundreds of years before the discovery of radioactivity enabled us to measure the age of a rock in years.

Small changes can have big results when they are repeated over enormous spans of time, as shown by Exercise 1.9. And at the very slow rates at which some geologic processes take place, familiar materials may behave in unfamiliar ways. For example, if you drop an ice cube on your kitchen floor, it behaves brittlely and shatters into pieces. But given time and the weight resulting from centuries of accumulation, ice in a glacier can flow plastically at tens of meters a year (**FIG. 1.10a**). Under geologic conditions and over long enough periods, even layers of solid rock can be bent into folds like those seen in **FIGURE 1.10b**.

FIGURE 1.10 Solid Earth materials change shape over long periods under appropriate conditions.

(a) Athabasca Glacier, Alberta, Canada.

(b) Folded sedimentary rocks in eastern Ireland.

1.6.3 "Life Spans" of Mountains and Oceans

You will learn later that mountains and oceans are not permanent landscape features. Mountains form by uplift or intense folding, but as soon as land rises above the sea, running water, ice, and wind begin to erode it away. When the forces that cause the uplift cease, the mountains are gradually leveled by the forces of erosion. Oceans are also temporary features. They form when continents split and the pieces move apart from one another, and they disappear when the continents on their margins collide. Exercise 1.10 explores the rates at which mountains and oceans form and are then destroyed.

EXERCISE 1.10 | **Rates of Mountain and Ocean Formation**

Name: _____ Section: _____

Course: _____ Date: _____

(a) **Rates of uplift and erosion.** The following questions give you a sense of the rates at which uplift and erosion take place. We will assume that uplift and erosion do not occur at the same time—that the mountains are first uplifted and only then does erosion begin—whereas the two processes actually operate simultaneously.
- If mountains rose by 1 mm/yr, how high would they be (in meters) after 1,000 years? _____ m 10,000,000 years? _____ m   50 million years? _____ m
- The Himalayas now reach an elevation of 8.8 km, and radiometric dating suggests that their uplift began about 45 Ma. Assuming a constant rate of uplift, how fast did the Himalayas rise? _____ km/yr _____ m/yr   _____ mm/yr
- Evidence shows that there were once Himalayan-scale mountains in northern Canada, an area now eroded nearly flat. If Earth were only 6,000 years old as was once believed, how fast would the rate of erosion have had to be for these mountains to be eroded to sea level in 6,000 years? _____ m/yr _____ mm/yr
- Observations of modern mountain belts suggest that ranges erode at rates of 2 mm per 10 years. At this rate, how long would it take to erode the Himalayas down to sea level? _____ years

(b) **Rates of seafloor spreading.** Today the Atlantic Ocean is about 5,700 km wide at the latitude of Boston. At one time, however, there was no Atlantic Ocean because the east coast of the United States and the northwest coast of Africa were joined in a huge supercontinent. The Atlantic Ocean started to form "only" 185,000,000 years ago, as modern North America split from Africa and the two slowly drifted apart in a process called *seafloor spreading*.
- Assuming that the rate of seafloor spreading has been constant, at what rate has North America been moving away from Africa? _____ mm per year   _____ km per million years

---

1.7 Applying the Basics to Interpreting the Earth

The concepts presented in this chapter are the foundations for understanding the topics that will be covered throughout this course. This section shows how you can apply these concepts with a little geologic reasoning to arrive at significant conclusions about the Earth.

1.7.1 Pressure in the Earth

The pull of Earth's gravity produces **pressure**. For example, the pull of gravity on the atmosphere creates a pressure of 1.03 kg/cm^2 (14.7 lb/in^2) at sea level. This means that every square centimeter of the ocean, the land, or your body is affected by a weight of 1.03 kg. We call this amount of pressure 1 **atmosphere** (atm). Scientists commonly specify pressures using a unit called the *bar* (from the Greek *barros*, meaning weight), where 1 bar ≈1 atm. Two kinds of pressure play important roles in the hydrosphere and geosphere. **Hydrostatic pressure**, pressure caused by the weight of water, increases as you descend into the ocean and can crush a submarine in the deep ocean. **Lithostatic pressure**, pressure caused by the weight of overlying rock, increases as you go deeper in the geosphere and is great enough in the upper mantle to change the graphite in a pencil into diamond. The rate at which lithostatic pressure increases is called the **geobaric gradient**, which along with the geothermal gradient plays a major role in determining what materials can exist at depth. Lithostatic pressures are so great that we must measure them in **kilobars** (Kbar), where 1 Kbar = 1,000 bars. Exercise 1.11 shows how pressure varies in the Earth.

EXERCISE 1.11   **Thinking about Pressure in the Earth**

Name: _____   Section: _____
Course: _____   Date: _____

Hydrostatic and lithostatic pressures are caused by the weight of the overlying water and rock, respectively, and these pressures, in turn, depend on the density of the overlying material. You calculated earlier that water has a density of 1 g/cm^3, whereas rock in the crust beneath continents has an average density of about 2.8 g/cm^3.

(a) Why does lithostatic pressure increase more rapidly with depth than hydrostatic pressure?

_____

_____

_____

(b) The geobaric gradient beneath continents is about 1 Kbar for every 3.3 km of depth.
  • On the top of the diagram used in Exercise 1.4 for the *geothermal* gradient, construct a scale showing pressure from 1 to 3 Kbar and draw the *geobaric* gradient in a different color.
  • What is the lithostatic pressure at a depth of 8 km below the surface of a continent? _____ Kbar
  • What are the temperature and pressure conditions in the Earth at a depth of 3.5 km?
    _____ °C   at 5.6 km? _____ °C
    _____ Kbar   _____ Kbar

Metric–English Conversion Chart

To convert U.S. Customary units to metric units	To convert metric units to U.S. Customary units
Length or distance inches × 2.54 = centimeters feet × 0.3048 = meters yards × 0.9144 = meters miles × 1.6093 = kilometers	centimeters × 0.3937 = inches meters × 3.2808 = feet meters × 1.0936 = yards kilometers × 0.6214 = miles
Area in^2 × 6.452 = cm^2 ft^2 × 0.929 = m^2 mi^2 × 2.590 = km^2	cm^2 × 0.1550 = in^2 m^2 × 10.764 = ft^2 km^2 × 0.3861 = mi^2
Volume in^3 × 16.3872 = cm^3 ft^3 × 0.02832 = m^3 U.S. gallons × 3.7853 = liters	cm^3 × 0.0610 = in^3 m^3 × 35.314 = ft^3 liters × 0.2642 = U.S. gallons
Mass ounces × 28.3495 = grams pounds × 0.45359 = kilograms	grams × 0.03527 = ounces kilograms × 2.20462 = pounds
Density lb/ft^3 × 0.01602 = g/cm^3	g/cm^3 × 62.4280 = lb/ft^3
Velocity ft/s × 0.3048 = m/s mph × 1.6093 = km/h	m/s × 3.2804 = ft/s km/h × 0.6214 = mph
Temperature 0.55 × (°F − 32) = °C	(1.8 × °C) + 32 = °F
Pressure lb/in^2 × 0.0703 = kg/cm^2	kg/cm^2 × 14.2233 = lb/in^2
For U.S. Customary units 1 foot (ft) = 12 inches (in) 1 yard (yd) = 3 feet 1 mile (mi) = 5,280 feet	**For metric units** 1 centimeter = 10 millimeters (mm) 1 meter (m) = 100 centimeters 1 kilometer (km) = 1,000 meters 1 milliliter (ml) = 1 cm^3

2

The Way the Earth Works: Examining Plate Tectonics

A path through the Þingvellir Valley in Iceland, the location of a rift valley that marks the crest of the Mid-Atlantic Ridge.

LEARNING
OBJECTIVES

■ Become familiar with the
geographic, geologic, and
geophysical evidence that
led geologists to develop the
plate tectonic theory

■ Explore plate tectonic
processes by observing
modern Earth features

■ Use twenty-first-century
technology to measure the
direction and rate at which
plates are moving today

MATERIALS
NEEDED

■ Tracing paper

■ Colored pencils

■ Ruler with divisions in tenths
of an inch or millimeters

■ Protractor

■ Calculator or pencil and
paper for simple arithmetic

2.1 Introduction

Earthquakes and volcanic eruptions show that Earth is a dynamic planet with enough energy beneath its surface to cause disasters for those who live on its surface. For thousands of years humans wondered about the causes of these events, until the 1960s and 1970s when geologists developed the **theory of plate tectonics**— a unifying theory that answered geologic questions that had puzzled us for many years. Plate tectonics explains the outer layer of the Earth as a group of separate plates that move with respect to each other and change the Earth's surface as they move. At first it was difficult to accept the concept that Earth's oceans, continents, and mountains are only temporary features that move and change over time, as the proposed changes are so slow that they could not be detected. Yet, according to the plate tectonics theory, planet-wide processes break continents apart, open and close oceans, and build and shrink great mountain chains. Local earthquakes and volcanoes are simply results of energy released as these processes occur.

No one ridicules plate tectonics now because evidence proved these processes are happening today, and geologists showed that these processes have been operating for billions of years. In this chapter, we will explore the evidence and the geologic reasoning that led to the plate tectonics theory.

2.2 The Plate Tectonics Theory

Plate tectonics is based on many kinds of information about the Earth that you will examine during this course, including the origin and distribution of different rock types, the topography of the continents and ocean basins, and the geographic distribution of earthquakes and volcanic eruptions. The basic concepts of the theory include the following:

■ Earth's crust and the uppermost part of the layer below it, called the **mantle**, form a relatively rigid outermost layer called the **lithosphere** that extends to a depth of 100 to 150 km.

■ The lithosphere is not a single shell, but consists of several large pieces called **lithosphere plates** or simply, **plates** (**FIG. 2.1**). There are about 12 major

FIGURE 2.1 Earth's major lithosphere plates.

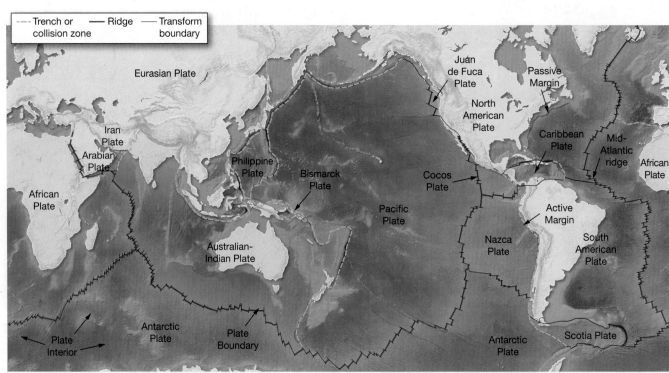

plates that are thousands of kilometers wide and several minor plates that are hundreds of kilometers wide.

- Continental lithosphere is thicker than oceanic lithosphere because continental crust alone (without the mantle component) is 25 to 70 km thick, whereas the oceanic crust is only about 7 km thick.

- The plates rest on the **asthenosphere**, a zone in the upper mantle (**FIG. 2.2**) that, although solid, has such low rigidity that it can flow like soft plastic. The asthenosphere acts as a lubricant, permitting the plates above it to move. Plates move relative to one another at 1 to 15 cm/yr, roughly the rate at which fingernails grow.

FIGURE 2.2 **Cross section showing activity at convergent, divergent, and transform plate boundaries.** At divergent boundaries (ocean ridges), new lithosphere is created; at convergent boundaries (subduction zones or areas of continent-continent-collision), old lithosphere is destroyed; and at transform faults (neutral plate boundaries), lithosphere is neither created nor destroyed.

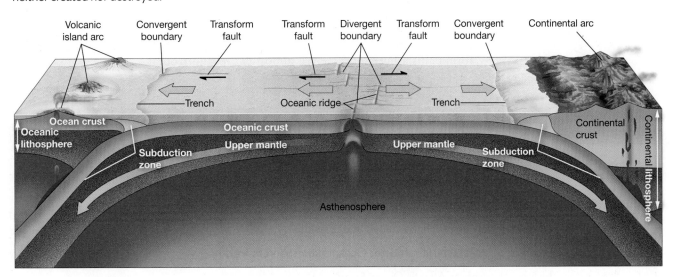

- A place where two plates make contact is called a **plate boundary**, and in some places *three* plates come together at a **triple junction**. There are three different kinds of plate boundaries, defined by the relative motions of the adjacent plates (Figs. 2.1 and 2.2):

1. At a **divergent plate boundary**, plates move away from one another at the axis of submarine mountain ranges called **mid-ocean ridges**, or oceanic ridges. Molten material rises from the asthenosphere to form new oceanic lithosphere at the ridge axis. The ocean grows wider through this process, called **seafloor spreading**, during which the new lithosphere moves outward from the axis to the flanks of the ridge.

2. At a **convergent plate boundary**, two plates move toward each other, and the oceanic lithosphere of one plate sinks (the subducted plate) into the mantle below the other (the overriding plate) in **subduction zones**. The lithosphere of the overriding plate may be oceanic or continental. The boundary between the two plates is a deep-ocean **trench**. At a depth of about 150 km, gases (mostly steam) released from the heated, subducted plate rise into the lower lithosphere. These gases help melt the lower mantle component of the lithosphere, and the resulting magma rises to the surface to produce volcanoes as either a **volcanic island arc** (like the Japanese Islands), where the overriding plate is made of oceanic lithosphere, or as a **continental arc** (like the Andes Mountains), where the overriding plate is made of continental lithosphere.

3. At a **transform boundary**, two plates slide past one another along a vertical zone of fracturing called a **transform fault**. Most transform faults break ocean ridges into segments and are also called *oceanic fracture zones*. A few, however, such as the San Andreas Fault in California, the Alpine Fault in New Zealand, and the Great Anatolian Fault in Turkey, cut through continental plates.

■ Continental crust cannot be subducted because it is too buoyant to "sink" into the mantle. When subduction completely consumes an oceanic plate between two continents, **continental collision** occurs, forming a **collisional mountain belt** like the Himalayas, Alps, or Appalachians. Folding during the collision thickens the crust to the extent that the thickest continental crust is found in these mountains.

■ **Continental rifts** are places where continental lithosphere is stretched and pulled apart in the process of breaking apart at a new divergent margin. If rifting is "successful," a continent splits into two pieces separated by a new oceanic plate, which gradually widens by seafloor spreading.

■ In the **tectonic cycle**, new oceanic lithosphere is created at the oceanic ridges, moves away from the ridges during seafloor spreading, and returns to the mantle in subduction zones. Oceanic lithosphere is neither created nor destroyed at transform faults, where movement is almost entirely horizontal.

EXERCISE 2.1 | **Recognizing Plates and Plate Boundaries**

Name: _____ Section: _____
Course: _____ Date: _____

Using Figure 2.1 as a reference, answer the following questions:

(a) What is the name of the plate on which the United States resides? _____

(b) Does this plate consist of continental lithosphere? Oceanic lithosphere? Or both? _____

(c) Where does the lithosphere of the Atlantic Ocean form? _____

(d) What kind of plate boundary occurs along the west coast of South America? _____

(e) Is the west coast of Africa a plate boundary? _____

2.3 Early Evidence for Plate Tectonics

A simple problem of scale and this low rate of plate movement delayed the discovery of plate tectonics until the late 1960s: lithosphere plates are so big and move so slowly that we didn't realize they were moving. Today there is no question because global positioning satellites and sensitive instruments can measure their directions and rates. We will look at some of the evidence that led geologists to accept the hypothesis that plates move and then see how we deduce the nature and rates of processes at the three types of plate boundaries.

2.3.1 Evidence from the Fit of the Continents

As far back as 500 years ago, mapmakers drawing the coastlines of South America and Africa noted that the two continents looked as if they might have fit together once in a larger continent. Those foolish enough to say it out loud were ridiculed, but today this fit is considered one of the most obvious lines of evidence for plate tectonics. You will explore this in the following exercise.

Name: _____　　Section: _____

Course: _____　　Date: _____

In this exercise, you will examine the shorelines of South America and Africa for evidence of the theory of plate tectonics.

(a) Sketch the shorelines of South America and Africa from the figure below on separate pieces of tracing paper. Rearrange the continents so that they fit as well as possible without overlapping or leaving large gaps. How well do the continents fit? Where are the problem areas?

_____

_____

(b) Are the current shorelines of South America and Africa accurate representations of those continents when they were rifted apart? What factors other than rifting and seafloor spreading could have modified the shape of the current shorelines?

_____

_____

Physiographic map of the South Atlantic Ocean floor and adjacent continents.

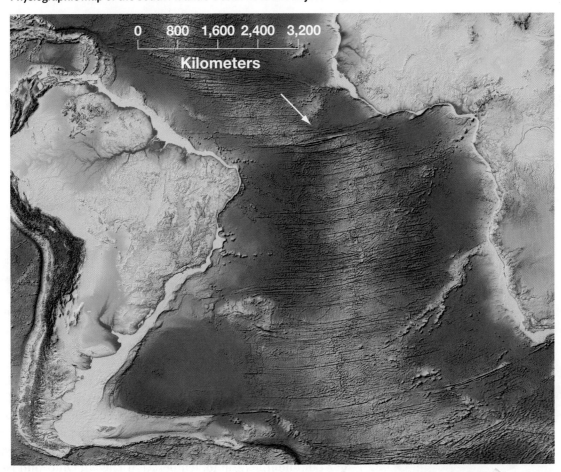

(continued)

Name: _____   Section: _____
Course: _____   Date: _____

(c) Trace the outlines of the continents again, this time using the edges of their continental shelves (the shallow, flat areas adjacent to the land) rather than the shoreline, and attempt to join them. Which reconstruction produces the best fit? In what ways is it better than the other?

_____

_____

_____

(d) Based on this evidence, what is the true edge of a continent?

_____

(e) When you fit the continents together, you could rotate them however you wished. However, this figure also contains clues that show exactly how Africa and South America spread apart. Place the *best-fit* tracings of South America and Africa over those continents on the figure. Now bring them closer to one another until they join, *using the oceanic fracture zones to guide the direction in which you move the two plates.* Do the continents fit well when moved this way?

_____

(f) What does this suggest about the age and origin of the fracture zones?

_____

_____

_____

2.3.2 Evidence from Reconstructing Paleoclimate Belts

FIGURE 2.3 Earth's major climate zones.

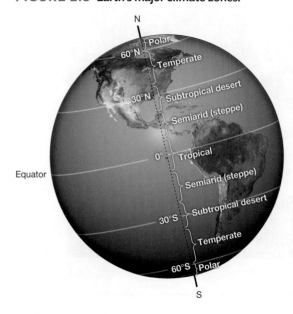

Earth's climate zones are distributed symmetrically about the equator today (**FIG. 2.3**), and the tropical, temperate, and polar zones support animals and plants unique to each environment. For example, walruses live in Alaska but palm trees don't, and large coral reefs grow in the Caribbean Sea but not in the Arctic Ocean. Climate zones also produce distinctive rock types: regional sand dunes and salt deposits form in arid regions, rocks made of debris deposited by continental glaciers accumulate in polar regions, and coal-producing plants grow in temperate and tropical forests.

However, geologists have found many rocks and fossils far outside the modern climate belts where we expect them to form. For example, 390-million-year-old (390-Ma) limestone containing reef-building organisms crops out along the entire Appalachian Mountain chain, some far north of where reefs exist today; 420-Ma salt underlies humid, temperate Michigan, Ohio, and New York; and deposits of 260- to 280-Ma continental glaciers are found in near-equatorial Africa, South America, India, and Madagascar.

According to the principle of uniformitarianism, these ancient salt deposits, reefs, and glacial deposits should have formed in locations similar to those where they form today. Either uniformitarianism doesn't apply to these phenomena or the landmasses have moved from their original climate zone into another. Observations like these led Alfred Wegener, a German meteorologist, to suggest in the 1920s that the continents had moved—a process he called *continental drift*. Modern geologists interpret these anomalies as the result of plate motion—when continents change position on the globe by moving apart during seafloor spreading or by coming together as subduction closes an ocean.

2.3.3 Geographic Distribution of Earthquakes and Active Volcanoes

By the mid-1800s, scientists realized that Earth's active volcanoes are not distributed randomly. Most are concentrated in narrow belts near the edges of continents, like the chain called the Ring of Fire surrounding the Pacific Ocean, while the centers of most continents have none (**FIG. 2.4**). In the late 1900s, technological advances provided new insights into the nature of the ocean floors, and we learned that submarine volcanoes are found in every major ocean basin.

More clues soon came from seismologists (geologists who study earthquakes). Records of where earthquakes occurred over a certain X period of time revealed a pattern. While this pattern (**FIG. 2.5**) was more complex than that for volcanoes, particularly in the deep ocean and continental interiors, it demonstrated important similarities. Something unique was happening along the volcanic chains and where earthquakes occurred, but geologists couldn't agree on what that was. Exercise 2.3 follows the reasoning geologists used to build the basic framework of the plate tectonic theory.

FIGURE 2.4 Worldwide distribution of active volcanoes. The solid, orange line surrounds the Ring of Fire.

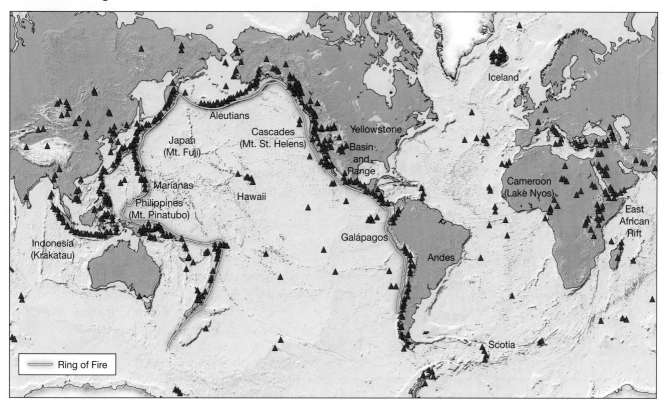

FIGURE 2.5 Worldwide distribution of earthquakes.

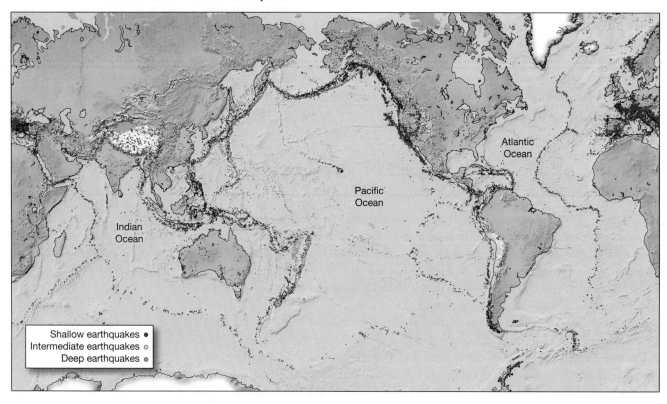

Shallow earthquakes ●
Intermediate earthquakes ○
Deep earthquakes ◉

EXERCISE 2.3	Putting the Early Evidence Together

Name: _____   Section: _____
Course: _____   Date: _____

In this exercise, you will study Figures 2.4 and 2.5 in order to understand more of the initial reasoning for plate tectonics theory.

(a) Look at the volcanic and earthquake activity that occurred on the west coast of North America. How do these compare? Examine the earthquake and volcanic material in the center of the United States. How do these compare?

_____

_____

(b) Examine the chain of volcanoes and line of earthquakes in the middle of the Atlantic Ocean in Figures 2.4 and 2.5. From reading the past sections, describe what you believe is causing these. _____

_____

_____

(c) Geologists have found rocks and fossils of species native to Africa in South America and rocks and fossils native to South America in Africa. Considering your response to (b), explain what they theorized had happened.

_____

_____

2.4 Modern Evidence for Plate Tectonics

The geographic fit and paleoclimate evidence convinced some geologists that plate tectonics was a reasonable hypothesis, but more information was needed to convince the rest. That evidence came from an improved understanding of Earth's magnetic field, the ability to date ocean-floor rocks, careful examination of earthquake waves, and direct measurements of plate motion using global positioning satellites and other exciting new technologies. The full body of evidence has converted nearly all doubters to ardent supporters.

2.4.1 Evidence for Seafloor Spreading: Oceanic Magnetic Anomalies

Earth has a magnetic field that can be thought of as having "north" and "south" poles like a bar magnet (**FIG. 2.6**). Navigational compasses are aligned by magnetic lines of force that emanate from one pole and reenter the Earth at the other. The magnetic field has been known for centuries, but two discoveries about the field in the mid-twentieth century provided new insights about how the field works and, soon afterward, the evidence that confirmed the plate tectonic theory.

First, geologists learned that when grains of magnetite or hematite crystallize, they are aligned magnetically parallel to Earth's lines of force. Some rocks that contain magnetite or hematite therefore preserve a weak record of Earth's ancient magnetic field, a record called **paleomagnetism**. Then geologists learned that the magnetic field reverses polarity from time to time, so that what is now the north magnetic pole becomes the south magnetic pole and vice versa. During periods of **normal polarity**, the field is the same as it is today, but during periods of **reversed polarity**, a compass needle that points to today's north magnetic pole would swing around and point south. So by finding and determining the polarities of rocks throughout the world, geologists have accurately learned the dates of the magnetic reversals back to 4.0 Ma (**FIG. 2.7**).

This was interesting and certainly surprising, but how does it support plate tectonics? Earth's magnetic field varies irregularly on the continents, with a complex pattern of areas where the field is anomalously stronger or weaker than average (positive and negative **magnetic anomalies**). In contrast, research in the late 1960s discovered that the pattern of magnetic anomalies in the oceans is much simpler and more regular than that on the continents—parallel linear belts of positive and negative anomalies

FIGURE 2.6 Earth's magnetic field is defined by magnetic lines of force shown by the arrows.

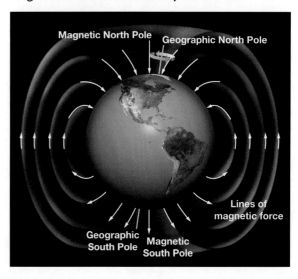

FIGURE 2.7 Radiometric dating of lava flows shows magnetic reversals for only the past 4 Ma.

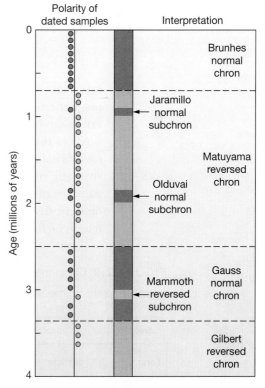

Major intervals of positive or negative polarity are called *chrons* and are named after scientists who contributed to the understanding of the magnetic field. Short-duration reversals are called *subchrons*.

FIGURE 2.8 Magnetic anomaly stripes in the Atlantic and Pacific oceans.

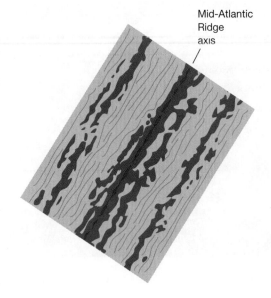

Mid-Atlantic Ridge axis

The pattern of anomalies is symmetrical, relative to mid-ocean ridges.

Canada

United States

Crest of Juan de Fuca Ridge

Crest of Gorda Ridge

dark gray = positive (+) anomaly
light gray = negative (–) anomaly

dark gray = positive (+) anomaly
light gray = negative (–) anomaly

(a) The Mid-Atlantic Ridge southwest of Iceland.

(b) The Juan de Fuca and Gorda ridges in the North Pacific off the state of Washington and the province of British Columbia.

informally called **magnetic stripes** (**FIG. 2.8a, b**). The measured strength of the magnetic field in the oceans is the result of two components: (1) Earth's modern magnetic field strength and (2) the paleomagnetism (remanent magnetic field) of the oceanic crust. If the paleomagnetic polarity of a rock is the same as today's magnetic field, the rock's weak paleomagnetism *adds* to the modern field strength, resulting in an observed magnetic field *stronger* than today's field. If the paleomagnetic polarity is reversed, the rock's paleomagnetism *subtracts* from the modern field, and the result is a measurement *weaker* than the average modern magnetic field (**FIG. 2.9**).

FIGURE 2.9 Components of Earth's magnetic field strength in the oceans.

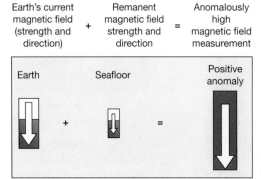

Earth's current magnetic field (strength and direction) + Remanent magnetic field strength and direction = Anomalously high magnetic field measurement

Earth's current magnetic field + Remanent magnetic field strength (reversed polarity) = Anomalously low magnetic field measurement

Earth   Seafloor   Positive anomaly

Earth   Seafloor   Negative anomaly

This pattern of reversals has been found at every oceanic ridge, proving that the reversals are truly worldwide events, and this paleomagnetism was convincing evidence for seafloor spreading. As lava erupting at ocean ridges cools, it records the magnetic field polarity in effect at that time. If Earth's polarity reverses, new lava will adopt the new polarity, and the older lava will yield a negative magnetic anomaly as shown in Figure 2.9.

Using this information, Exercise 2.4 shows the reasoning by which geologists connected magnetic anomaly stripes in the oceans with the plate tectonic model.

EXERCISE 2.4   **Interpreting Ocean Ridge Magnetic Stripes**

Name: _____   Section: _____
Course: _____   Date: _____

In Figure 2.8a, compare the orientation of the magnetic anomaly stripes for the Mid-Atlantic Ridge with the orientation of the ridge crest (illustrated with a red line.) In Figure 2.8b, do the same for the Juan de Fuca Ridge and its anomalies.

(a) Are the individual anomalies oriented randomly? Are they parallel to the ridge crests? Oblique to the ridge crests?

_____

_____

_____

(b) Explain how the process of seafloor spreading can produce these orientations and relationships.

_____

_____

_____

(c) Some magnetic stripes are wider than others. Knowing what you do about seafloor spreading and magnetic reversals, suggest an explanation. _____

_____

_____

2.4.2 Direct Measurement of Plate Motion

Skeptics can no longer argue that Earth's major features are fixed in place. Satellite instruments can measure Earth's features with precision not even dreamed of 10 years ago, and they make it possible to measure the directions and rates of plate motion. Data for the major plates are shown in **FIG 2.10**. The length of each arrow indicates the relative rate of plate motion caused by seafloor spreading. We will see later how geologists were able to deduce the same information using other data.

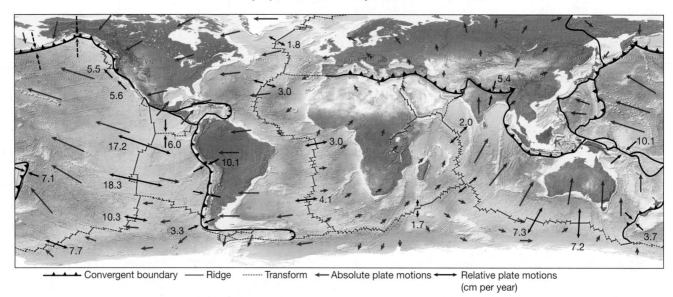

——▲—▲—▲ Convergent boundary —— Ridge ------- Transform ◄—— Absolute plate motions ◄——► Relative plate motions
(cm per year)

2.5 Processes at Plate Boundaries Revealed in Earth Features

The next few exercises examine the three kinds of plate boundaries and show how geologists deduce details of their geometry, the rates of plate motion involved, and their histories. Let's start with information that we can gather about seafloor spreading.

EXERCISE 2.5   **Estimating Seafloor Spreading Rates**

Name: _____   Section: _____
Course: _____   Date: _____

The South Atlantic Ocean formed by seafloor spreading at the Mid-Atlantic Ridge. Geologists can get a rough estimate of the spreading rate (i.e., the relative motion of South America with respect to Africa) by measuring the distance between the two continents in a direction parallel to the fracture zones and determining the time over which the spreading occurred.

(a) Measure the distance between South America and Africa along the fracture zone (indicated by the arrow) on the map in Exercise 2.2 (see p.29) _____ km

The oldest rocks in the South Atlantic Ocean, immediately adjacent to the African and South American continental shelves, are 120,000,000 years old.

(b) Calculate the average rate of seafloor spreading for the South Atlantic Ocean over its entire existence. Express your answer in _____ km/million years = _____ km/yr = _____ cm/yr = _____ mm/yr.

(c) Assuming someone born today lives to the age of 100, how much wider will the Atlantic Ocean become during his or her lifetime? _____ cm

---

2.5.1 Seafloor Spreading and Continental Rifting

These results dramatically demonstrate how slowly the South Atlantic Ocean is spreading and why plate tectonics met widespread disbelief initially. They also reinforce the importance of understanding the vast expanse of geologic time discussed in Chapter 1. Even extremely slow processes can have great impact given enough time to operate!

A new ocean forms when rifting takes place beneath a continent. The continental crust first thins and then breaks into two pieces separated by an oceanic ridge. As seafloor spreading proceeds, an ocean basin grows between the fragments of the original continent. This process is in an early stage today in eastern Africa.

Exercises 2.6, 2.7, and 2.8 examine these concepts and processes further.

EXERCISE 2.6	Comparing Seafloor Spreading Rates of Different Ocean Ridges

Name: _____   Section: _____

Course: _____   Date: _____

Magnetic reversals are found worldwide, so magnetic stripes should be the same width in every ocean *if the rate of seafloor spreading is the same at all ridges*. If a particular anomaly is wider in one ocean than another, however, it must result from faster spreading. The figure on the right shows simplified magnetic stripes from the South Atlantic and Pacific oceans, the ages of the rocks, and the distance from the spreading center (the red line). For simplicity, only the most recent 80 million years of data are shown for the two oceans, and we will only estimate the spreading rate for that time span.

(a) Measure the width of the South Atlantic Ocean _____ km

(b) Estimate the average rate at which the South Atlantic Ridge has been opening over the 80 million years for which data are provided. _____cm/yr.

(c) Now look at the data for the South Pacific Ocean and its spreading center, the East Pacific Rise. Considering the width of this ocean, will the spreading rate be the same, greater, or less than that of the South Atlantic? _____ Explain.

_____

_____

_____

(d) Now get the details. Measure the width of the South Pacific Ocean. _____ km

(e) What is the spreading rate of the East Pacific Rise? _____ km/million years

These spreading rates are typical of the range measured throughout the world's oceans and represent "fast spreaders" and "slow spreaders."

Map view of magnetic anomaly stripes in two oceans.

(red line = axis of ocean ridges; black = positive anomaly; white = negative anomaly)

The figures below show bathymetric maps of parts of the East Pacific Rise and Mid-Atlantic Ridge at the same scale.

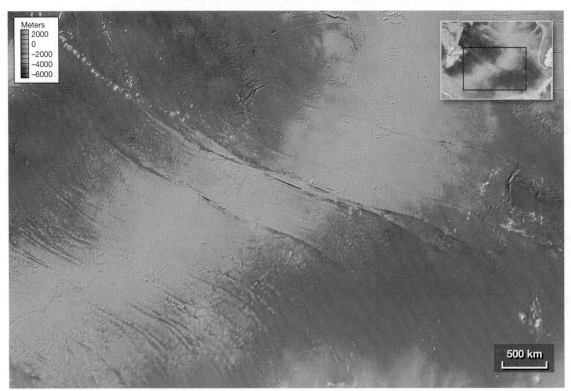

(a) The East Pacific Rise in the South Pacific Ocean.

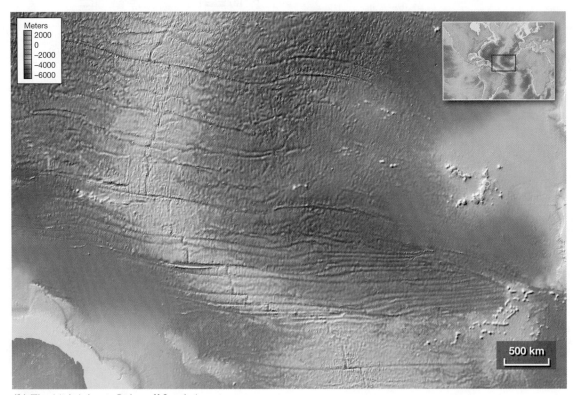

(b) The Mid-Atlantic Ridge off South America.

(continued)

Name: _____   Section: _____
Course: _____   Date: _____

(a) Ocean ridges typically have a rift valley at their axes—a valley created when two continents split. Which of the two ridges in the previous figures has the deepest and longest rift valley?

_____

_____

_____

_____

Different spreading rates cause variations in shape because of the way the cooling lithosphere behaves. Lithosphere near the ridge axis is young, thin, still hot, and therefore has a lower density than older, colder lithosphere far from the axis. As a result, the ridge axis area floats relatively high on the underlying asthenosphere, and the water above it is relatively shallow. As seafloor spreading moves the oceanic lithosphere away from the ridge axis, the rocks cool and get thicker and denser. The farther it is from the ridge axis, the lower the oceanic lithosphere sits on the asthenosphere and the deeper the water will be. This concept is known as the **age-versus-depth relationship**.

(b) Keeping in mind the age-versus-depth relationship, why is the belt of shallow sea wider over the East Pacific Rise than over the Mid-Atlantic Ridge?

_____

_____

_____

_____

(c) On the graph provided, plot depth (on the vertical axis) against distance from the ridge axis (on the horizontal axis) for both the East Pacific Rise and the Mid-Atlantic Ridge. Use five to ten points for each ridge, marking the points you use on the maps on page 38. Connect the dots for the East Pacific Rise with red pencil and those for the Mid-Atlantic Ridge with green pencil to make cross sections of each ridge.

(d) Does the rate at which the depth increases with distance from the ridge stay the same over time, decrease over time, or increase over time?

_____

_____

_____

_____

Name: _____ Section: _____
Course: _____ Date: _____

Examine the bathymetric/physiographic map of the region that includes eastern Africa, the Red Sea, and the western Indian Ocean below. The Arabian Peninsula was split from Africa by the Red Sea rift, which continues to widen today. The East African rift was an earlier attempt that failed to split off a piece of the continent.

Red Sea rift zone.

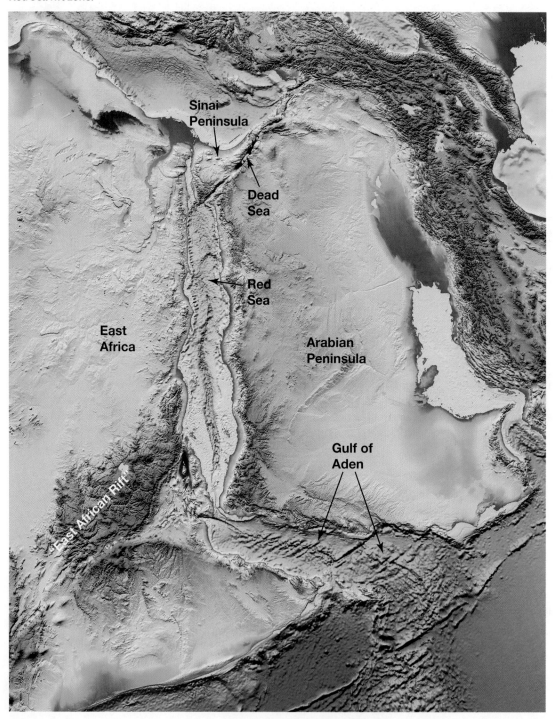

(continued)

Name: _____   Section: _____
Course: _____   Date: _____

Rifting has split Africa from the Arabian Peninsula to form the Red Sea. A new ocean ridge and ocean lithosphere have formed in the southern two-thirds of the Red Sea. A narrow belt of deeper water defines the trace of this ridge. At the northern end of the Red Sea, the ridge/rift axis is cut by a transform fault that runs along the eastern side of the Sinai Peninsula and through the Dead Sea.

(a) Use a red line to show the trace of the Red Sea rift/ridge axis. Use a purple line to show the trace of the Dead Sea transform fault.

(b) The narrow ocean bordering the southeast edge of the Arabian Peninsula is the Gulf of Aden. Use red and purple lines to trace the ridge segments and transform fault in this narrow sea.

(c) Based on the geometry of the ridges and transform faults in the Red Sea and Gulf of Aden, draw an arrow showing the motion of the Arabian Peninsula (the Arabian Plate) relative to Africa.

2.5.2 Subduction Zones: Deducing the Steepness of Subduction

When two oceanic plates collide head-on, one is subducted beneath the other and returns to the mantle, completing the tectonic cycle begun when the ocean crust initially erupted at a mid-ocean ridge. Melting occurs above the subducted plate when it reaches a depth of about 150 km. As a result, the volcanic arc forms at some distance from the ocean trench. The area between the volcanic arc and trench typically contains an accretionary prism, which is composed of highly deformed sediment scraped off the sinking plate, and a forearc basin, in which debris eroded from the arc accumulates (**FIG. 2.11**).

No two subduction zones are identical. Differences can include the width of the accretionary prism, the rate of subduction, and the steepness of the subducted plate as it moves into the mantle. Earthquakes occur in the subducted plate as it moves, concentrated in what geologists call a *Wadati-Benioff zone*. We can track the plate as it is subducted by locating the depth and location of the earthquake foci (points where the energy is released).

FIGURE 2.11 Anatomy of an island arc–trench system.

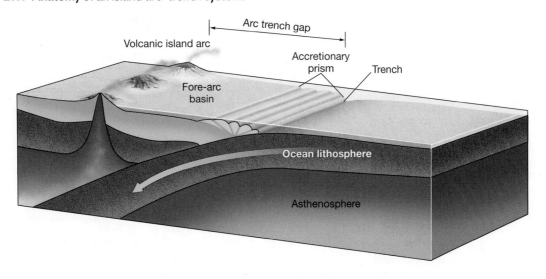

Even without any earthquake information, it is possible to estimate the steepness of a subduction zone. All you need is an accurate map of the seafloor and a little geologic reasoning based on knowledge of its arc-trench system. We will see how this is done in Exercise 2.9.

EXERCISE 2.9   **Estimating the Steepness of Subduction Zones**

Name: _____   Section: _____

Course: _____   Date: _____

(a) Based on island arc–trench geometry (Fig. 2.11) and the fact that melting typically occurs at about the same depth in subduction zones, what is the major factor that controls the width of the arc-trench gap? Explain. _____

_____

_____

(b) Based on your answer to (a), sketch two island arc–trench systems, one with a wider arc-trench gap than the other.

Narrow arc-trench gap   Wide arc-trench gap

The Aleutian Island arc extends westward from mainland Alaska into the Pacific Ocean. Profiles across four segments of the Aleutian arc are given in the illustration that follows, which shows the positions of the volcanic arc and trench for each segment. Gray dots in the profile of the Amchitka segment show the earthquakes in the Wadati-Benioff zone.

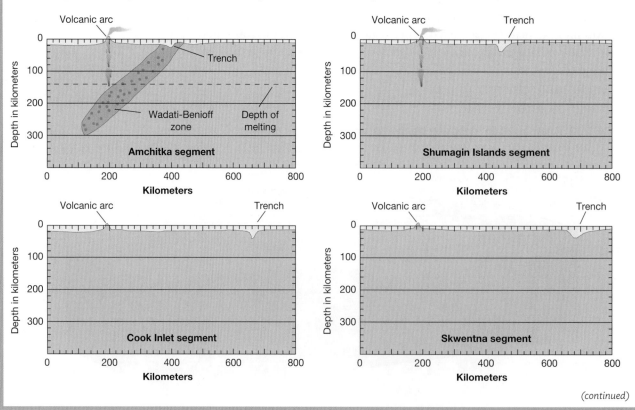

Name: _____   Section: _____

Course: _____   Date: _____

(c) Using this earthquake information, you will see the boundaries of the subducted plate sketched on the Amchitka profile. Assuming that the deepest part of the Aleutian Trench is at the middle of the trench, explain the reason for these boundaries.

_____

_____

_____

(d) The volcanic arc sits directly above the point where the subducted plate begins to melt. At what depth does melting apparently begin beneath the Amchitka segment of the arc? _____ km. How does this point relate to the position of the volcanic arc?

_____

_____

_____

The positions of volcanic arcs and trenches are shown for three other segments of the Aleutian arc. Although earthquake information is not available for these segments, you have enough information to sketch the subducted plates in all three areas.

(e) Sketch the outlines of the subducted plate for each segment in the profiles.

Hints:
   • Assume that melting occurs at the same depth in each segment.
   • Draw a vertical line from the volcanic arc to the melting depth.
   • Draw a horizontal line from the melting depth.
   • Draw the subducted slab so that its upper surface passes through the intersection point.
   • Using your protractor, measure the angle of subduction downward from the horizontal.

(f) From the profiles, record the following data in the table provided:
   • The width of the arc-trench gap
   • The angle of subduction

	Amchitka	Shumagin Islands	Cook Inlet	Skwentna
Arc-trench gap				
Subduction angle				

(g) Based on this information, what is the relationship between the steepness of subduction and the width of the gap between the volcanic island arc and the trench?

_____

_____

_____

2.5.3 Transform Faults

We have thus far looked at transform faults only in the oceans—the ocean fracture systems—but some also cut continental lithosphere. Continental transform faults can be found in New Zealand (the Alpine Fault), Turkey (the Great Anatolian Fault), and Haiti (Enriquillo-Plantain Garden Fault). However, the most famous of these is the San Andreas Fault system of California, an active continental transform fault that has caused major damage and loss of life over the past 100 years. The San Andreas Fault *system* (shown in Exercise 2.10) is not a single fault but rather a zone containing several faults. It extends for more than 1,000 km, connecting segments of the Juan de Fuca Ridge and Cascade Trench at its northern end to an unnamed ridge segment in the Gulf of California to the south.

EXERCISE 2.10 **Estimating the Amount and Rate of Motion in a Continental Transform Fault**

Name: _____   Section: _____
Course: _____   Date: _____

The more we know about the history of a continental transform fault close to heavily populated regions, the better we can prepare for its next pulse of activity. Geologists try to find out how long a continental transform fault has been active, how much it offsets the plates it separates, and how fast it has moved in the past.

Geologists estimate that the San Andreas Fault system has been active for about 20 million years. This exercise shows how geologists use geologic markers cut by faults to measure the amount and rate of motion along a transform fault.

(a) Simplified geologic setting of the San Andres Fault system.

(b) Amount of fault movement indicated by offset bodies of identical rock.

(continued)

Name: _____ Section: _____

Course: _____ Date: _____

Field geologists have mapped an active continental transform fault (red line in the figure below) for several hundred kilometers. A 50-million-year-old (50-Ma) body of granite and a 30-Ma vertical layer of marble have been offset by the fault as shown.

(a) Draw arrows to indicate the direction in which the plate on the northeast side of the fault has moved relative to the plate on the southwest side.

(b) Measure the amount of offset of the 50-Ma granite body. _____ km

(c) The geologists have proved that faulting began almost immediately after the granite formed and continues today. If the fault blocks moved at a constant rate for the past 50 million years, calculate the rate of offset. _____ km/million years; mm/year _____

(d) Was the rate constant for 50 million years? Measure the offset of the 30-Ma marble. _____ km

Geologic markers displaced by a continental transform fault.

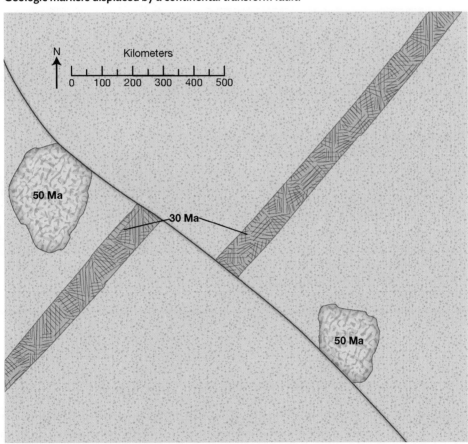

(e) Assume that the fault blocks moved at a constant rate for the past 30 million years. Offset rate: _____ km/30 million years; _____ km/million years; _____ mm/year

(continued)

Name: _____   Section: _____

Course: _____   Date: _____

(f) Compare the two rates. Has the rate of fault offset been constant or has it increased or decreased over time? Explain.

_____

_____

_____

Now let's look at the San Andreas Fault system again (the maps at the beginning of this exercise):

(g) On map (a), draw arrows to indicate the direction along the San Andreas Fault in which the Pacific and North American Plates have been offset.

(h) San Francisco and Los Angeles are on opposite sides of the San Andreas Fault. If the San Andreas Fault is undergoing offset at the rates that you measured above, how many years will it take before the two cities are directly opposite one another? _____

? What Do You Think Because of the earthquake threats from the San Andreas Fault, the city of San Francisco has rigorous building codes requiring that buildings be designed to withstand earthquakes in the area. On the other side of the country, however, the codes are less rigorous. An earthquake in Virginia in August 2011 was felt along most of the east coast of the United States, prompting questions about whether existing building codes should be changed to meet rigorous San Francisco standards. Meeting San Francisco standards makes construction a lot more expensive than that in New York. Imagine you had to make a recommendation to the New York City Council about whether they should or should not adopt San Francisco building codes. Using the information you have in this chapter, on a separate sheet of paper, note what would your recommendation be and why?

2.5.4 Hotspots and Hot-Spot Tracks

Paleoclimate anomalies show that plates have moved, magnetic stripes help measure the rate of seafloor spreading, and satellites measure the rates and directions of plate motion today. But how can we determine if a plate always moved at its current rate? Or if it has changed direction over time? The answers come from the study of hot-spot volcanic island chains.

FIGURE 2.12 shows how hot-spot island chains form. Each volcano in the chain forms at a **hot spot**—an area of unusual volcanic activity not associated with processes at plate boundaries. The cause is still controversial, but many geologists propose that hot spots form above a narrowly focused source of heat called a **mantle plume**—a column of very hot rock that rises by slow plastic flow from deep in the mantle.

FIGURE 2.12 Origin of hot-spot island chains and seamounts.

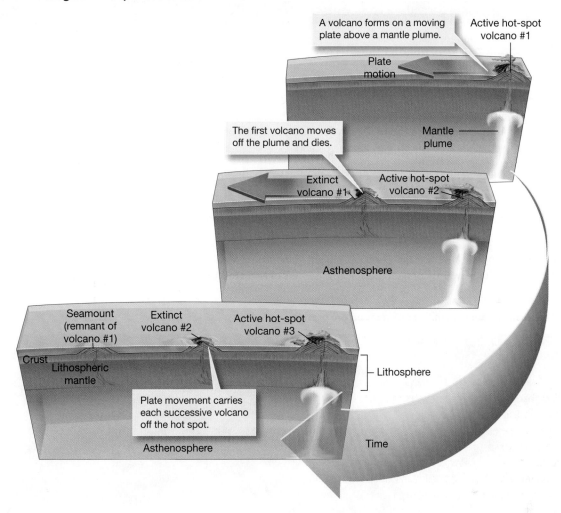

When the plume reaches the base of the lithosphere, it melts the lithosphere rock and produces magma that rises to the surface, erupts, and builds a volcano. The plume is thought to be relatively motionless. If the plate above it moves, the volcano is carried away from its magma source and becomes extinct. A new volcano then forms above the hot spot until it, too, is carried away from the hot spot.

Over millions of years, a chain of volcanic islands forms, the youngest at the hot spot, the oldest farthest from it. As the volcanoes cool, they become denser, subside, and are eroded by streams and ocean waves. Eventually, old volcanoes sink below the ocean surface, forming **seamounts**. The chain of islands and seamounts traces plate motion above the hot spot, just as footprints track the movement of animals.

For Example, the Hawaiian Islands are the youngest volcanoes in the Hawaiian–Emperor seamount chain. Most of the older volcanoes are seamounts detected by underwater oceanographic surveys. The Hawaiian–Emperor seamount chain tracks the motion of the Pacific Plate and lets us interpret Pacific Plate motion for a longer time span than that recorded by the Hawaiian Islands alone.

In Exercises 2.11 and 2.12, you will examine both the Hawaiian Islands (as hot-spot volcanic islands) and the Hawaiian-Emperor seamount chain.

Name: _____   Section: _____
Course: _____   Date: _____

The Hawaiian Islands, located in the Pacific Ocean far from the nearest oceanic ridge, are an excellent example of hot-spot volcanic islands (see figure below). Volcanoes on Kauai, Oahu, and Maui haven't erupted for millions of years, but the island of Hawaii hosts five huge volcanoes, one of which (Kilauea) has been active for the past 31 years. In addition, a new volcano, Loihi, is growing on the Pacific Ocean floor just southeast of Kilauea. As the Pacific Plate moves, Kilauea will become extinct and Loihi will be the primary active volcano.

(a) Where is the Hawaiian hot-spot plume located today relative to the Hawaiian Islands? Explain your reasoning.

_____

_____

(b) Draw a line connecting the volcanic center (highlighted in red) on Maui to the spot between Kilauea and Loihi (the volcanic center for Hawaii). Connect the volcanic centers of Maui to Molokai, Molokai to Oahu, and Oahu to Kauai as well.

Ages of Hawaiian volcanoes in millions of years before present.

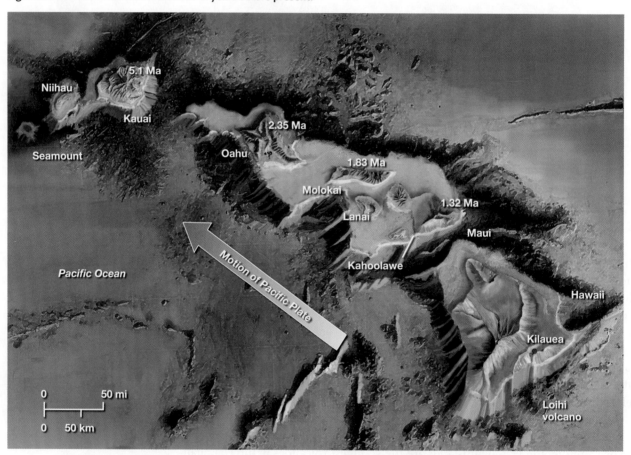

(c) Geologists use the *azimuth system* to describe direction precisely. In the azimuth system, north is 0°, east is 090°, south is 180°, and west is 270°, as shown in the following figure. The green arrow points to 052° (northeast). To practice using the azimuth system, estimate and then measure the directions shown by arrows A through D. Use the protractor in your toolkit.

Name: _____   Section: _____
Course: _____   Date: _____

The azimuth system.

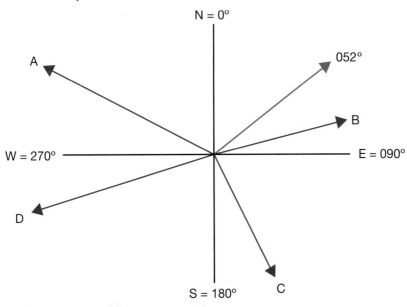

(d) Measure the direction and distance between the volcanic centers of the Hawaiian Islands using a ruler, a protractor, and the map scale. Calculate the rate of plate motion (distance between volcanoes divided by the time interval between eruption ages) and fill in **Table 2.1.** Express the rates in millimeters per year.

TABLE 2.1 Movement of the Pacific Plate over the Hawaiian hot spot.

	Distance between volcanic centers (km)	Number of years of plate motion	Rate of plate motion (mm/yr)	Azimuth direction of plate motion (e.g., 325°)
Hawaii to Maui				
Maui to Molokai				
Molokai to Oahu				
Oahu to Kauai				

Name: _____   Section: _____
Course: _____   Date: _____

The oldest volcano of the Hawaiian–Emperor seamount chain was once directly above the hot spot but is now in the northern Pacific, thousands of kilometers away (see the figure below). Seamount ages show that the hot spot has been active for a long time and reveal the direction and rate at which the Pacific Plate has moved.

(a) What evidence is there that the Pacific Plate has not always moved in the same direction?

_____

_____

_____

(b) How many years ago did the Pacific Plate change direction? Explain your reasoning.

_____

_____

_____

Physiography of the Pacific Ocean floor, showing ages of volcanoes in the Hawaiian–Emperor seamount chain.

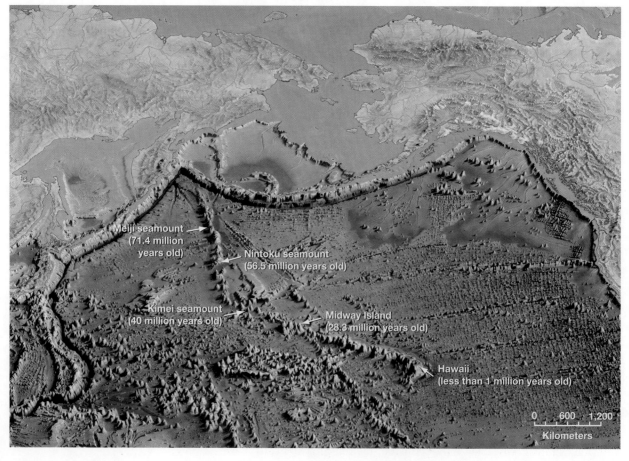

Meiji seamount
(71.4 million
years old)

Nintoku seamount
(56.5 million years old)

Kimei seamount
(40 million years old)

Midway Island
(28.3 million years old)

Hawaii
(less than 1 million years old)

0   600   1,200
Kilometers

(continued)

Name: _____   Section: _____
Course: _____   Date: _____

(c) Based on the information in the map on page 50, in what direction did the Pacific Plate move originally?

_____

(d) How far has the Meiji seamount moved from the hot spot? Explain your reasoning.

_____

_____

_____

(e) At what rate has the Pacific Plate moved
   (i)   based on data from the Hawaii-Midway segment? _____ km/Ma
   (ii) based on data from the Hawaii-Kimei segment? _____ km/Ma
   (iii) based on data from the Kimei-Meiji segment? _____ km/Ma

(f) Has the Meiji seamount moved at a constant rate? Explain your reasoning.

_____

_____

_____

(g) In what direction is the Pacific Plate moving today? Explain your reasoning.

_____

_____

_____

(h) Assuming that the current direction of motion continues, what will be the eventual fate of the Meiji seamount? Explain in as much detail as possible.

_____

_____

_____

_____

2.6 Active versus Passive Continental Margins

Earthquakes and volcanic eruptions are common on the west coast of North America, but there are no active volcanoes on the east coast, and earthquakes there are rare. Geologists call the west coast a (tectonically) **active continental margin** and the east coast a **passive continental margin**. These differences can be found on several continents. Most passive continental margins have broad continental shelves, whereas active continental margins typically have narrow continental shelves.

In Exercise 2.13, you will examine why this phenomena occurs.

Name: _____ **Section:** _____
Course: _____ **Date:** _____

Based on what you know about plate tectonics, explain why some continental margins are active and others passive.

(a) Are all continental coastlines plate boundaries? Explain.

_____

_____

_____

(b) Compare the west coast of South America with the west coast of Africa. Which has a broad continental shelf? A narrow shelf? Which is close to an ocean trench? Which coast would you expect to have the most earthquakes? Explain.

_____

_____

_____

Physiographic map of the South Atlantic Ocean floor and adjacent continents.

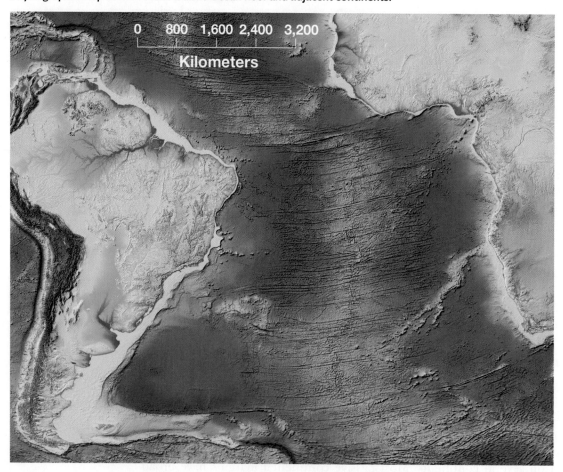

(continued)

EXERCISE 2.13   Why Are There More Earthquakes and Volcanoes on One Side of Some Continents than on the Other? (continued)

Name: _____   Section: _____

Course: _____   Date: _____

(c) On **FIGURE 2.13**, a plate tectonic map of the world, label active continental margins with a red letter A and passive margins with a blue letter P. Two have been identified to get you started.

FIGURE 2.13 Earth's major lithosphere plates.

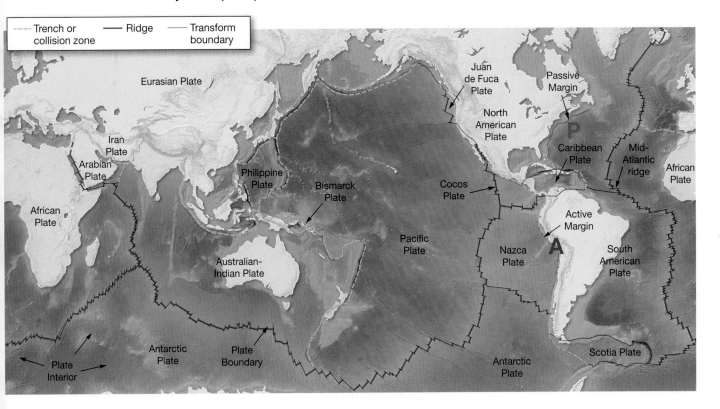

3 Minerals

Giant gypsum crystals in the Cave of the Crystals, Chihuahua, Mexico

- Understand what makes minerals different from other materials
- Become familiar with and learn to describe the physical properties of minerals
- Identify the most common rock-forming minerals using a systematic approach
- Recognize the economic value of minerals
- Learn what minerals can reveal about Earth processes and Earth history

MATERIALS NEEDED

- Sets of mineral and rock specimens
- Hand lens, streak plate, glass plate, knife or steel nail, and a penny to determine the physical properties of minerals
- Dilute hydrochloric acid (HCl) for a simple chemical test
- Small magnet

3.1 Introduction

This chapter begins our study of Earth materials. It starts by examining the different kinds of materials in the geosphere and then focuses on minerals, the basic building blocks of most of the Earth. You will learn how minerals are different from other substances, how to study their physical properties, and how to use those properties to identify common minerals.

3.2 Classifying Earth Materials

Imagine that an octopus is swimming in the ocean when a container falls off a freighter overhead. The container breaks up and spills its entire cargo of sneakers, sandals, flip-flops, shoes, moccasins, and boots into the sea. The octopus is curious about these objects and wants to learn about them. How would it classify them? Remember, an octopus doesn't have heels or toes, has eight legs, doesn't understand "left" and "right," and doesn't wear clothes. One system might be to separate items that are mostly enclosed (shoes, boots, sneakers, moccasins) from those that are open (sandals, flip-flops). Another might be to separate objects made of leather from those made of cloth; or brown objects from black ones; or big ones from small ones. There are many ways to classify footwear, some of which might lead our octopus to a deeper understanding of the reasons for these differences.

Early geologists faced a similar task in the seventeenth century when they began to study Earth materials systematically by classifying them. Why classify things? Because classification shows us relationships between things that lead to understanding them and the processes by which they were made. Biologists classify organisms, art historians classify paintings, and geologists classify Earth materials. Exercise 3.1 introduces the thought processes involved in developing a classification scheme.

EXERCISE 3.1	Classifying Earth Materials

Name: _____   Section: _____

Course: _____   Date: _____

(a) Examine the specimens of Earth materials provided by your instructor. Group them into categories you believe are justified by your observations, and explain the criteria you used to set up the groups.

Group	Defining criteria for each group	Specimens in group

(continued)

Name: _____   Section: _____
Course: _____   Date: _____

(b) Compare your results with those of others in the class. Did you all use the same criteria? Are your classmates' specimens in the same groups as yours?

_____

_____

(c) What does your comparison tell you about the process of classification?

_____

_____

_____

3.3 What Is a Mineral and What Isn't?

Most people know that the Earth is made of minerals and rocks, but they don't know the difference between them. The words *mineral* and *rock* have very specific meanings to geologists, often much more precise than those used in everyday language. For example, what a dietitian calls a mineral is not a mineral to a geologist. To most geologists, a **mineral** is a naturally occurring homogeneous solid formed by geologic processes with an ordered internal arrangement of atoms, ions, or molecules, and a composition definable by a chemical formula. There has traditionally been another criterion: a mineral must be inorganic, not produced by any animal or plant. Today, some geologists have dropped this requirement in recognition that bones and teeth are identical to the mineral apatite, many clam shells are identical to the mineral calcite, and many microscopic creatures build shells from material identical to the mineral quartz. Let's look at the definition more closely.

■ *Naturally occurring* means that a mineral forms by natural Earth processes. Thus, human-made materials like steel and plastic are not minerals.

■ *Homogeneous* means that a piece of a mineral contains the same pure material throughout.

■ *Formed by geologic processes* traditionally implied processes such as solidification or precipitation, which did not involve living organisms. As noted above, however, many geologists now consider solid, crystalline materials produced by organisms to be minerals too.

■ *Solid* means that minerals retain their shape indefinitely under normal conditions. Therefore, liquids like oil and water and gases like air and propane cannot be minerals.

■ An *ordered internal arrangement of atoms* is an important characteristic that separates minerals from substances that may fit all other parts of the definition. Atoms in minerals occupy positions in a grid called a *crystalline structure.* Solids in which atoms occur in random clusters rather than in a crystalline structure are called *glasses.*

■ *Definable chemical composition* means that the elements present in a mineral and the proportions of their atoms can be expressed by a simple formula—for example, quartz is SiO_2 and calcite is $CaCO_3$—or by one that is more complex: the mineral muscovite is $KAl_2(AlSi_3O_{10})(OH)_2$.

When a mineral grows without interference from other minerals, it develops smooth flat surfaces and a symmetrical geometric shape that we call a **crystal**. When a mineral forms in an environment where other minerals interfere with its growth, it has an irregular shape but still has the appropriate crystalline structure for that mineral, as would a piece broken off a crystal during erosion. An irregular or fragmented piece of mineral is a **grain**, and a single piece of a mineral, either crystal or grain, is called a **specimen.**

In the geosphere, most minerals occur as parts of rocks. It is important to know the difference between a mineral specimen and a rock. A **rock** is *a coherent, naturally occurring, inorganic solid consisting of an aggregate of mineral grains, pieces of older rocks, or a mass of natural glass.* Some rocks, like granite, contain grains of several different minerals; and some, like rock salt, are made of many grains of a single mineral. Others are made of fragments of previously existing rock that are cemented together. And a few kinds of rock are natural glasses, cooled so rapidly from a molten state that their atoms did not have time to form the gridlike crystalline structures required for minerals. Exercise 3.2 helps you practice this terminology with specimens provided by your instructor.

EXERCISE 3.2   **Is It a Mineral or a Rock?**

Name: _____   Section: _____
Course: _____   Date: _____

(a) Based on the definitions of *mineral* and *rock*, determine which specimens you used in Exercise 3.1 are minerals, which are rocks, and which are *other*—neither minerals nor rocks. Write the specimen number in the appropriate column.

Minerals	Rocks	Other

(b) Choose and look carefully at one of the *rock* specimens. How many different minerals are in this rock? _____

(c) How do you know? What visual or other clues did you use to determine how each mineral is different from its neighbors?

_____

_____

_____

(d) Describe each of the minerals in this rock in your own words. (Up to four, if possible.)

Mineral 1

_____

_____

_____

(continued)

Name: _____   Section: _____
Course: _____   Date: _____

Mineral 2

_____

_____

_____

Mineral 3

_____

_____

_____

Mineral 4

_____

_____

_____

3.4 Physical Properties of Minerals

Mineralogists (geologists who specialize in the study of minerals) have named more than four thousand minerals that differ from one another in composition and crystalline structure. These characteristics, in turn, determine a mineral's **physical properties**, which include how it looks (color and luster), breaks, feels, smells, and even tastes. Some minerals are colorless and nearly transparent; others are opaque, dark colored, and shiny. Some are hard, others soft. Some form long, needlelike crystals, others blocky cubes. You probably instinctively used some of these physical properties in Exercise 3.2 to decide how many minerals were in your rocks and then to describe them. We discuss the major physical properties of minerals below so you can use them to identify common minerals—in class, at home, or while on vacation.

3.4.1 Diagnostic versus Ambiguous Properties

Geologists use physical properties to identify minerals, much as detectives use physical descriptions to identify suspects. And, as with people, some physical properties are *diagnostic properties*—they immediately help identify an unknown mineral or rule it out as a possibility. Other properties are *ambiguous properties* because they may vary in different specimens of the same mineral. For example, color is a notoriously ambiguous property in many minerals (**FIG. 3.1**). Size doesn't really matter either; a large specimen of quartz has the same properties as a small one. Exercise 3.3 shows how diagnostic and ambiguous properties affect everyday life.

FIGURE 3.1 Varieties of the mineral fluorite, showing its range in color.

3.4.2 Luster

One of the first things we notice when we pick up a mineral is its luster—the way light interacts with its surface. For mineral identification, we distinguish minerals that have a metallic luster from those that are nonmetallic. Something with a *metallic* luster is shiny and opaque like an untarnished piece of metal. Materials with a *nonmetallic* luster look earthy (dull and powdery like dirt), glassy (vitreous), waxy, silky, or pearly—terms relating luster to familiar materials. Luster is a diagnostic property for many minerals, but be careful: some minerals may tarnish, and their metallic luster may be dulled.

3.4.3 Color

The color we see when we look at a mineral is controlled by how the different wavelengths of visible light are absorbed or reflected by the mineral's atoms. Color is generally a diagnostic property for minerals with a metallic luster and for *some* with a nonmetallic luster. Specimens of some nonmetallic minerals, like the fluorite in Figure 3.1, have such a wide range of colors that they were once thought to be different minerals. We now know that the colors are caused by impurities. For example, rose quartz contains a very small amount of titanium.

3.4.4 Streak

The streak of a mineral is the color of its powder, which we can determine by rubbing a mineral against an unglazed porcelain plate. Streak and color are the same for most minerals, but for some they are different. In these cases, the difference between streak and color is an important diagnostic property. A mineral's color may vary widely, as in Figure 3.1; but its streak is generally similar for all specimens regardless of their color, as seen in **FIGURE 3.2**.

FIGURE 3.2 Color and streak.

(a) Red hematite has a reddish-brown streak.

(b) But this dark, metallic-looking specular hematite also has a reddish-brown streak.

EXERCISE 3.3	People Have Diagnostic Properties Too

Name: _____ Section: _____

Course: _____ Date: _____

Your father has asked you to pick up his old college roommate at the airport. You've never met him, but your father gave you a yearbook photo and described what he looked like 30 years ago: height, weight, hair color, beard, eye color. Which of these features would still be diagnostic today? Which, considering the passage of time, might be ambiguous? What other properties might also be diagnostic? Indicate in the following chart which properties, considering the passage of time, might be diagnostic and which might be ambiguous. Then suggest two others that would be diagnostic despite the years that have passed.

Property	Diagnostic (explain)	Ambiguous (explain)
Height		
Weight		
Hair color		
Beard		
Eye color		
Others		

3.4.5 Hardness

The hardness of a mineral is a measure of how easily it can scratch or be scratched by other substances. A nineteenth-century mineralogist, Friedrich Mohs, created a mineral hardness scale that we still use today, using ten familiar minerals. He assigned a hardness of 10 to the hardest mineral and a hardness of 1 to the softest (**TABLE. 3.1**). This is a **relative scale**, meaning that a mineral can scratch those lower in the scale but cannot scratch those that are higher. It is not an *absolute scale* in which diamond would be ten times harder than talc, and corundum would be three times harder than calcite. The hardness of common materials, such as the testing materials listed in **FIGURE 3.3**, can also be described using the Mohs hardness scale. To determine the hardness of a mineral, see which of these materials it can scratch and which ones can scratch it. You will practice this in Exercise 3.4.

TABLE 3.1 Mohs hardness scale and its relationship to common testing materials.

Mineral	Mohs hardness number (H)	Mohs hardness of testing materials	
Diamond	10		HARD
Corundum	9		
Topaz	8		
Quartz	7		
Orthoclase	6	Streak plate (6.5–7)	MODERATE
Apatite	5	Window glass; steel cut nail (5.5)	
Fluorite	4	Common nail or pocket knife (5.0-5.5)	
Calcite	3	U.S. penny (3.0)	
Gypsum	2	Fingernail (2.5)	SOFT
Talc	1		

EXERCISE 3.4   **Constructing and Using a Relative Hardness Scale**

Name: _____   Section: _____
Course: _____   Date: _____

Your instructor will tell you which specimens from your mineral set to use for this exercise. Arrange them in order of increasing hardness by seeing which can scratch the others and which are most easily scratched.

Softest ————————————————————————➤ *Hardest*

Specimen no.:   _____   _____   _____   _____   _____

Now use the testing materials listed in Table 3.1 to determine the Mohs hardness of minerals in your set.

Mohs hardness: _____   _____   _____   _____   _____

FIGURE 3.3 **Testing for hardness.**

(a) A fingernail (H=2.5) can scratch gypsum (H=2.0) but not calcite (H=3.0).

(b) A knife blade (H=5.0) can scratch fluorite (H=4.0) but not quartz (H=7.0).

3.4.6 Crystal Habit

Crystal shapes found in the mineral kingdom range from simple cubes with six faces to complex 12-, 24-, or 48-sided (or more) crystals (**FIG. 3.4a–h**). Some crystals are flat like a knife blade, others are needlelike. Each mineral has its own diagnostic **crystal habit**, a preferred crystal shape that forms when it grows unimpeded by other grains. For example, the habit of halite is a cube, and that

FIGURE 3.4 **Crystal habits of some common minerals with diagrams of perfect crystal shapes.**

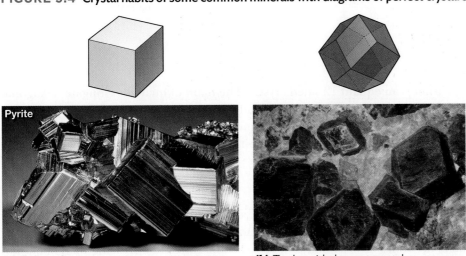

(a) Cubes of pyrite

(b) Twelve-sided garnet crystals.

FIGURE 3.4 Crystal habits of some common minerals (*cont.*).

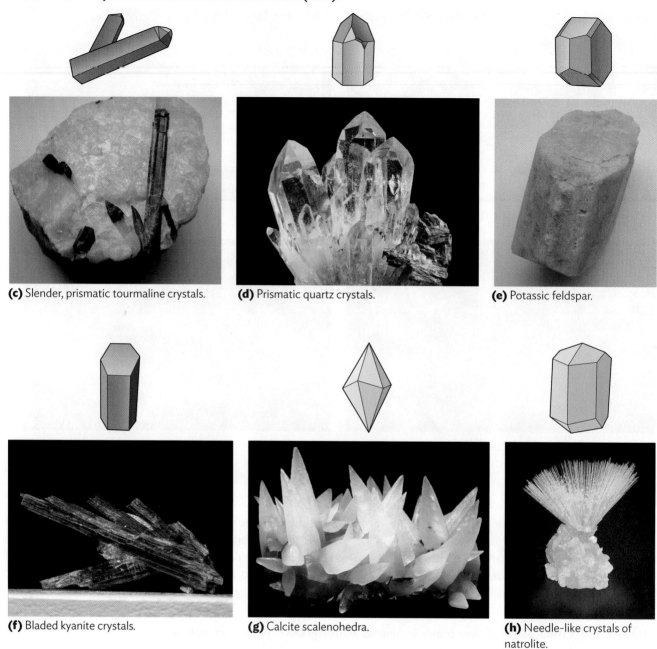

(c) Slender, prismatic tourmaline crystals.

(d) Prismatic quartz crystals.

(e) Potassic feldspar.

(f) Bladed kyanite crystals.

(g) Calcite scalenohedra.

(h) Needle-like crystals of natrolite.

of garnet is an equant 12-sided crystal. The habit of quartz is elongate hexagonal crystals topped by a six-sided pyramid. *Remember:* Crystal growth requires very special conditions. As a result, most mineral specimens are irregular grains; few display their characteristic crystal habit.

3.4.7 Breakage

Some minerals break along one or more smooth planes, others along curved surfaces, and still others in irregular shapes. The way a mineral breaks is controlled by whether there are zones of weak bonds in its structure. Instead of breaking the minerals in your sets with a hammer, you can examine specimens to see how they have already broken—using a microscope or magnifying glass to see more clearly. Two kinds of breakage are important: fracture and cleavage.

Fracture occurs when there are no zones of particularly weak bonding within a mineral. When such a mineral breaks, either irregular (**irregular fracture**) or curved surfaces (*conchoidal fracture*) form (**FIG. 3.5**). Conchoidal fracture surfaces are common in thick glass and in minerals whose bond strength is nearly equal in all directions (e.g., quartz).

FIGURE 3.5 Types of fractures in minerals.

Irregular fracture

Crystal face

(a) Irregular fracture in garnet.

(b) Conchoidal fracture in quartz.

Cleavage occurs when bonds holding atoms together are weaker in some directions than in others. The mineral breaks along these zones of weakness, producing flat, smooth surfaces. Some minerals have a single zone of weakness, but others may have two, three, four, or six (**FIG. 3.6**). If there is more than one zone of weakness, a mineral cleaves in more than one direction. It is important to note *how many* directions there are and *the angles between those directions*. Two different minerals might each have two directions of cleavage, but those directions might be at 90° in one mineral but not in the other (see Fig. 3.6). For example, amphiboles and pyroxenes (two important groups of minerals) are similar in most other properties and have two directions of cleavage, but amphiboles cleave at 56° and 124° whereas pyroxenes cleave at 90°.

Note that in Figure 3.6 there may be many cleavage *surfaces*, but several of those surfaces are parallel to one another, as shown for halite. All of these parallel surfaces define a single *cleavage direction*. To help observe a mineral's cleavage, hold it up to the light and rotate it. Parallel cleavage surfaces reflect light at the same time, making different cleavage directions easy to see.

Both crystal faces and cleavage surfaces are smooth, flat planes and might be mistaken for one another. If you can see many small, parallel faces, these are cleavage faces because crystal faces are not repeated. In addition, breakage occurs after a crystal has grown, thus cleavage or fracture surfaces generally look less tarnished or altered than crystal faces. Exercise 3.5 will help you recognize the difference between cleavage and fracture.

FIGURE 3.6 Cleavage in common minerals. Some minerals cleave in four or six directions, but all of the directions are rarely visible in a single specimen.

Types of cleavage	Diagram	Visual	Examples
1 Direction			Muscovite
2 Directions at 90°	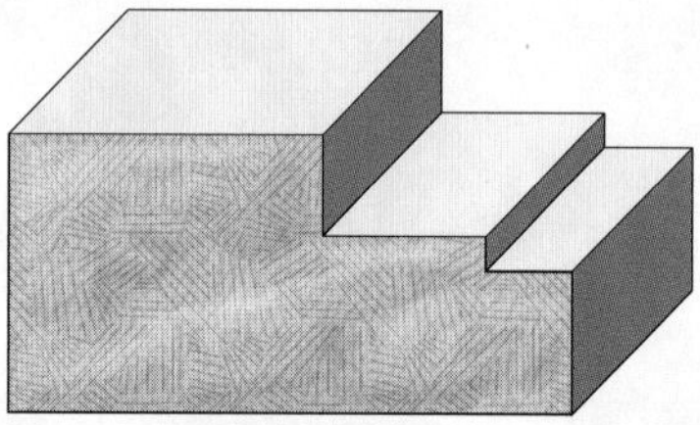		Pyroxene
2 Directions not at 90°	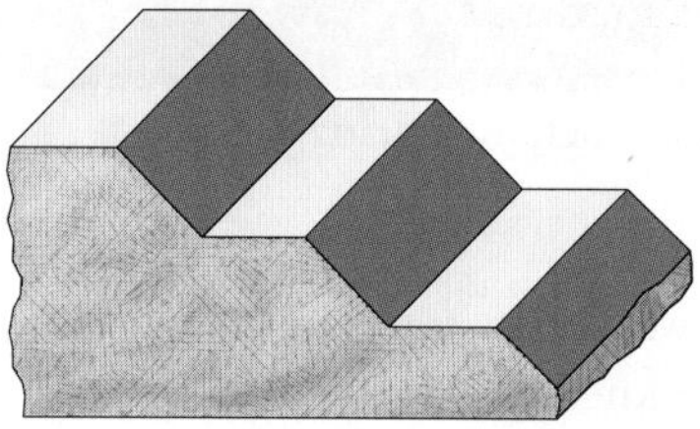		Amphibole
3 Directions at 90°	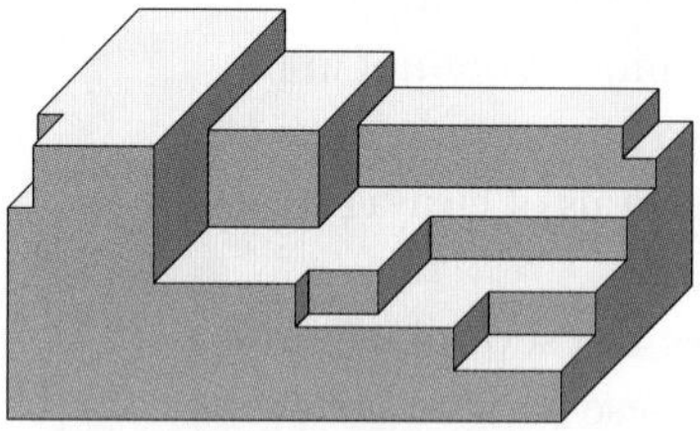		Halite
3 Directions not at 90°	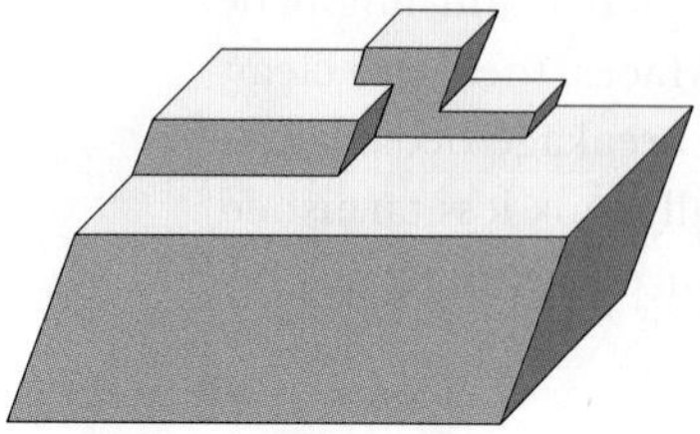	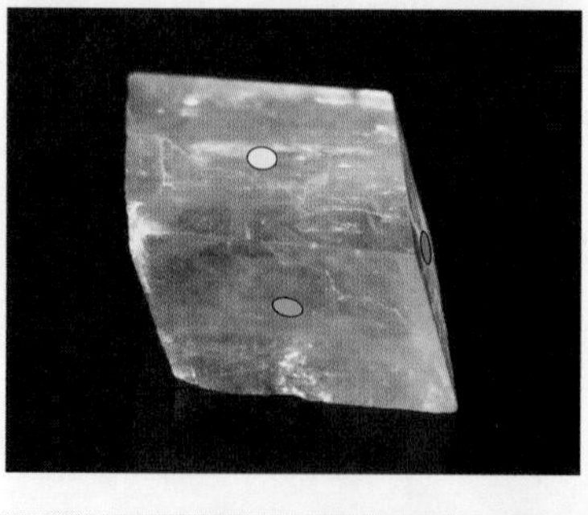	Calcite

Name: _____   **Section:** _____
Course: _____   **Date:** _____

Examine the specimens indicated by your instructor. Which have cleaved and which have fractured? For those with cleavage, indicate the number of directions and the angles between them.

_____

_____

_____

_____

_____

3.4.8 Specific Gravity

The specific gravity (**SpG**) of a mineral is a comparison of its density with the density of water. The density of pure water is 1 g/cm^3, so if a mineral has a density of 4.68 g/cm^3, its specific gravity is 4.68.

$\dfrac{4.68\ \cancel{g/cm^3}}{1.00\ \cancel{g/cm^3}}$ The units cancel, so **SpG = 4.68**.
        This means that the mineral is 4.68 times denser than water.

You can measure specific gravity by calculating the density of a specimen (density = mass ÷ volume). But geologists generally estimate specific gravity by *hefting* a specimen and determining if it seems heavy or light. To compare the specific gravities of two minerals, pick up similar-sized specimens to get a general feeling for their densities. You will feel the difference—just as you would feel the difference between a box of Styrofoam packing material and a box of marbles. In Exercise 3.6 you will practice estimating specific gravity by heft, and then measure it precisely.

| EXERCISE 3.6 | Heft and Specific Gravity |

Name: _____   **Section:** _____
Course: _____   **Date:** _____

Separate the minerals provided by your instructor into those with relatively high specific gravity and those with relatively low specific gravity by hefting them.

(a) What luster do most of the minerals in the high specific gravity group have? _____ This is not a coincidence. In general, minerals with this luster have higher specific gravities than minerals with other lusters.

(b) To become familiar with the range of specific gravity in common minerals, select the most dense and least dense specimens based on their heft. Measure their specific gravities. To calculate density, measure the mass of the specimen by submerging it in a graduated cylinder and calculating the change in volume (see Chapter 1 for a detailed procedure). Density = mass ÷ volume.

Specific gravity of   Specific gravity of
most dense specimen _____   least dense specimen _____

(c) Do not try this procedure with halite (rock salt). Why wouldn't it work?

_____

3.4.9 Magnetism

A few minerals are attracted to a magnet or act like a magnet and attract metallic objects like nails or paper clips. The most common example is, appropriately, called *magnetite*. Because so few minerals are magnetic, this is a diagnostic property.

3.4.10 Feel

Some minerals feel greasy or slippery when you rub your fingers over them. They are greasy because their chemical bonds are so weak in one direction that the pressure of your fingers is enough to break them and to slide planes of atoms past one another. Talc and graphite are common examples.

3.4.11 Taste

Yes, geologists sometimes taste minerals. Taste is a *chemical* property, determined by the presence of certain elements. The most common example is halite (common salt), which tastes salty because of the chloride ion (Cl^-). **Do not taste minerals in your set unless instructed to do so!** We taste minerals only *after* we have narrowed the possibilities down to a few for which taste would be the diagnostic property. Why not taste every mineral? Because some taste bitter (like sylvite—KCl), some are poisonous, and you don't want to get other students' germs!

3.4.12 Odor

As geologists we use all of our senses to identify minerals. A few minerals, and the streak of a few others, have a distinctive odor. For example, the streak of minerals containing sulfur smells like rotten eggs, and the streak of some arsenic minerals smells like garlic.

3.4.13 Reaction with Dilute Hydrochloric Acid

Many minerals containing the carbonate anion (CO_3^{2-}) effervesce (fizz) when dilute hydrochloric acid is dropped on them. The acid frees carbon dioxide from the mineral, and the bubbles of gas escaping through the acid produce the fizz.

3.4.14 Tenacity

Tenacity refers to the way in which materials respond to being pushed, pulled, bent, or sheared. Most adjectives used to describe tenacity are probably familiar: *malleable* materials can be bent or hammered into a new shape; *ductile* materials can be pulled into wires; *brittle* materials shatter when hit hard; and *flexible* materials can bend. Flexibility is a diagnostic property for some minerals. After being bent, thin sheets of *elastic* minerals return to their original unbent shape, but sheets of *flexible* minerals retain the new shape.

3.5 Identifying Mineral Specimens

You are now ready to use these physical properties to identify minerals. Although there are more than 4,000 minerals, only about 30 occur commonly and an even smaller number make up most of the Earth's crust—the part of the geosphere we are most familiar with. Identification is easier if you follow the systematic approach used by geologists:

STEP 1 Assemble the equipment available in most geology classrooms to study minerals.

- A glass plate, penny, and knife or steel nail to test hardness
- A ceramic streak plate to test streak
- A magnifying glass, hand lens, or microscope to help determine cleavage
- Dilute hydrochloric acid to identify carbonate minerals
- A magnet to identify magnetic minerals

STEP 2 Observe or measure the specimen's physical properties. Profile the properties on a standardized data sheet like the one at the end of this chapter, as shown in the example given in **TABLE 3.2**.

TABLE 3.2 **Profile of a mineral's physical properties.**

Specimen number	Luster	Color	Hardness	Breakage	Other diagnostic properties
1	Metallic	Dark gray	Less than a fingernail	Excellent cleavage in one direction	Leaves a mark on a sheet of paper

STEP 3 Eliminate from consideration all minerals that do not have the properties you have recorded. This can be done systematically by using a **flowchart** (**FIG. 3.7**) that asks key questions in a logical sequence, so that each answer eliminates entire groups of minerals until only a few remain (one, if you're lucky). Or you can use a **determinative table** in which each column answers the same questions, like those at the branches of a flowchart. Appendices 3.1, 3.2, and 3.3 at the end of this chapter provide flowcharts and determinative tables. Experiment to find out which tool works best for you.

Use these steps to help you complete Exercise 3.7.

FIGURE 3.7 **Flowchart showing the steps used to identify minerals.**

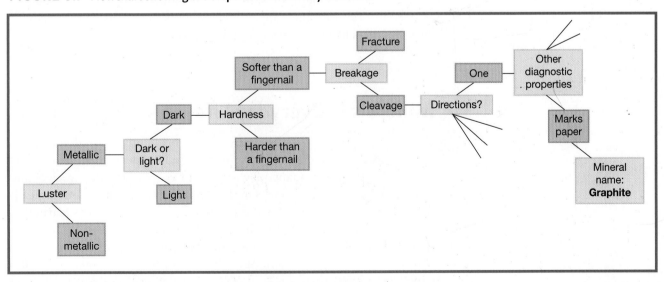

Note: For simplicity, only the path for the unknown mineral is shown.

Name: _____ Section: _____

Course: _____ Date: _____

Your instructor will provide a set of minerals to identify. Record the profile of physical properties for each specimen on the data sheets at the end of this chapter. Then use either the flowcharts (Appendix 3.1) or determinative tables (Appendix 3.2) to identify each mineral. If these lead to more than one possibility, look at Appendix 3.3 for additional information.

3.6 Mineral Classification

The minerals in your set were chosen because they are important rock-forming minerals that make up most of the geosphere, are economically valuable resources, or illustrate the physical properties used to study minerals. Geologists classify all minerals into a small number of groups based on their chemical composition. These groups include

- *silicates,* such as quartz, feldspars, amphiboles, and pyroxenes, which contain silicon and oxygen. Silicates are divided into *ferromagnesian* minerals, which contain iron and magnesium, and *nonferromagnesian* minerals, which do not contain those elements.
- *oxides,* such as magnetite (Fe_3O_4) and hematite (Fe_2O_3), in which a cation is bonded to oxygen anions.
- *sulfides,* such as pyrite (FeS_2) and sphalerite (ZnS), in which a cation is bonded to sulfur anions.
- *sulfates,* such as gypsum ($CaSO_4 \cdot 2H_2O$), in which cations are bonded to the sulfate complex (SO_4^{2-}).
- *halides,* such as halite (NaCl) and fluorite (CaF_2), in which cations are bonded to halogen anions (elements in the second column from the right in the periodic table).
- *carbonates,* such as calcite ($CaCO_3$) and dolomite ($CaMg[CO_3]_2$), containing the carbonate complex (CO_3)$^{2-}$.
- *native elements,* which are minerals that consist of atoms of a single element. The native elements most likely to be found in your mineral sets are graphite (carbon) and copper. You are unlikely to find more valuable native elements like gold, silver, and diamond (which is carbon, just like graphite).

3.7 Minerals in Everyday Life

When people think about minerals, many picture brilliantly colored gemstones like diamond, ruby, and emerald. Most minerals are less spectacular, but despite their "ordinary" appearance many play important roles in modern society and are extremely valuable. Indeed, stages in the development of our technologically advanced civilization are named for the resources that our ancestors learned to obtain from rocks and minerals: the Stone Age, when rocks were the major resource, was followed by the Copper Age, Bronze Age (copper plus tin melted from ores of these metals), and Iron Age (the iron initially from meteorites and later smelted from iron ores).

Ore minerals are those containing metals that can be separated from the rest of the elements in the minerals, usually by melting. These, mostly oxides and sulfides, are important resources, but so too are minerals whose physical properties make them valuable. Our ancestors were the first economic geologists, learning how to use minerals based on their physical properties and then figuring out where to find them. In Exercise 3.8, let's look at some of the physical properties that make minerals useful.

Everyday Uses of Minerals

Name: _____   Section: _____

Course: _____   Date: _____

In following table, indicate what physical property you think would make a mineral appropriate for the use indicated, and name a mineral from your set that could be used for this purpose. In some instances, more than one mineral will meet the requirements and more than one property is required.

Economic use	Physical property or properties needed	Minerals
Abrasives (e.g., sandpaper)		
Old-time window coverings before glass was widely available		
Modern window panes		
Writing on paper		
Lubricant for locks		
After-bath powder		
Bright eye shadow		
Pigment for paints		
Navigation with a compass		

? **What Do You Think** Copper is used widely for electrical wires, pipes, construction (and pennies), and is finding new uses today in high-tech electronics. Exploration geologists recently discovered a very large deposit of copper minerals in a remote part of the woods in northern Maine shown in the accompanying illustration. Imagine that you own land in the area and that a mining company has made an offer to buy or lease the property so it can extract the copper ore. On a separate sheet of paper, explain what you think the pros and cons are to deciding whether to sell or not to sell.

In these appendices, we present two ways of identifying or determining minerals. First, in Appendix 3.1, we provide flowcharts for students who may be more visually oriented. Then, in Appendix 3.2, we present the same information in the more standard determinative tables. Because both students and geologists think and work in different ways, we felt it was important to provide both options. Use whichever works best for you!

Mineral Identification Flowcharts

A. Minerals with Metallic Luster

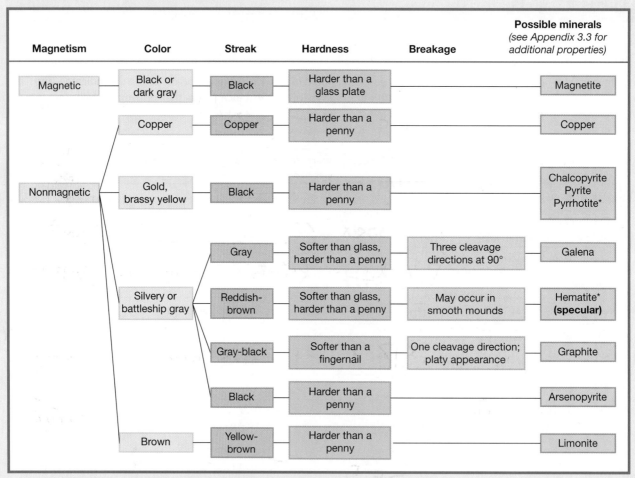

* Pyrrhotite and hematite are sometimes weakly magnetic.

Mineral Identification Flowcharts

B. Minerals with Nonmetallic Luster, Dark Colored

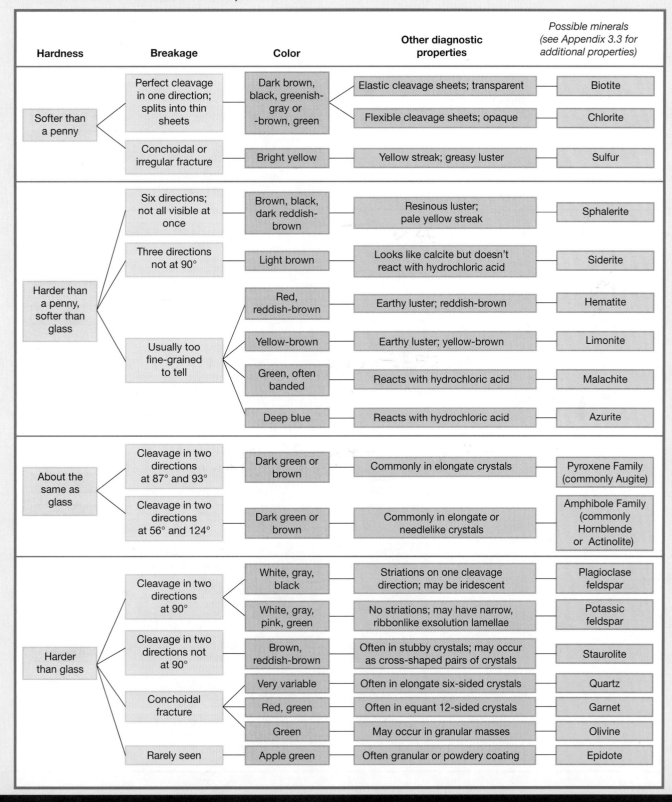

Hardness	Breakage	Color	Other diagnostic properties	Possible minerals (see Appendix 3.3 for additional properties)
Softer than a penny	Perfect cleavage in one direction; splits into thin sheets	Dark brown, black, greenish-gray or -brown, green	Elastic cleavage sheets; transparent	Biotite
			Flexible cleavage sheets; opaque	Chlorite
	Conchoidal or irregular fracture	Bright yellow	Yellow streak; greasy luster	Sulfur
Harder than a penny, softer than glass	Six directions; not all visible at once	Brown, black, dark reddish-brown	Resinous luster; pale yellow streak	Sphalerite
	Three directions not at 90°	Light brown	Looks like calcite but doesn't react with hydrochloric acid	Siderite
	Usually too fine-grained to tell	Red, reddish-brown	Earthy luster; reddish-brown	Hematite
		Yellow-brown	Earthy luster; yellow-brown	Limonite
		Green, often banded	Reacts with hydrochloric acid	Malachite
		Deep blue	Reacts with hydrochloric acid	Azurite
About the same as glass	Cleavage in two directions at 87° and 93°	Dark green or brown	Commonly in elongate crystals	Pyroxene Family (commonly Augite)
	Cleavage in two directions at 56° and 124°	Dark green or brown	Commonly in elongate or needlelike crystals	Amphibole Family (commonly Hornblende or Actinolite)
Harder than glass	Cleavage in two directions at 90°	White, gray, black	Striations on one cleavage direction; may be iridescent	Plagioclase feldspar
		White, gray, pink, green	No striations; may have narrow, ribbonlike exsolution lamellae	Potassic feldspar
	Cleavage in two directions not at 90°	Brown, reddish-brown	Often in stubby crystals; may occur as cross-shaped pairs of crystals	Staurolite
	Conchoidal fracture	Very variable	Often in elongate six-sided crystals	Quartz
		Red, green	Often in equant 12-sided crystals	Garnet
		Green	May occur in granular masses	Olivine
	Rarely seen	Apple green	Often granular or powdery coating	Epidote

Mineral Identification Flowcharts

C. Minerals with Nonmetallic Luster, Light Colored

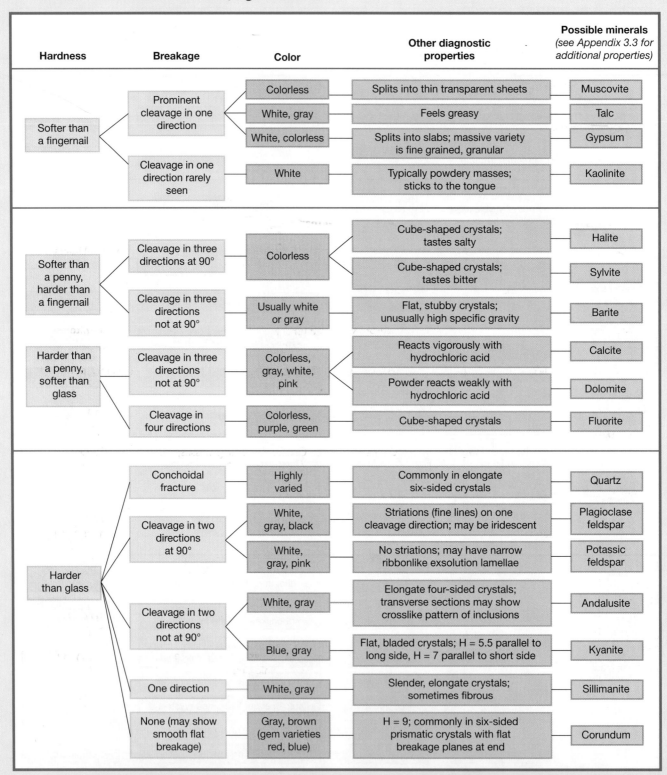

Hardness	Breakage	Color	Other diagnostic properties	Possible minerals (see Appendix 3.3 for additional properties)
Softer than a fingernail	Prominent cleavage in one direction	Colorless	Splits into thin transparent sheets	Muscovite
		White, gray	Feels greasy	Talc
		White, colorless	Splits into slabs; massive variety is fine grained, granular	Gypsum
	Cleavage in one direction rarely seen	White	Typically powdery masses; sticks to the tongue	Kaolinite
Softer than a penny, harder than a fingernail	Cleavage in three directions at 90°	Colorless	Cube-shaped crystals; tastes salty	Halite
			Cube-shaped crystals; tastes bitter	Sylvite
	Cleavage in three directions not at 90°	Usually white or gray	Flat, stubby crystals; unusually high specific gravity	Barite
Harder than a penny, softer than glass	Cleavage in three directions not at 90°	Colorless, gray, white, pink	Reacts vigorously with hydrochloric acid	Calcite
			Powder reacts weakly with hydrochloric acid	Dolomite
	Cleavage in four directions	Colorless, purple, green	Cube-shaped crystals	Fluorite
Harder than glass	Conchoidal fracture	Highly varied	Commonly in elongate six-sided crystals	Quartz
	Cleavage in two directions at 90°	White, gray, black	Striations (fine lines) on one cleavage direction; may be iridescent	Plagioclase feldspar
		White, gray, pink	No striations; may have narrow ribbonlike exsolution lamellae	Potassic feldspar
	Cleavage in two directions not at 90°	White, gray	Elongate four-sided crystals; transverse sections may show crosslike pattern of inclusions	Andalusite
		Blue, gray	Flat, bladed crystals; H = 5.5 parallel to long side, H = 7 parallel to short side	Kyanite
	One direction	White, gray	Slender, elongate crystals; sometimes fibrous	Sillimanite
	None (may show smooth flat breakage)	Gray, brown (gem varieties red, blue)	H = 9; commonly in six-sided prismatic crystals with flat breakage planes at end	Corundum

Determinative Tables for Systematic Mineral Identification

Sequence of questions: Luster? Approximate hardness? Streak? Breakage? Color? Other?

TABLE 1 Minerals with metallic luster

(a) Hardness less than 2.5 (softer than a fingernail)

Streak	Cleavage or fracture	H	Color	Other diagnostic properties	Mineral name (composition)
Black	Perfect cleavage in one direction	1	Dark gray-black	Greasy feel; leaves a mark on paper; SpG = 2.23	**Graphite** C
Yellow-brown	Diffcult to see	—	Yellow-brown	Very rarely in masses with metallic luster; more commonly dull, earthy; SpG = 3.6–4	**Limonite** $FeO(OH) \cdot nH_2O$

(b) Hardness between 2.5 and 5.5 (harder than a fingernail; softer than glass)

Streak	Cleavage or fracture	H	Color	Other diagnostic properties	Mineral name (composition)
Gray	Three directions at 90° angles	2.5	Lead gray	Commonly in cubic crystals; SpG = 7.4–7.6	**Galena** PbS
Black	Rarely seen	3	Bronze-brown when fresh	Commonly with purplish, iridescent tarnish; SpG = 5.06–5.08	**Bornite** Cu_5FeS_4
	Rarely seen	3.5–4	Brassy yellow	Often tarnished; similar to pyrite but not in cubes	**Chalcopyrite** $CuFeS_2$
	Rarely seen	4	Brown-bronze	Slightly magnetic; SpG = 4.62	**Pyrrhotite** $Fe_{1-x}S$
Copper-red	Rarely seen	2.5–3	Copper	Often in branching masses; SpG = 8.9	**Copper** (Cu)

(c) Hardness greater than 5.5 (harder than glass; cannot be scratched by a knife)

Streak	Cleavage or fracture	H	Color	Other diagnostic properties	Mineral name (composition)
Black	Conchoidal fracture	6–6.5	Brassy yellow	Commonly in 12-sided crystals or cubes with striated faces; SpG = 5.02	**Pyrite** FeS_2
	Rarely seen	6	Iron black	Strongly magnetic; SpG = 5.18	**Magnetite** Fe_3O_4
	Rarely seen	5.5–6	Silver white	Streak smells like garlic because of arsenic; SpG = 6.07	**Arsenopyrite** FeAsS
Reddish-brown	Rarely seen	5.5–6.5	Black, red	Black variety is metallic; red variety is more common and has nonmetallic, earthy luster	**Hematite** Fe_2O_3

Determinative Tables for Systematic Mineral Identification

TABLE 2 Minerals with nonmetallic luster

(a) Hardness less than 2.5 (softer than a fingernail)

Streak	Cleavage or fracture	H	Color	Other diagnostic properties	Mineral name (composition)
Yellow	Conchoidal or uneven fracture	1.5–2.5	Bright yellow	Resinous luster; SpG = 2.05–2.09	**Sulfur** S
White or colorless	Perfect cleavage in one direction	2–2.5	Colorless, light tan, yellow	Can be peeled into transparent, elastic sheets; SpG = 2.76–2.88	**Muscovite** $KAl_2(AlSi_3O_{10})(OH)_2$
	Perfect cleavage in one direction	1	Green, gray, white	Greasy feel; may occur in irregular masses (soapstone); SpG = 2.7–2.8	**Talc** $Mg_3Si_4O_{10}(OH)_2$
	Perfect cleavage in one direction; may show two other directions not at 90°	2	Colorless, white, gray	Occurs in clear crystals or gray or white, earthy masses (alabaster); SpG = 2.32	**Gypsum** $CaSO_4 \cdot 2H_2O$
	Three directions at 90°	2	Colorless, white	Cubic crystals like halite but has very bitter taste; SpG = 1.99	**Sylvite** KCl
	Perfect in one direction, but rarely seen	2–2.5	White	Usually in dull, powdery masses that stick to the tongue; SpG = 2.6	**Kaolinite** $Al_2Si_2O_5(OH)_4$
	Perfect in one direction, but not always visible	2–5	Green, white	Platy and fibrous (asbestos) varieties; greasy luster; SpG = 2.5–2.6	**Serpentine** $Mg_3Si_2O_5(OH)_4$
	—	—	White, brown, gray	Not really a mineral; rock often made of small, spherical particles containing several clay minerals; SpG = 2–2.55	**Bauxite** Mixture of aluminum hydroxides
Brown or green	Perfect cleavage in one direction	2.5–3	Brown, black, green	Can be peeled into thin, elastic sheets; SpG = 2.8–3.2	**Biotite** $K(Fe,Mg)_3 AlSi_3O_{10} (OH)_2$
	Perfect cleavage in direction	2–2.5	Green, dark green	A mica-like mineral, but sheets are flexible not elastic; SpG = 2.6–3.3	**Chlorite** Complex Fe-Mg sheet silicate

(b) Hardness between 2.5 and 5.5 (harder than a fingernail; softer than glass)

Streak	Cleavage or fracture	H	Color	Other diagnostic properties	Mineral name (composition)
Green	—	3.5–4	Bright green	Occurs in globular or elongate masses; reacts with HCl; SpG = 3.9–4.03	**Malachite** $Cu_2CO_3(OH)_2$

Determinative Tables for Systematic Mineral Identification

TABLE 2 Minerals with nonmetallic luster

Streak	Cleavage or fracture	H	Color	Other diagnostic properties	Mineral name (composition)
Blue	—	3.5	Intense blue	Often in platy crystals or spherical masses; reacts with HCl; SpG = 3.77	**Azurite** $Cu_3(CO_3)_2(OH)_2$
Reddish-brown	—	—	Reddish brown	Usually in earthy masses; also occurs as black, metallic crystals; SpG = 5.5–6.5	**Hematite** Fe_2O_3
Yellow-brown	—	—	Brown, tan	Earthy, powdery masses and coatings on other minerals; SpG = 3.6–4	**Limonite** $FeO(OH) \cdot nH_2O$
	Three directions not at 90°	3.5–4	Light to dark brown	Often in rhombic crystals; reacts with hot HCl; SpG = 3.96	**Siderite** $FeCO_3$
	Six directions, few of which are usually visible	3.5	Brown, white, yellow, black, colorless	Resinous luster; SpG = 3.9–4.1	**Sphalerite** ZnS
White or colorless	Three directions at 90°	2.5	Colorless, white	Cubic crystals or massive (rock salt); salty taste; SpG = 2.5	**Halite** $NaCl$
	Three directions not at 90°	3	Varied; usually white or colorless	Rhombic or elongated crystals; reacts with HCl; SpG = 2.71	**Calcite** $CaCO_3$
	Three directions not at 90°	3.5–4	Varied; commonly white or pink	Rhombic crystals; *powder* reacts with HCl but crystals may not; SpG = 2.85	**Dolomite** $CaMg(CO_3)_2$
	Three directions at 90°	3–3.5	Colorless, white	SpG = 4.5 (unusually high for a nonmetallic mineral)	**Barite** $BaSO_4$
	Four directions	4	Colorless, purple, yellow, blue, green	Often in cubic crystals; SpG = 3.18	**Fluorite** CaF_2
	One direction, poor	5	Usually green or brown	Elongate six-sided crystals; may be purple, blue, colorless; SpG = 3.15–3.20	**Apatite** $Ca_5(PO_4)_3(OH,Cl,F)$

(c) Hardness between 5.5 and 9 (harder than glass or a knife; softer than a streak plate)

Streak	Cleavage or fracture	H	Color	Other diagnostic properties	Mineral name (composition)
	Conchoidal fracture	7	Red, green, brown, black, colorless, pink, orange	Equant 12-sided crystals; SpG = 3.5–4.3	**Garnet family** Complex Ca, Fe, Mg, Al, Cr, Mn silicate
	Two directions at 90°	6	Colorless, salmon, gray, green, white	Stubby prismatic crystals; three polymorphs: orthoclase, microcline, sanidine; may show exsolution lamellae; SpG = 2.54–2.62	**Potassic feldspar** $KAlSi_3O_8$

Determinative Tables for Systematic Mineral Identification

TABLE 2 Minerals with nonmetallic luster *(continued)*

Streak	Cleavage or fracture	H	Color	Other diagnostic properties	Mineral name (composition)
White or colorless	Two directions at 90°	6	Colorless, white, gray, black	Striations (fine lines) on one of the two cleavage directions; solid solution between sodium (albite) and calcium (anorthite) plagioclase; SpG = 2.62–2.76	**Plagioclase feldspar** $CaAl_2Si_2O_8$ – $NaAlSi_3O_8$
	Conchoidal fracture	7	Colorless, pink, purple, gray, black, green, yellow	Elongate six-sided crystals; SpG = 2.65	**Quartz** SiO_2
	Rarely seen	7.5	Gray, white, brown	Elongate four-sided crystals; SpG = 3.16–3.23	**Andalusite** Al_2SiO_5
	One direction	6–7	White, rarely green	Long, slender crystals, often fibrous; SpG = 3.23	**Sillimanite** Al_2SiO_5
	One direction	5 *and* 7	Blue, blue-gray to white	Bladed crystals; *two hardnesses:* H = 5 parallel to long direction of crystal, H = 7 across the long direction	**Kyanite** Al_2SiO_5
Colorless to light green	Conchoidal fracture	6.5–7	Most commonly green	Stubby crystals and granular masses; solid solution between Fe (fayalite) and Mg (forsterite)	**Olivine family** Fe_2SiO_4 Mg_2SiO_4
	Two directions at 56° and 124°	5–6	Dark green to black	An amphibole with elongate crystals; SpG = 3–3.4	**Hornblende** Complex double-chain silicate with Ca, Na, Fe, Mg
	Two directions at 56° and 124°	5–6	Pale to dark green	An amphibole with elongate crystals; SpG = 3–3.3	**Actinolite** Double-chain silicate with Ca, Fe, Mg
	Two directions at 87° and 93°	5–6	Dark green to black	A pyroxene with elongate crystals; SpG = 3.2–3.3	**Augite** Single-chain silicate with Ca, Na, Mg, Fe, Al
	One direction perfect, one poor; not at 90°	6–7	Apple green to black	Elongate crystals and fine-grained masses; SpG = 3.25–3.45	**Epidote** Complex twin silicate with Ca, Al, Fe, Mg
No streak; mineral scratches streak plate	One direction, imperfect	7.5–8	Blue-green, emerald green, yellow, pink, white	Six-sided crystals with flat ends; SpG = 2.65–2.8; gem variety: emerald (green)	**Beryl** $Be_3Al_2Si_6O_{18}$
	—	7–7.5	Dark brown, Reddish-brown, brownish-black	Stubby or cross-shaped crystals; SpG = 3.65–3.75	**Staurolite** Hydrous Fe, Al silicate
	No cleavage	9	Gray, light brown; gem varieties red (ruby), blue (sapphire)	Six-sided crystals with flat ends	**Corundum** Al_2O_3

Common Minerals and Their Properties

Mineral	Additional diagnostic properties and occurrences
Actinolite	Elongate green crystals; cleavage at 56° and 124°; H = 5.5–6. An amphibole found in metamorphic rocks.
Amphibole*	Stubby rod-shaped crystals common in igneous rocks; slender crystals common in metamorphic rocks; two cleavage directions at 56° and 124°.
Andalusite	Elongate gray crystals with rectangular cross sections.
Apatite	H = 5; pale to dark green, brown, white; white streak; six-sided crystals.
Augite	H = 5.5–6; green to black rod-shaped crystals; cleavage at 87° and 93°. A pyroxene common in igneous rocks.
Azurite	Deep blue; reacts with HCl. Copper will plate out on a steel nail dipped into a drop of HCl and placed on this mineral.
Barite	H = 3–3.5; SpG is unusually high for nonmetallic mineral.
Bauxite	Gray-brown earthy *rock* commonly containing spherical masses of clay minerals and mineraloids.
Beryl	Six-sided crystals; H = 7.5–8.
Biotite	Dark-colored mica; one perfect cleavage into flexible sheets.
Bornite	High SpG; iridescent coating on surface gives it "peacock ore" nickname.
Calcite	Reacts with HCl. Produces double image from text viewed through transparent cleavage fragments.
Chalcopyrite	Similar to pyrite, but softer and typically has iridescent tarnish.
Chlorite*	Similar to biotite, but does not break into thin, flexible sheets; forms in metamorphic rocks.
Copper	Copper-red color and high specific gravity are diagnostic.
Corundum	Six-sided prismatic crystals with flat ends; hardness of 9 is diagnostic. Most lab specimens have dull luster and are gray, brown.
Dolomite	Similar to calcite, but only weak or no reaction with HCl placed on a grain of the mineral; *powder* reacts strongly. Slightly curved rhombohedral crystals.
Epidote	Small crystals and thin, granular coatings form in some metamorphic rocks and [by alteration of] some igneous rocks.
Fluorite	Commonly in cube-shaped crystals with four cleavage directions cutting corners of the cubes.
Galena	Commonly in cube-shaped crystals with three perfect cleavages at 90°.
Garnet*	Most commonly dark red; 12- or 24-sided crystals in metamorphic rocks.
Graphite	Greasy feel; leaves a mark on paper.
Gypsum	Two varieties: *selenite* is colorless and nearly transparent with perfect cleavage; *alabaster* is a rock—an aggregate of grains with an earthy luster.
Halite	Cubic crystals and taste are diagnostic.
Hematite	Two varieties: most common is reddish-brown masses with earthy luster; rare variety is black crystals with metallic luster.

* Indicates mineral family.

Common Minerals and Their Properties (continued)

Mineral	Additional diagnostic properties and occurrences
Hornblende	Dark green to black amphibole; two cleavages at 56° and 124°.
Kaolinite	Earthy, powdery white to gray masses; sticks to tongue.
Kyanite	Bladed blue or blue-green crystals; H = 5 parallel to blade, H = 7 across blade.
Limonite	Earthy, yellow-brown masses, sometimes powdery; forms by the "rusting" (oxidation) of iron-bearing minerals.
Magnetite	Gray-black; H = 6; magnetic.
Malachite	Bright green. Copper will plate out on a steel nail dipped into a drop of HCl and placed on this mineral.
Muscovite	A colorless mica; one perfect cleavage; peels into flexible sheets.
Olivine	Commonly as aggregates of green granular crystals.
Plagioclase feldspar*	Play of colors and striations distinguish plagioclase feldspars from potassic feldspars, which do not show these properties.
Potassic feldspar*	
Pyrite	Brassy gold color, hardness, and black streak are diagnostic. Cubic crystals with striations on their faces or in 12-sided crystals with five-sided faces.
Pyrrhotite	Brownish-bronze color; black streak; may be slightly magnetic.
Pyroxene*	Two cleavages at 87° and 93°; major constituent of mafic and ultramafic igneous rocks.
Quartz	Wide range of colors; six-sided crystal shape, high hardness (7) and conchoidal fracture are diagnostic.
Serpentine	Dull white, gray, or green masses; sometimes fibrous (asbestos).
Siderite	Three cleavages not at 90°; looks like brown calcite; powder may react to HCl.
Sillimanite	Gray, white, brown; slender crystals, sometimes needlelike.
Sphalerite	Wide variety of colors (including colorless); distinctive pale yellow streak.
Staurolite	Reddish-brown to dark brown stubby crystals in metamorphic rocks.
Sulfur	Bright yellow with yellow streak; greasy luster.
Sylvite	Looks like halite but has very bitter taste.
Talc	Greasy feel; H = 1.

*Indicates mineral family.

MINERAL PROFILE DATA SHEET

Sample	Luster	Hardness	Streak	Color	Cleavage or fracture (describe)	Other properties	Mineral name and composition

Name _____

MINERAL PROFILE DATA SHEET

Sample	Luster	Hardness	Streak	Color	Cleavage or fracture (describe)	Other properties	Mineral name and composition

MINERAL PROFILE DATA SHEET

Name _____

Sample	Luster	Hardness	Streak	Color	Cleavage or fracture (describe)	Other properties	Mineral name and composition

MINERAL PROFILE DATA SHEET   Name _____

Sample	Luster	Hardness	Streak	Color	Cleavage or fracture (describe)	Other properties	Mineral name and composition

4

Minerals, Rocks, and the Rock Cycle

From left to right: sedimentary (conglomerate), metamorphic (gneiss), and igneous (granite) rocks are the three main classes of rock.

LEARNING OBJECTIVES

- Understand the differences between minerals and rocks
- Become familiar with the three main classes of rock: igneous, sedimentary, metamorphic
- Understand the different processes that form rocks. Learn how the grains in a rock act as records of geologic processes
- Use textures and mineral content to determine if rocks are igneous, sedimentary, or metamorphic
- Understand the rock cycle: how rocks are recycled to make new ones

MATERIALS NEEDED

- An assortment of igneous, sedimentary, and metamorphic rocks
- Supplies to make artificial igneous, sedimentary, and metamorphic rocks; tongs for moving hot petri dishes, hot plate, sugar, thymol, Na-acetate in a dropper bottle, sand grains, glass petri dishes, $Ca(OH)_2$ solution, straws, Play-Doh, plastic coffee stirring rods, plastic chips
- Magnifying glass or hand lens
- Mineral-testing supplies: streak plate, steel nail, and so on

4.1 Introduction

A rock is an aggregate of mineral grains, fragments of previously existing rock, or a mass of natural glass. Rocks can form in a variety of ways—through cooling and solidification of a melt, by cementation of loose grains, by precipitation from water solutions, or from changes that happen in response to temperature and pressure underground. Geologists can identify rocks and interpret aspects of Earth's history by studying two principal characteristics of rocks: **rock composition** (the proportions of different chemicals and thus the identity of minerals or glass that make up a rock) and **rock texture** (the dimensions and shape of grains, and the ways that grains are arranged, oriented, and held together). On our dynamic planet, rocks don't survive forever—nature recycles materials, using those in one rock to form new rocks through a series of steps called the *rock cycle*.

In this chapter, we first introduce the three basic classes of rocks. Then we describe how to observe the characteristics of rocks and how to use these characteristics to classify rocks. Then you will make your own "rocks" in the classroom and see how different processes produce different textures. By combining your mineral identification skills and knowledge of rock textures, you can face the challenge of determining how specimens of common rocks have formed.

4.2 The Three Classes of Rocks

Geologists struggled for centuries with the question of how to classify rocks. They finally concluded that rocks can best be classified on the basis of how they formed. Rock composition and texture provide the basis for this classification, for these characteristics reflect the process of formation. In modern terminology, we distinguish three classes of rocks: igneous, sedimentary, and metamorphic.

■ **Igneous rocks** form through the cooling and solidification of molten rock, which is created by the melting of preexisting rock in the mantle or lower crust. We refer to molten rock below Earth's surface as **magma** and to molten rock that has erupted onto the surface as **lava**. Some volcanoes erupt explosively, blasting rock fragments into the air, and when these fragments fall back to Earth, coalesce, and solidify, the resulting rock is also considered to be igneous.

■ **Sedimentary rocks** form at or near the surface of the Earth in two basic ways: (1) when grains of preexisting rocks accumulate, are buried, and then are cemented together by minerals precipitating out of groundwater; and (2) when minerals precipitate out of water near the Earth's surface, either directly or through the life function of an organism, and either form a solid mass or are cemented together later. The grains that become incorporated in sedimentary rocks form when preexisting rocks are broken down into their smallest pieces by processes and interactions with air, water, and living organisms. These interactions are called **weathering**. Some weathering simply involves the physical fragmentation of rock. Other kinds of weathering involve chemical reactions that produce new minerals, most notably clay. The products of weathering can be transported (eroded) by water, ice, or wind to the site where they are deposited (settle out), buried, and transformed into new rock.

■ **Metamorphic rocks** form when preexisting rocks are subjected to physical and chemical conditions within the Earth, such as the increase in pressure and temperature and/or shearing at elevated temperature. For example, when buried very deeply, rock is warmed to high temperature and squeezed by high pressure. The texture and/or mineral content of the initial rock changes in response to the new conditions in *the solid state*. Metamorphic rock thus forms without the melting or weathering that makes igneous and sedimentary rocks.

Environments of Rock Formation

Name: _____ Section: _____
Course: _____ Date: _____

Three settings are described below in which three different classes of rocks form. Fill in the blanks by indicating which class of rocks is the result.

(a) Along the coast, waves carry sand out into deeper, quieter water, where it accumulates and gradually becomes buried. A rock formed from sand grains cemented together is _____.

(b) Thick flows of glowing red lava engulf farms and villages on the flanks of Mt. Etna, a volcano in Sicily. The cold, black rock formed when these flows are cool enough to walk on is _____.

(c) The broad tundra plains of northern Canada expose gray, massive rock that formed many kilometers below a mountain belt. This rock, exposed only after the overlying rock was stripped away, is _____.

4.2.1 The Rock Cycle

An igneous rock that has been exposed at the surface of the Earth will not last forever. Minerals that crystallized from magma to form the rock can be broken apart at the surface by weathering, undergo **erosion** (the grinding effects of moving ice, water, or air), be transported by streams, be deposited in the ocean, and be cemented together to form a sedimentary rock. These same minerals, now in a sedimentary rock, can later be buried so deeply beneath other sedimentary rocks that they are heated and squeezed to form a metamorphic rock with new minerals. Erosion may eventually expose the metamorphic rock at the Earth's surface, where it can be weathered to form new sediment and eventually a different sedimentary rock. Or the metamorphic rock may be buried so deeply that it melts to produce magma and eventually becomes a new igneous rock with still different minerals.

This flux (flow) of material from one rock type to another over geologic time is called the **rock cycle** (**FIG. 4.1**). The rock cycle involves the reuse of mineral grains or the breakdown of minerals into their constituent atoms and the reuse of those atoms to make other minerals. Rocks formed at each step of the rock cycle look very different from their predecessors because of the different processes by which they formed.

4.3 A Rock Is More than the Sum of Its Minerals

The first thing a geologist wants to determine about a rock is whether it is igneous, sedimentary, or metamorphic. Composition *alone* (i.e., the component minerals of the rock) is rarely enough to define a rock's origin, because some of the most common minerals (quartz, potassic feldspar, sodium-rich plagioclase feldspar) are found in all three rock classes. For example, a rock made of these minerals could be igneous (granite), sedimentary (sandstone), or metamorphic (gneiss). Texture *alone* can identify which type of process was involved for many rocks, but some textures can also develop in all three types of rock. The best clue to the origin of a rock is its unique combination of texture *and* mineralogy.

4.3.1 Describing Texture

Geologists begin characterizing a rock's texture by asking: Does the rock consist of a mass of glass or does it consist of grains? If it consists of glass, it will be shiny and develop conchoidal fractures (curving, ridged surfaces). Rocks composed of glass

FIGURE 4.1 The rock cycle.

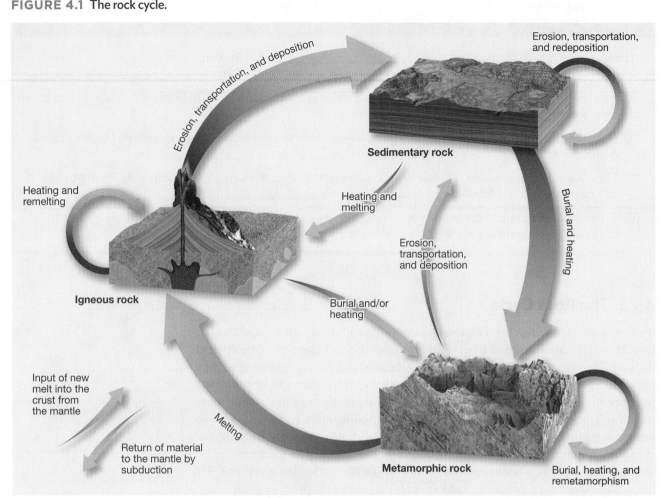

are relatively rare—they form only at certain kinds of volcanoes. In most cases, we jump to the next question: What are the grains like, and how are they connected? To answer this question, we first describe the size and shape of grains and then determine if the grains are interlocking, like pieces of a jigsaw puzzle, or if they are stuck together by cement (mineral "glue").

Grain Size A rock description should include the size(s) of grains and whether the rock has a single (homogeneous) grain size, a wide range of sizes, or two distinctly different sizes. Are the grains **coarse** enough to be identified easily;

TABLE 4.1 Common grain size terminology

Grain size term	Practical definition	Approximate size range
Coarse grained	Grains are large enough so you can identify the minerals or grains present.	Larger than 5 mm
Medium grained	Individual grains can be seen with the naked eye but are too small to be identified.	0.1–4.9 mm
Fine grained	Individual grains are too small to be seen with the naked eye or identified with a hand lens; feels "gritty" to the edge of a fingernail.	Smaller than 0.1 mm
Very fine grained	Individual grains cannot be seen with the naked eye *or* a hand lens; feels smooth to the edge of a fingernail.	Much smaller than 0.1 mm
Glassy	No grains at all; rock is a homogeneous mass of glass.	—

FIGURE 4.2 Grain sizes in sedimentary rocks (a–d) and an igneous rock (e).

(a) Very fine grained.

(b) Fine grained.

1 cm

(c) Medium grained.

(d) Coarse grained.

(e) Glassy.

medium sized, so you can see that there are separate grains but can't easily identify them; **fine**, so that they can barely be recognized; or **very fine**, so small that
they cannot be seen even with a magnifying glass? Special terms are used for
grain size in igneous and sedimentary rocks, but here we use common words that
work just as well (**TABLE 4.1** and **FIG. 4.2**).

Note that some rocks contain grains of different sizes, as in Figure 4.2d where very
coarse grains are embedded in a very-fine-grained matrix. As we will see later, these
mixed grain sizes give important information about the rock-forming processes.

Grain Shape When we describe the shape of grains, we ask: Are the grains **equant**, meaning they have the same dimension in all directions, or **inequant**, meaning that the dimension in different directions is different? Grains that resemble spheres or cubes are equant, like the garnets in **FIGURE 4.3a** or the halite crystals in **FIGURE 4.3b**. Those that resemble rods, sheets, ovals, or have an irregular form are inequant, like the shapes in **FIGURE 4.3c, d**. Next, we look at the grains to see if they are rounded, angular, or some combination of these shapes. Finally we ask: Are *all* grains the same shape, or do they vary depending on mineral type or on position within the rock? Each of these characteristics tells us something about the formation of the rock.

Relationships among Grains During most igneous- and metamorphic-rock-forming processes, and during some sedimentary-rock-forming processes, crystallizing minerals interfere with one another and interlock like a three-dimensional jigsaw puzzle (**FIG. 4.4a, b**). We refer to the result as a **crystalline texture**. A very different texture results from sedimentary processes that deposit grains and then cement them together (**FIG. 4.4c, d**). Such sedimentary rocks are said to have a **clastic texture**, and individual grains in these rocks are called **clasts**.

Grain Orientation and Alignment The orientation of grains relative to one another often provides a key clue to the classification of a rock. For example, in most igneous rocks, the inequant grains are randomly oriented in that they point in a variety of directions. In many metamorphic rocks, however, inequant minerals,

FIGURE 4.3 Typical grain shapes.

(a) Garnet crystals in metamorphic rock.

(b) Halite crystals in sedimentary rock.

(c) Rounded grains.

(d) Angular grains. (Photo courtesy of Michael C. Rygel.)

FIGURE 4.4 Grain relationships.

(a) Crystalline texture: interlocking quartz and feldspar grains in an igneous rock. The dark grains are biotite and hornblende.

(b) Crystalline texture: photomicrograph of interlocking calcite grains in a metamorphic rock.

(c) Clastic texture: cemented grains in a sedimentary rock.

(d) Clastic texture: cemented fossils.

such as platy micas or elongate amphiboles, are oriented parallel to one another. Thus, when looking at a rock, you should ask: Are the inequant grains aligned parallel to one another or are they randomly oriented throughout the rock? If the grains are aligned parallel to one another, the rock develops a type of layering called **foliation** (**FIG. 4.5**).

Now that we've developed a basic vocabulary for important characteristics of rocks, we can begin to describe a rock the way a geologist would. You will try this in Exercise 4.2.

FIGURE 4.5 A foliated rock (schist).

Name: _____   Section: _____
Course: _____   Date: _____

Describe the *texture* of the rocks provided by your instructor. Be sure to consider the size and shape of grains, the relation-ships among grains, and whether grains are randomly oriented or aligned. Use your own words—the formal geologic terms for the textures will be introduced later.

(a) Rock 1:

_____

_____

_____

(b) Rock 2:

_____

_____

_____

(c) Rock 3:

_____

_____

_____

(d) What textural features are common to all three rock samples, and which are not?

_____

_____

_____

(e) Did all three rocks form by the same process? Explain your reasoning.

_____

_____

_____

_____

4.4 The Processes That Produce Textures

The best way to understand how a rock's texture forms is to observe the rock-forming process in action. You can't go inside a lava flow to watch an igneous rock crystallize, or deep below a mountain belt to see rock metamorphose, or under the ocean floor to watch sediment become a sedimentary rock—but you can *model* the formation of igneous, sedimentary, and metamorphic rocks well enough in the classroom to understand how textures form. You will use textural features more fully later in detailed interpretations of igneous, sedimentary, and metamorphic rocks. In Exercise 4.3, you will conduct simple experiments to understand how igneous, sedimentary, and metamorphic processes create different and distinctive textures.

EXERCISE 4.3 **Understanding the Origin of Rock Textures**

Name: _____   Section: _____
Course: _____   Date: _____

In this exercise, you will conduct simple experiments to better understand rock textures.

(a) **Crystalline igneous rock.** Partially fill ($\frac{1}{2}$ inch) a glass petri dish with the powder provided by your instructor. Heat on a hot plate until the material melts completely. Carefully remove the dish from the hot plate using forceps, and allow the liquid to cool. Observe the crystallization process closely. You may have to add a crystal seed if crystallization does not begin in a few minutes. Congratulations—you have just made a "magma" and then an "igneous rock" with an interlocking crystalline texture.
  • Describe the crystallization process. How and where did individual grains grow, and how did they eventually join with neighboring grains?

_____

_____

_____

Sketch and describe the texture of the cooled "rock."

Sketch: Describe: _____

_____

_____

_____

_____

_____

_____

_____

_____

_____

This crystalline texture formed when crystals in the cooling "magma" grew until they interfered with one another and eventually interlocked to form a cohesive solid. Crystals grow only when the magma cools slowly enough for atoms to have time to fit into a crystalline lattice.

(continued)

Name: _____   Section: _____
Course: _____   Date: _____

(b) Glassy igneous texture. Add sugar ($\frac{1}{2}$ inch) to another petri dish, melt it on a hot plate, and allow it to cool.

Describe the cooling process and sketch the resulting texture.

Sketch: Describe: _____

_____

_____

_____

_____

_____

_____

_____

This glassy texture is typical of igneous rocks that cool so quickly that atoms do not have time to arrange into crystalline lattices. In some cooling lavas, the bubbles that you saw may be preserved as the lava freezes.

(c) Clastic sedimentary texture. Cover the bottom of another petri dish with a mixture of sand grains and small pebbles. Add a small amount of the liquid provided by your instructor, and allow the mixture to sit until the liquid has evaporated. Now turn the dish upside down.
  • Why don't the grains fall out of the dish?

_____

_____

Examine the "rock" with your hand lens. Sketch and describe the texture.

Sketch: Describe: _____

_____

_____

_____

_____

_____

_____

_____

You have just made a clastic sedimentary rock composed of individual grains cemented to one another.

(continued)

Name: _____   Section: _____

Course: _____   Date: _____

(d) **Fine-grained chemical sedimentary texture.** Using a straw, blow *gently* into the beaker of clear liquid provided by your instructor until you notice a change in the water. Continue blowing for another minute and then observe what happens.

- Describe the change that occurred in the liquid as you were blowing into it.

_____

_____

_____

- Describe what happened to the liquid after you stopped blowing.

_____

_____

_____

- Pour off the liquid and add a drop of dilute hydrochloric acid to the mineral that you formed. What is the mineral?

_____

You have just made a chemical sediment, one that forms when minerals precipitate (grow and settle out) from a solution. This sediment would be compacted and solidified to make a very fine-grained sedimentary rock.

(e) **Crystalline chemical sedimentary texture.** Suspend a piece of cotton twine in a supersaturated salt solution, and allow the salt to evaporate over a few days.

Sketch and describe the texture of the salt crystals attached to the twine; compare it with the interlocking igneous texture from Exercise 4.3a.

Sketch: Describe: _____

_____

_____

_____

_____

_____

_____

_____

You've now seen that crystalline textures can form in both sedimentary and igneous rocks. How might you distinguish the rocks? (*Hint:* Think about whether the chemical composition of lava is the same as that of seawater.) By the way, crystalline textures also form in metamorphic rocks.

_____

_____

_____

(continued)

Name: _____   Section: _____
Course: _____   Date: _____

(f) Metamorphic foliation. Take a handful of small plastic chips and push them with random orientation into a mass of Play-Doh. Then flatten the Play-Doh with a book.

Sketch and describe the orientation of the plastic chips in the "rock." Are they randomly oriented, as they were originally? How is their alignment related to the pressure you applied?

Sketch: Describe: _____

_____

_____

_____

_____

_____

_____

_____

The parallel alignment of platy minerals is called *foliation* and is found in metamorphic rocks that have been strongly squeezed, as at a convergent boundary between colliding lithosphere plates.

• What real minerals would you expect to behave the way the plastic chips did?

_____

_____

EXERCISE 4.4   **Interpreting the Texture of Real Rocks**

Name: _____   Section: _____
Course: _____   Date: _____

Now look at the rocks used in Exercise 4.2. Compare their textures with those that you made in Exercise 4.3, and suggest whether each is igneous, sedimentary, or metamorphic. Explain your reasoning.

Rock 1:

_____

_____

_____

_____

(continued)

Name: _____ Section: _____
Course: _____ Date: _____

Rock 2:

_____

_____

_____

_____

Rock 3:

_____

_____

_____

_____

_____

4.5 Clues about a Rock's Origin from the Minerals It Contains

Some minerals can form in only one class of rocks, so their presence in a sample tells us immediately whether the rock is igneous, sedimentary, or metamorphic. For example, halite forms only by the evaporation of salt water, and staurolite forms only by the metamorphism of aluminum-rich sedimentary rocks at high temperature. However, as mentioned earlier, several common minerals can form in more than one class of rocks. For example, quartz crystallizes in many magmas; but it also forms during metamorphic reactions and can precipitate from water to form a fine-grained chemical sedimentary rock or to cement grains together in clastic sedimentary rocks.

TABLE 4.2 shows which common minerals indicate a unique rock-forming process and which can form in two or more ways. Remember the rock cycle: a grain of quartz that originally *formed* by cooling from magma may now be *found* in a sedimentary or metamorphic rock.

The presence of some minerals in a rock is evidence that the rock can only be igneous, sedimentary, or metamorphic. Thus, a rock containing gypsum *must be* sedimentary. The presence of sillimanite in a rock means the rock *is* metamorphic. The blue triangles in Table 4.2 highlight minerals whose presence gives less definite evidence. These can survive the breakup of the igneous or metamorphic rocks in which they formed originally, be deposited along with other clasts, and be found in small amounts in sedimentary rocks. Other minerals, like quartz, can form in igneous, sedimentary, *or* metamorphic rocks, so their presence doesn't help at all in determining a rock's origin.

TABLE 4.2 Occurrence of common rock-forming minerals.

Mineral	Igneous	Sedimentary	Metamorphic
Plagioclase feldspar	▲	▼ ■	▲
Potassic feldspar	▲	▼ ■	▲
Quartz	▲	▼ ▲	▲
Quartz can form (▲) in any rock. It can also form in an igneous or metamorphic rock and be deposited and found (▼) when the rocks break up in a sedimentary rock.			
Hornblende	▲		▲
Actinolite			▲
Augite	▲		
Muscovite	▲	▼	▲
Biotite	▲		▲
Chlorite			▲
Olivine	▲		▲
Garnet	▲	▼	▲
Andalusite	■		▲
Kyanite			▲
Sillimanite			▲
A rock containing sillimanite must be metamorphic.			
Epidote	■		▲
Halite		▲	
Calcite		▲	▲
Dolomite		▲	▲
Gypsum		▲	
A rock containing gypsum must be sedimentary.			
Talc			▲
Serpentine	■		▲

▲ Commonly *forms* in these rocks ■ Rarely *forms* in these rocks
▼ Commonly *found* in these rocks as clasts

4.6 Identifying Minerals in Rocks

The techniques you used to identify individual specimens of minerals are also used when minerals are combined with others in rocks, but the relationships among grains may make the process more difficult, especially when the grains are small. For example, it is usually easy to determine how many minerals are in coarse-grained rocks because color and luster are as easy to determine in rocks as in minerals. But it is difficult to determine hardness and cleavage of small grains without interference from their neighbors. And when a rock contains many mineral grains cemented together, it may be difficult to distinguish between the hardness of individual minerals and the strength of the cement holding them together.

Extra care must be taken with rocks to be sure you are measuring the mineral properties you think you are measuring. In very-fine-grained rocks, geologists have to use microscopes and even more sophisticated methods (e.g., X-rays) to identify the minerals. Keep the following tips in mind as you determine the properties of minerals in rocks in Exercises 4.5 and 4.6.

■ *Color:* Whenever possible, look at a rock's weathered outer surface *and* a freshly broken surface, because weathering often produces a surface color different from the color of unaltered minerals. This difference can be helpful for identification. For example, regardless of whether plagioclase feldspar is dark gray, light gray, or colorless, weathering typically produces a very fine white coating of clay minerals. Weathering of a fine-grained rock can help distinguish dark gray plagioclase (white weathering) from dark gray pyroxene (brown weathering) that might otherwise be hard to tell apart.

A rock's color depends on its grain size as well as the color of its minerals. All other things being equal, a fine-grained rock appears darker than a coarse-grained rock made of the same minerals.

■ *Luster:* Using a hand lens, rotate the rock in the light to determine how many different kinds of luster (and therefore how many different minerals) are present.

■ *Hardness:* Use a steel safety pin or the tip of a knife blade and a magnifying glass to determine mineral hardness in fine-grained rocks. Be sure to scratch a single mineral grain or crystal; and when using a glass plate, be sure a single grain is scratching it. Only then are you testing the mineral's hardness. Otherwise, you might be dislodging grains from the rock and demonstrating how strongly the rock is cemented together rather than measuring the hardness of any of its minerals.

■ *Streak:* Similarly, be very careful when using a streak plate, because it may not be obvious which mineral in a fine-grained rock is leaving the streak. Some minerals are also harder than the plate. As a result, streak is generally not useful for identifying single grains in a rock that contains many minerals.

■ *Crystal habit:* Crystal habit is valuable in identifying minerals in a rock when it has well-shaped crystals. But when many grains interfere with one another during growth, the result is an interlocking mass of irregular grains rather than crystals. Weathering and transportation break off corners of grains and round their edges, destroying whatever crystal forms were present.

■ *Breakage:* Use a hand lens to observe cleavage and fracture. Rotate the rock in the light while looking at a single grain. Remember that multiple *parallel* shiny surfaces represent a single cleavage direction. Breakage can be a valuable property in distinguishing light-colored feldspars (two directions of cleavage at 90°) from quartz (conchoidal fracture), and amphibole (two directions of cleavage not at 90°) from pyroxene (two directions of cleavage at 90°).

■ *Specific gravity:* A mineral's specific gravity can be measured only from a pure sample. When two or more minerals are present in a rock, you are measuring the *rock's* specific gravity, and that includes contributions from all of the minerals present. However, the heft of a very-fine-grained rock can be very helpful in interpreting what combination of minerals might be in it.

■ *Hydrochloric acid test for carbonates:* Put a *small* drop of acid on a fresh, unaltered surface. Be careful: a thin film of soil or weathering products may contain calcite and make a noncarbonate mineral appear to fizz. Use a hand lens to determine exactly where the carbon dioxide is coming from. Are all of the grains reacting with the acid or is gas coming only from the cement that holds the grains together?

Name: _____   Section: _____
Course: _____   Date: _____

I.   Your instructor will provide similar-sized samples of granite and basalt. Granite has grains that are large enough for you to identify. It forms by slow cooling of molten rock deep underground. Basalt also forms from cooling molten material, but it cools faster and has much smaller grains.

(a) Examine the granite. Does it have well-shaped crystals or irregular grains? Suggest an explanation for their shapes.

_____

_____

_____

(b) Identify three or four minerals in the granite. Use a hand lens or magnifying glass if necessary.

_____

_____

(c) Now examine the basalt. Identify the minerals present if you can, using a hand lens or magnifying glass. What problems did you have?

_____

_____

_____

(d) Heft the granite and basalt. Are their specific gravities similar or is one denser than the other? If so, which is denser?

_____

_____

(e) Based on specific gravity, do you think that the most abundant minerals in the basalt are the same as those in the granite? Explain.

_____

_____

_____

(continued)

Name: _____　**Section:** _____

Course: _____　**Date:** _____

II. Your instructor will provide samples of sandstone (made mostly of quartz) and limestone (made mostly of calcite).

(f) Suggest two tests that will enable you to tell which sample is which. Explain.

_____

_____

_____

_____

_____

_____

(g) Okay: Which is the sandstone and which is the limestone?

_____

_____

_____

_____

_____

_____

4.7　Interpreting the Origin of Rocks

You are now ready to begin your examination of rocks. In Exercise 4.6, the goal is to determine which of the three types of rock-forming processes was involved in the origin of each specimen. Later in the book, you will examine each group separately and be more specific about details of those processes. The flowchart on p. 100 combines textural and mineralogical features to make your task easier. Then Exercise 4.7 asks you to apply your knowledge to make a recommendation.

EXERCISE 4.6　　Classifying Rocks: Igneous, Sedimentary, or Metamorphic?

Name: _____　**Section:** _____

Course: _____　**Date:** _____

Use the flowchart on page 103 to separate the rocks provided by your instructor into igneous, sedimentary, and metamorphic piles based on their textures, mineral content, and the information in Table 4.2. You will look at these rocks again in more detail in the next three chapters.

(continued)

Name: _____   Section: _____
Course: _____   Date: _____

Igneous rocks	
#	Reasons for classifying as igneous

Sedimentary rocks	
#	Reasons for classifying as sedimentary

Metamorphic rocks	
#	Reasons for classifying as metamorphic

(continued)

Name: _____   Section: _____
Course: _____   Date: _____

What is the rock's texture?

1. Rock has **glassy** texture ⟶ Rock is **IGNEOUS**.
   (smooth, shiny, no grains).

2. Rock has **porous**   Look at material   No minerals (glass) ⟶ Rock is **IGNEOUS**.
   texture (numerous   between grains.   Nonsilicate minerals ⟶ Rock is **SEDIMENTARY**.
   holes).   Silicate minerals ⟶ Rock is **IGNEOUS**.

3. Rock has **interlocking**
   grains.

   Identify minerals.

   Nonsilicates — Halite, gypsum ⟶ Rock is **SEDIMENTARY**.
   Calcite, dolomite ⟶ Rock may be — **SEDIMENTARY***.
   METAMORPHIC*.

   Silicates

   Foliated ⟶ Rock is **METAMORPHIC**.

   NONFOLIATED
   Metamorphic minerals (garnet, staurolite, etc.) — Rock is **METAMORPHIC**.
   No metamorphic minerals ⟶ Rock is **IGNEOUS**.

   *** Look for metamorphic minerals. Are interlayered rocks in the field sedimentary or metamorphic?**

4. Rock consists of **visible grains separated by finer-grained material.**

   (a) Matrix is **glassy.** ⟶ Rock is **IGNEOUS**.

   (b) Matrix is **interlocking crystals.**   Identify matrix minerals.
   Silicate mineral matrix ⟶ Rock is **IGNEOUS** or **METAMORPHIC***.
   Nonsilicate mineral matrix ⟶ Rock is **SEDIMENTARY**.
   *If metamorphic minerals are present, then the rock is metamorphic.

   (c) Matrix is **small, rounded grains that do not interlock.** ⟶ Rock is **SEDIMENTARY**.

   (d) Matrix is **too fine grained to identify.** ⟶ Rock could be **SEDIMENTARY** or **IGNEOUS**. Use microscope.

5. Rock is **very fine grained:** i. Use microscope to observe relations between grains if thin section is available.
   ii. Use field observations to determine relations with rocks of known origin.
   iii. Estimate specific gravity to get an indication of what minerals are present.

EXERCISE 4.7 What Do You Think?

Name: ______________________________ Section: ____________
Course: ______________________________ Date: ____________

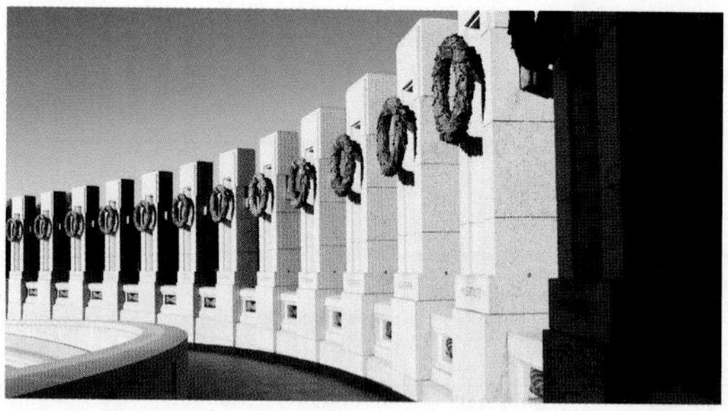

What Do You Think A veterans' group has contacted you about the best stone to use for memorials honoring the men and women who have served our country. They want to use local stone whenever possible and they have found a nearby source of marble and granite. Assuming this is taking place near your institution, what factors should you consider before making a recommendation? (Consider the ease of carving and polishing, the response of the materials to weathering, etc.) What factors would you consider if you lived in a very wet and humid climate? What factors would you consider if you lived in a hot and arid climate?

5

Using Igneous Rocks to Interpret Earth History

Solidified pahoehoe lava engulfs an intersection in Hawaii—an example of extrusive igneous rock.

- Become familiar with igneous textures and mineral assemblages
- Use texture and mineral content to interpret the history of igneous rocks
- Understand the classification of igneous rocks
- Associate igneous rocks with different plate tectonic settings
- Recognize the kinds of volcanoes and why they have different shapes

MATERIALS
NEEDED

- Set of igneous rocks
- Magnifying glass or hand lens and, ideally, a microscope and thin sections of igneous rocks
- Standard supplies for identifying minerals (streak plate, glass plate, etc.)

5.1 Introduction

Every rock has a story to tell. The story of an igneous rock begins when rock in the lower crust or upper mantle melts to form molten material called **magma**, which rises up through the crust. Some magma flows or spatters out on the surface as **lava** or it explodes—often through a crack or vent in the Earth—into the air as tiny particles of **volcanic ash** or larger blocks. Igneous rock that forms from solidified lava or ash is called **extrusive igneous rock** because it comes out (*extrudes*) onto the surface. Other magma never reaches the surface and solidifies underground to form **intrusive igneous rock**, so called because it squeezes into (*intrudes*) the surrounding rocks. Intrusions come in many shapes. Massive blobs are called **plutons**, and the largest of these—called **batholiths**—are generally composites of several plutons. Other intrusions that form thin sheets cutting across layering in the wall rock (the rock around the intrusion) are called **dikes**, and those that form thin sheets parallel to the layers of wall rock are called **sills** (**FIG. 5.1**).

When looking at an igneous rock, geologists want to know: Where did it cool (was it intrusive or extrusive)? Where in the Earth did the rock's parent magma form? In what tectonic setting—ocean ridge, mid-continent, subduction zone, hot spot—did it form? In this chapter, you will learn how to answer these questions and how to identify common types of igneous rock. Few new skills are needed—just observe carefully and apply some geologic reasoning. Exercise 5.1 shows how easy the process is.

Most of the questions geologists ask about an igneous rock can be answered with three simple observations: grain size, color, and specific gravity. The following sections provide the additional information you need to interpret the history of an igneous rock.

FIGURE 5.1 Intrusive and extrusive igneous rock bodies.

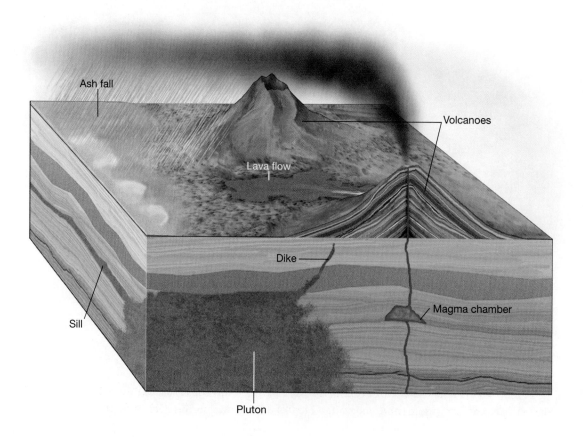

Ash fall

Volcanoes

Lava flow

Dike

Magma chamber

Sill

Pluton

Name: _____   Section: _____

Course: _____   Date: _____

There are many ways to classify igneous rocks, but for now let's use three easily observable criteria: grain size, color, and specific gravity. Using the set of igneous rocks provided by your instructor, first group the specimens by grain size and record the specimen numbers in the appropriate column of the following table. Then group the specimens by color and specific gravity.

Grain size		Color		Specific gravity (heft)	
Coarse	Fine	Light colored	Dark colored	Relatively high	Relatively low

(a) Which two properties seem to be related to one another? Which is different?

_____

_____

(b) Consider which magma properties and rock-forming processes control the grain size, color, and specific gravity of an igneous rock. Which two of these properties have a common cause and are related to one another? Explain. (This connection will be helpful later when you are identifying igneous rocks.)

_____

_____

_____

_____

_____

5.2 Interpreting the Cooling Histories of Igneous Rocks

Imagine that you are looking at an ancient igneous rock in an outcrop. That rock might have formed from volcanic debris blasted into the air, lava that frothed out of a volcano or flowed smoothly across the ground, magma that cooled just below the surface, or magma that solidified many kilometers below the surface. But millions of years have passed since the rock formed, and if a volcano had been involved, it has long since been eroded away. If the rock was intrusive, kilometers of overlying rock would have been removed to expose it at the surface. How can you determine which of these possibilities is the right one?

    The key to understanding the cooling history of an igneous rock is its **texture**—the size, shape, and arrangement of its grains. The texture of an igneous rock, is formed by the settling of ash and other volcanic fragments, looks very different from that of a rock formed by magma cooling deep underground or by lava cooling at the surface.

Specimens composed of interlocking grains—whether large enough to be identified or too small to be identified—are called **crystalline** rocks. Those that are shiny and contain no grains are called **glasses**; sponge-like masses are said to be **porous**; and those that appear to have pieces cemented together are said to be **fragmental**. Each of these textures indicates a unique cooling history—if you understand how to "read" the textural information. Exercise 5.2 will help you start this process.

5.2.1 Grain Size in Crystalline Igneous Rock

Grain size is the key to understanding the cooling history of most igneous rocks that solidified underground or on the surface (you will experiment with this in Exercise 5.3 and 5.4). When magma or lava begins to cool, small crystal seeds form (a process called *nucleation*), and crystals grow outward from the seeds until they interfere with one another. The result is a three-dimensional interlocking texture found in *most* igneous rocks. But why do some igneous rocks have coarser (larger) grains than others?

Crystals grow as ions migrate through magma to the crystal seeds, so anything that assists ionic migration increases grain size in an igneous rock. **Cooling rate** is the most important factor controlling grain size. The slower a magma cools, the more time ions have to migrate to crystal seeds; the faster it cools, the less time there is and the smaller the grains will be. Another factor is a magma's **viscosity** (the resistance of material to flow). The *less* viscous a magma is (i.e., the more *fluid*), the easier it is for ions to migrate and the larger the crystals can become.

Cooling rate and viscosity cause some igneous rocks to have coarse grains (sometimes called a **phaneritic texture**) and others to have fine grains (**aphanitic texture**). **FIGURE 5.2** illustrates the difference between course and fine grains.

EXERCISE 5.2	A First Look at Igneous Rock Textures

Name: _____   Section: _____
Course: _____   Date: _____

We look first at rocks that have the same composition, because their different textures could only be caused by the different ways in which they cooled. For example, the rocks in Figure 5.2 have the same minerals and thus similar compositions. Separate the light-colored igneous rocks in your set. Describe their textures, paying careful attention to the sizes and shapes of the grains and the relationships among adjacent grains. Use everyday language—the appropriate geologic terms will be introduced later.

Specimen	Textural description

Some igneous rocks have grains of two different sizes, one much larger than the other (**FIG. 5.3**). This is called a **porphyritic texture**. The larger grains are called *phenocrysts* and the finer grains are called, collectively, the rock's *groundmass*.

·GURE 5.2 Grain sizes in light-colored igneous rocks.

  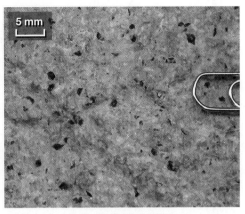

y coarse-grained (pegmatitic crystals).   **(b)** Coarse-grained (phaneritic crystals).   **(c)** Fine-grained (most of the aphanitic grains are too small to see with the naked eye).

EXERCISE 5.3	Interpreting Igneous Cooling Histories from Grain Size

Name: _____   Section: _____
Course: _____   Date: _____

It is a short step from understanding how cooling rate affects grain size to deriving the basic rules of how magma and lava cool in nature. Three simple thought experiments will help you understand the cooling process.

(a) Imagine you have two balls of pizza dough 20 cm in diameter. As shown in the figure on the right, one is rolled out to form a crust 1 cm thick and 50 cm in diameter. Both pieces of dough still have the same mass and volume, even though their shapes are very different. Both are baked in a 450°F oven for 20 minutes and removed.

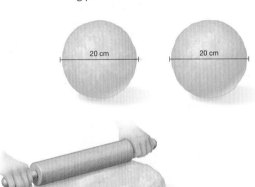

(i) Which will cool faster, the large cube or one of the small cubes? _____

Why? _____

_____

Rule 1 of magma cooling: *Circle the correct two choices in the following statement.* A thin sheet of magma loses heat (**faster/ slower**) than a blob containing the same amount of magma. This is because the amount of surface area available for cooling in the sheet is (**larger/smaller**) than the surface area available in a blob of the same volume.

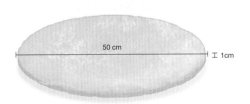

(b) Equal amounts of hot coffee are poured into thin plastic and Styrofoam cups.
(i) Which cools faster, the coffee in the plastic cup or in the Styrofoam cup? _____

(*Hint:* Which cup can you hold the longest before burning your fingers?) Explain. _____

_____

(continued)

Name: _____   Section: _____
Course: _____   Date: _____

Rule 2 of magma cooling: *Circle the correct choice.* Magma loses heat much (**faster/slower**) when exposed to air or water than it does when surrounded by other rock, because wall rock is a good insulator.

(c) Consider two cubes of steel measuring 1 m on a side. One is cut in half in each dimension to make eight smaller cubes.

1m
1m

  (i)   What is the surface area of the large cube? _____ cm² Of the eight small cubes? _____ cm²
  (ii) Imagine the large and small cubes are heated in a furnace to 500°C and removed. Which will cool faster, the large cube or the smaller ones? _____
  (iii) Explain why this happens.

  _____

  _____

  _____

Rule 3 of magma cooling: *Circle the correct choice.* A small mass of magma loses heat (**faster/slower**) than a large one.

(d) Now you can put the rules to use.
  (i)   In general, lava flows and shallow intrusive igneous rocks have (finer/coarser) grains than deep intrusive rocks.
  (ii) In general, thick lava flows, sills, and dikes have (finer/coarser) grains than thin ones.
  (iii) Which of the rocks in Figure 5.2 was probably extrusive?

  _____

  _____

  (iv) Which were probably intrusive?

  _____

  _____

(continued)

Name: _____   Section: _____
Course: _____   Date: _____

(v) Many dikes, sills, and plutons have *chilled margins*—smaller grains at the contact with their wall rock than in their interiors. Explain how this happens.

_____

_____

_____

_____

_____

FIGURE 5.3 Porphyritic texture (two different grain sizes).

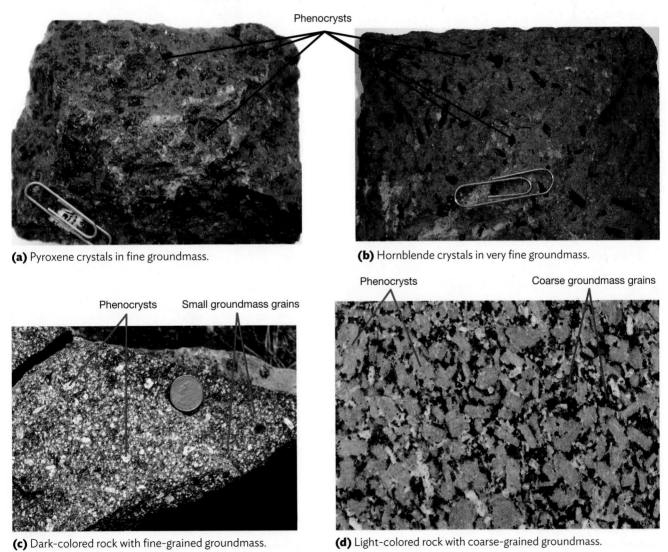

(a) Pyroxene crystals in fine groundmass.

(b) Hornblende crystals in very fine groundmass.

(c) Dark-colored rock with fine-grained groundmass.

(d) Light-colored rock with coarse-grained groundmass.

FIGURE 5.4 Glassy texture (no mineral grains).

5.2.2 Glassy Igneous Textures

Some magmas cool so fast or are so viscous that crystal seeds can't form. Instead, atoms in the melt are frozen in place, arranged haphazardly rather than in the orderly arrangement required for mineral grains. Because haphazard atomic arrangement is also found in window glass, we call a rock that has cooled this quickly **igneous glass**, and it is said to have a **glassy texture** (**FIG. 5.4**). Like thick glass bottles, volcanic glass fractures conchoidally. Volcanic glass commonly looks black, but impurities may cause it to be red-brown or streaked.

Most igneous glass forms at the Earth's surface when lava exposed to air or water cools very quickly, but some may form just below the surface in the throat of the volcano. In addition, much of the ash blasted into the air during explosive eruptions is made of volcanic glass, formed when tiny particles of magma freeze instantly in the air.

EXERCISE 5.4 **Interpreting Porphyritic Textures**

Name: _____ Section: _____
Course: _____ Date: _____

(a) Based on what you have deduced about magma cooling, how does a porphyritic texture form?

_____

_____

_____

(b) Describe the cooling histories of the rocks shown in Figure 5.3 in as much detail as you can.

5.3a

_____

_____

5.3b

_____

_____

5.3c

_____

_____

Name: _____   Section: _____
Course: _____   Date: _____

5.3d

_____

_____

_____

5.2.3 Porous (Vesicular) Textures

As magma rises toward the surface, the pressure on it decreases. This allows dissolved gases (H_2O, CO_2, SO_2) to come out of solution and form bubbles, like those that appear when you open a can of soda. If bubbles form just as the lava solidifies, their shapes are preserved in the lava (**FIG. 5.5**). The material between the vesicles may be fine-grained crystalline material (Fig. 5.5a) or volcanic glass (Fig. 5.5b). These rocks are said to have a **porous** or **vesicular texture**, and the individual bubbles are called **vesicles**. "Lava rock" used in outdoor grills and pumice used to smooth wood and remove calluses are common and useful examples of porous igneous rocks.

5.2.4 Fragmental Textures

Violent volcanic eruptions can blast enormous amounts of material into the air. This material is collectively called **pyroclastic debris** (from the Greek *pyro*, meaning fire, and *clast*, meaning broken). Pyroclastic debris includes volcanic glass, large blocks or bombs erupted as liquid and cooled as fine-grained crystalline rock (**FIG. 5.6a**), crystals formed in the magma before eruption (**FIG. 5.6b**), tiny ash particles, and pieces of rock broken from the walls of the volcanic vent or ripped from the ground surface during an eruption (**FIG. 5.6c**).

FIGURE 5.5 Porous (vesicular) igneous textures.

(a) Dark-colored porous igneous rock.

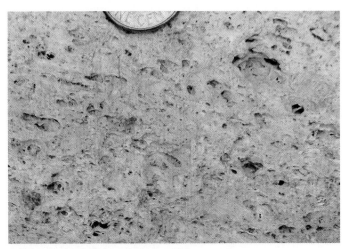

(b) Light-colored porous igneous rock.

FIGURE 5.6 Fragmental (pyroclastic) igneous rocks in a range of grain sizes and textures.

(a) Coarse volcanic bombs in finer-grained pyroclastic matrix at Kilauea in Hawaii.

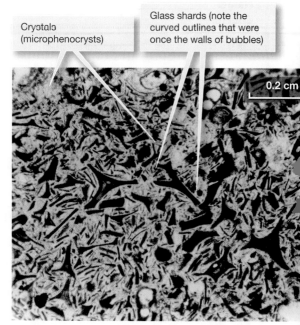

Crystals (microphenocrysts)

Glass shards (note the curved outlines that were once the walls of bubbles)

0.2 cm

(b) Photomicrograph of glass shards (black) welded together with a few tiny crystals.

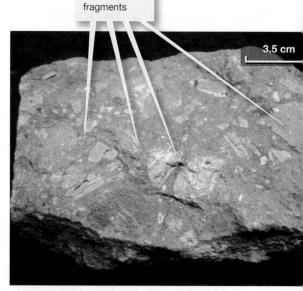

Volcanic rock fragments

3.5 cm

(c) Hand sample with volcanic rock fragments in a very fine-grained brown ash.

In an **ash fall**, fine-grained pyroclastic material falls quietly like hot snow and blankets the ground. The ash may be compacted and turned to rock by the pressure of ash from later eruptions. In more violent eruptions, avalanches of ash called **pyroclastic flows** rush down the side of a volcano while still so hot that the fragments weld together immediately to form rock.

Name: _____   Section: _____
Course: _____   Date: _____

Which rock in Figure 5.6 most likely formed in an ash fall? _____ In a pyroclastic flow? _____ Explain your reasoning.

_____

_____

_____

5.2.5 Grain Shape

Some grains in igneous rocks are well-shaped crystals (phenocrysts in Fig. 5.3a–c), but others are irregular (Fig. 5.2a, b; groundmass grains in Fig. 5.3c, d). When magma begins to cool, its grains start as well-shaped crystals; but they interfere with one another as they grow, resulting in irregular shapes. This process enables us to determine the sequence in which grains grew in an igneous rock: the well-shaped crystals formed early when there was no interference from others. Can you tell which grains grew first in Figure 5.3?

You are now able to read the story recorded by igneous textures and can do so without identifying a single mineral or naming the rocks. **TABLE 5.1** summarizes the origin of common igneous textures and the terms used by geologists to describe them. In Exercise 5.6 you will sharpen your skills by interpreting the history of igneous rocks provided by your instructor.

TABLE 5.1 Interpreting igneous rock textures

Texture	Pegmatitic	Coarse grained (phaneritic)	Fine grained (aphanitic)	Porphyritic	Glassy	Porous (vesicular)	Fragmental (pyroclastic)
Description	Very large grains (>2.5 cm)	Individual grains are visible with the naked eye	Individual grains cannot be seen without magnification	A few large grains (phenocrysts) set in a finer-grained groundmass	Smooth, shiny; looks like glass; no mineral grains present	Spongy; filled with large or small holes	Mineral grains, rock fragments, and glass shards welded together
Interpretation	Very slow cooling or cooling from an extremely fluid magma (usually the latter)	Slow cooling; generally *intrusive*	Rapid cooling; generally *extrusive*	Two cooling rates; slow at first to form the phenocrysts; then more rapid to form the groundmass	Extremely rapid cooling; generally *extrusive*	Rapid cooling accompanied by the release of gases	Explosive eruption of ash and rock into the air
Example							

Name: _____   Section: _____

Course: _____   Date: _____

Examine the igneous rocks provided by your instructor. Apply what you have learned about the origins of igneous textures to fill in the "cooling history" column in the study sheets at the end of the chapter. Use the following questions as a guide to your interpretation.

- Which specimens cooled quickly? Which cooled slowly?
- Which specimens cooled very rapidly at the Earth's surface (i.e., are extrusive)?
- Which specimens cooled slowly beneath the surface (i.e., are intrusive)?
- Which specimens cooled from a magma rich in gases?
- Which specimens experienced more than one cooling rate?

5.3 Igneous Rock Classification and Identification

You now know how to determine the conditions under which igneous rocks form by examining their textures. Igneous rock composition is the key to answering other questions, because it helps reveal how magma forms and why some igneous rocks occur in specific tectonic settings. We look first at how geologists use composition to classify igneous rocks and then at how igneous rocks are named by their composition and texture.

5.3.1 Igneous Rock Classification: The Four Major Compositional Groups

In Exercise 5.1, you saw that some of the igneous rocks in your set are relatively dark colored and others are light colored. Some have high specific gravities, others relatively low specific gravities. A rock's color and specific gravity are controlled mostly by its minerals, which are, in turn, determined by its chemical composition. Oxygen and silicon are by far the two most abundant elements in the lithosphere, so it is not surprising that nearly all igneous rocks are composed primarily of *silicate minerals* like quartz, feldspars, pyroxenes, amphiboles, micas, and olivine.

There are many different kinds of igneous rock, but all fit into one of four major compositional groups—felsic, intermediate, mafic, and ultramafic—defined by how much silicon and oxygen (*silica*) they contain and by which other elements are most abundant (**TABLE 5.2**).

Felsic igneous rocks (from *fel*dspar and *si*lica) have the most silica and the least iron and magnesium. They contain abundant potassic feldspar and sodic plagioclase, commonly quartz, and only sparse ferromagnesian minerals—usually biotite or hornblende. Like their most abundant minerals, felsic rocks are light colored and have low specific gravities.

Intermediate igneous rocks have chemical compositions, colors, specific gravities, and mineral assemblages between those of felsic and mafic rocks: plagioclase feldspar with nearly equal amounts of calcium and sodium, both amphibole and pyroxene, and only rarely quartz.

Mafic igneous rocks (from *ma*gnesium and the Latin *f*errum, meaning iron) have much less silica, potassium, and sodium than felsic rocks but much more calcium, iron, and magnesium. Their dominant calcium-rich plagioclase and ferromagnesian minerals are dark green or black and have higher specific gravities than minerals in felsic rocks. Even fine-grained mafic rocks can therefore be recognized by their dark color and relatively high specific gravity.

TABLE 5.2   The four major groups of igneous rocks.

Igneous rock group	Approximate weight % silica (SiO2)	Other major elements	Most abundant minerals
Felsic	>70	Aluminum (Al), potassium (K), sodium (Na)	K-feldspar, Na-plagioclase, quartz
Intermediate	~55–70	Al, Na, calcium (Ca), iron (Fe), magnesium (Mg)	Ca-Na plagioclase, amphibole, pyroxene
Mafic	~45–55	Al, Ca, Mg, Fe	Ca-plagioclase, pyroxene, olivine
Ultramafic	<45	Mg, Fe	Olivine, pyroxenes

Ultramafic igneous rocks have the least silica and the most iron and magnesium, but very little aluminum, potassium, sodium, or calcium. As a result, they contain mostly ferromagnesian minerals like olivine and pyroxene with very little, if any, plagioclase. Ultramafic rocks are very dark colored and have the highest specific gravities of the igneous rocks.

5.3.2 Identifying Igneous Rocks

The name of an igneous rock is based on its mineral content *and* texture (**FIG. 5.7**). Each of the four igneous rock groups is shown by a column, each of which has coarse, fine, porphyritic, glassy, porous, and fragmental varieties. Rocks in a column may have exactly the same minerals but look so different that they are given different names. For example, granite and rhyolite are both felsic but look different because of their different textures. Although gabbro and granite have the same *texture*, they contain different minerals and therefore look different.

Identifying an igneous rock requires no new skills, just a few simple observations and your ability to identify the common rock-forming minerals.

If you are wondering why there aren't pictures of all the rock types to help you identify the specimens in your rock set, it's because granite may be gray, red, white, or even purple or black depending on the color of its feldspars. A picture can thus help only if it is exactly the same as the rock in your set. But if you understand the combination of minerals that defines granite, you will get it right every time. You will try this in Exercise 5.7.

EXERCISE 5.7   **Identifying Igneous Rocks**

Name: _____   Section: _____
Course: _____   Date: _____

Name the specimens in your igneous rock set by following the steps below.

Step 1: Place the rock in the correct *column* in Figure 5.7a by noting its color and heft and estimating mineral abundance if its grains are coarse enough.

Step 2: Determine the texture of the rock and note which *row* in Figure 5.7b it corresponds to.

Step 3: The name of the rock is found in the box at the intersection of the column and row.

Record the names of the rocks on the study sheets given at the end of the chapter.

FIGURE 5.7 Classification of igneous rocks.

(a) Abundances of minerals in felsic, intermediate, mafic, and ultramafic igneus rocks.

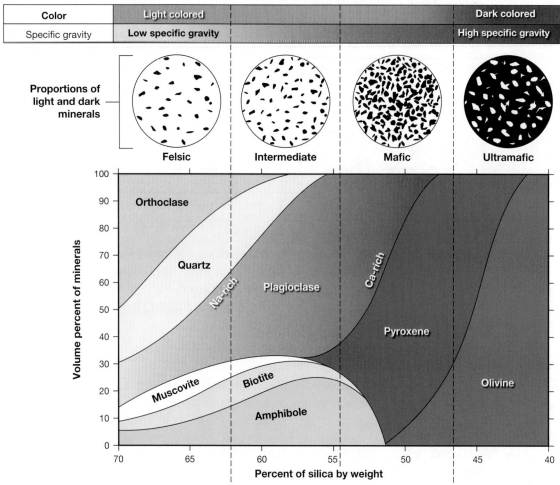

Color	Light colored			Dark colored
Specific gravity	Low specific gravity			High specific gravity

Proportions of light and dark minerals

Felsic | Intermediate | Mafic | Ultramafic

Orthoclase, Quartz, Na-rich, Plagioclase, Ca-rich, Pyroxene, Muscovite, Biotite, Amphibole, Olivine

Volume percent of minerals

Percent of silica by weight

(b) Choose a column based on the abundance of minerals, and select the rock name based on the appropriate texture.

		Silica (SiO₂) content	High silica content			Low silica content
Texture		Pegmatitic	Granitic pegmatite		Mafic pegmatite	**Dunite** (olivine only) **Pyroxenite** (pyroxene) **Peridotite** (olivine + pyroxene)
		Coarse-grained (Phaneritic)	Granite	Diorite	Gabbro	
		Fine-grained (aphanitic)	Rhyolite	Andesite	Basalt	*Rocks with these textures and compositions are very rare.*
		Porphyritic	Porphyritic granite or* Porphyritic rhyolite	Porphyritic diorite or* Porphyritic andesite	Porphyritic gabbro or* Porphyritic basalt	
		Glassy	Obsidian		Tachylite	
		Porous	Pumice		Scoria/Vesicular basalt	
	Fragmental	Fine	Rhyolite tuff	Andesite tuff	Basalt tuff	
		Coarse	Volcanic breccia			

*Porphyritic rocks are named for the size of the groundmass grains. For example, a felsic porphyry in which the groundmass grains are coarse is called *granite porphyry*. If the groundmass grains are small, it is called a *rhyolite porphyry*.

5.4 Origin and Evolution of Magmas

To understand why there are four basic kinds of igneous rocks and why they can be used to interpret ancient plate tectonic events, we need to know (1) *where* and *why* rocks and minerals melt in the Earth, (2) *how* rocks and minerals melt, and (3) how plate tectonic settings produce specific rock types. The next three sections explore these issues.

5.4.1 *Where* and *Why* Do Rocks and Minerals Melt?

Some science fiction movies and novels suggest that the rigid outer shell of the Earth floats on a sea of magma. In fact, most of the crust and mantle is solid rock. Melting occurs only in special places, generally by one of the following three processes.

Decompression melting: Stretching of the lithosphere at a divergent plate boundary (ocean ridge) or continental rift lowers the confining pressure on the asthenosphere below. The amount of heat in the asthenosphere couldn't originally overcome the chemical bonds *and* the confining pressure; but once the pressure decreases, the heat already present in the rock is sufficient to overcome the bonds and cause melting. This process is called **decompression melting** *and requires no additional input of heat.*

Flux melting: Subducted oceanic crust contains hydrous minerals, some formed when ocean-ridge basalt reacts with seawater shortly after erupting, others formed during weathering of island-arc volcanoes. Water and other volatiles are released from these minerals when the subducted plate reaches critical depth (about 150 km) and rise into the asthenosphere, where they help melt the asthenosphere in the process of **flux melting**. In flux melting, no additional heat is needed because the flux of water lowers the melting point of the rocks.

Heat transfer melting: Iron in a blast furnace melts when enough heat is added to break the bonds between atoms, and some magmas form the same way in continental crust. Some very hot mafic magmas that rise into continental crust from the mantle bring along enough heat to start melting the crust, much as hot fudge does when poured onto ice cream. This process, called **heat-transfer melting**, occurs mostly in four settings: (1) oceanic or continental hot spots where mantle plumes bring heat to shallower levels; (2) continental volcanic arcs where mafic magma transfers heat from the mantle to the continental crust of the arc; (3) during late stages of continent-continent collision when the base of the lithosphere peels off and the asthenosphere rises to fill that space, melts, and sends large amounts of basaltic magma upward; and (4) in continental rifts where mafic magma from the asthenosphere rises into the base of thinned continental crust.

5.4.2 *How* Do Rocks and Minerals Melt?

We can't directly observe melting in the mantle or a subduction zone, but we can study it in the laboratory. In the 1920s, N. L. Bowen and other pioneers melted minerals and rocks and chilled them at different stages of melting to learn how magma forms. They learned that magma melting (and crystallization) is very different from the way ice melts into water, and their results paved the way for understanding the origins of the different igneous rock groups.

- **Most rocks melt over a range of temperature**, because some of their minerals have lower melting points than others. The minerals with low melting points melt first while others remain solid. The entire rock melts only when the temperature rises enough to melt all the minerals—a range of several hundred degrees in some cases.

- **Some minerals, like quartz, melt simply, like ice.** When heated to their melting temperatures, these minerals melt completely to form a liquid with the same composition as the mineral. This is not surprising, but the next two findings are.

- **Plagioclase feldspars melt** *continuously*, **over a span of temperature.** When plagioclase containing equal amounts of calcium and sodium begins to melt, more sodium atoms are freed from the crystalline structure than calcium. The first liquid is thus more sodium-rich than the original plagioclase. More and more calcium is freed from the residual mineral as temperature rises, so the compositions of both the liquid and residual mineral change continuously. Only when the entire mineral has melted does the liquid have the same composition as the original mineral.

▪ **Ferromagnesian minerals melt** *discontinuously*, **and new minerals form in the process.** Melt ice and you get water; but melt pyroxene, for example, and when the first liquid forms, the remaining solid changes from pyroxene to olivine. Keep heating it and the olivine gradually melts. Similar things happen to the other common ferromagnesian minerals: biotite, amphibole, and olivine. As with plagioclase, the melt only has the same composition as the starting material when the process is complete.

Bowen summarized these findings graphically in what is now named Bowen's reaction series in his honor (**FIG. 5.8**). Note the relationship between the melting temperatures of the common rock-forming minerals and the assemblages that define the four major igneous rock groups. In Exercise 5.8, you will practice with results and insights learned from these experiments.

5.4.3 Origin of the Igneous Rock Groups: Factors Controlling Magma Composition

Laboratory melting and crystallization experiments last days, weeks, or months and are carried out in sealed containers to prevent contamination. Melting in the Earth is more complicated because it may take millions of years, and many things can happen to the magma. For example, material from the surrounding rocks can be absorbed by the melt, or magma can escape before melting is complete. Several factors control the composition of a magma and the rocks that can form from it, but four processes play the most important roles.

FIGURE 5.8 Bowen's reaction series showing the sequence and types of crystallization from magma.

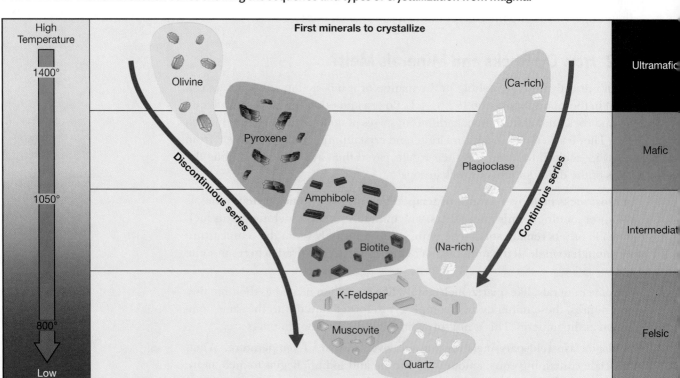

Partial melting: Magma forms by **partial melting** of preexisting rock, not *complete* melting. When ferromagnesian minerals and plagioclase start to melt, the initial liquid has lower density than the source rock and therefore rises, escaping from the melting zone before the rock has melted entirely. In general, partial melting produces a magma that is more felsic (i.e., contains more silica and less iron and magnesium) than its source rock.

Thus, partial melting of ultramafic rock normally produces basalt, but different kinds of basalt could form depending on whether 10%, 20%, or 30% of the source rock melted before the magma escaped. (We said it was more complicated than in the lab!)

EXERCISE 5.8   **Insights from Melting Experiments**

Name: _____   Section: _____
Course: _____   Date: _____

Results from melting experiments help explain several aspects of igneous rock texture and composition. Remember that *crystallization* is the exact opposite of melting. Answer the following questions.

(a) The figure below shows textural features displayed by plagioclase feldspar (left) and pyroxene and amphibole (right) found in many igneous rocks. Considering the continuous and discontinuous legs of Bowen's reaction series, suggest an explanation for the origin of these textures.

_____

_____

_____

_____

Photomicrographs of common mineral textures in igneous rocks.

Compositionally zoned plagioclase feldspar with Ca-rich core and progressively richer Na rims.

Pyroxene grain (px; dark green) with amphibole rim (am; light green).

(b) From Bowen's reaction series, is it likely that minerals with high melting points will occur in rocks with minerals with low melting points? Explain.

_____

_____

_____

(continued)

Name: _____   Section: _____
Course: _____   Date: _____

(c) List the four major igneous rock groups in compositional order from lowest magma temperature to highest.

1. _____

2. _____

3. _____

4. _____

When the initial magma escapes, it rises to cooler levels and begins to crystallize. Even then it may not "follow the rules" of Bowen's reaction, because three other processes can change its composition and thus the minerals that can crystallize from it.

Magmatic differentiation (also called **fractional crystallization**): Early-formed minerals may separate from a magma, usually by sinking, because they are denser than the liquid. If early-crystallized olivine sinks to the bottom of the magma, it can no longer react with residual liquid to make pyroxene as it should according to Bowen's reaction series. Instead, the results are (1) dunite (an ultramafic rock composed entirely of olivine) at the bottom of the magma chamber and (2) a magma more silicic than the original magma. This residual magma might then crystallize minerals that would not have formed from the original. In extreme cases, differentiation of mafic magma can actually produce very small amounts of felsic magma.

Assimilation: As a magma rises, it may add ions by melting some of the surrounding rocks. As the new material is incorporated, the magma composition may change enough locally to enable minerals to crystallize that could not otherwise have been produced.

Magma mixing: Field and chemical evidence suggests that some intermediate rocks did not crystallize from an intermediate magma but rather formed when felsic and mafic magmas mixed. Exercise 5.9 covers these concepts.

Name: _____   Section: _____
Course: _____   Date: _____

Use your knowledge of melting and magmatic processes to explain the origin of the following features.

(a) Most continental rift zones contain basalt and rhyolite, but it is also possible to find small amounts of andesite in this setting. Suggest an origin for the andesite.

_____

_____

_____

(continued)

EXERCISE 5.9 Explaining Igneous Features Found in the Field (continued)

Name: ______________________ Section: ____________
Course: ______________________ Date: ____________

(b) The Palisades Sill on the west bank of the Hudson River in New Jersey is a shallow basaltic intrusive. The sill varies texturally, as shown in the figure below (left). Explain how the observed mineralogic and textural variations were produced by crystallization of the sill.

____

____

____

____

Schematic cross section of the Palisades Sill

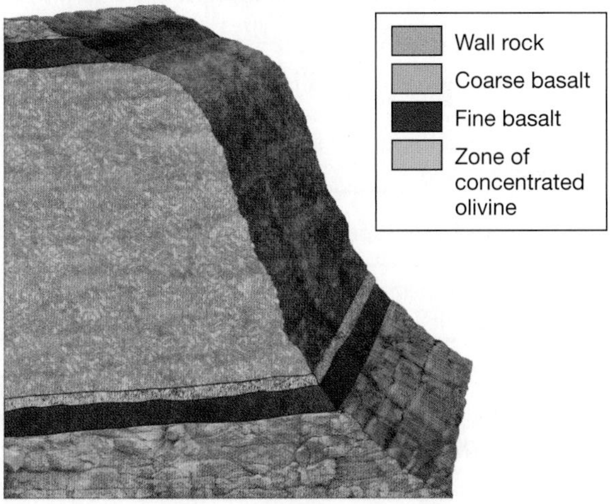

Wall rock
Coarse basalt
Fine basalt
Zone of concentrated olivine

Mafic inclusions (xenoliths) in a granitic pluton

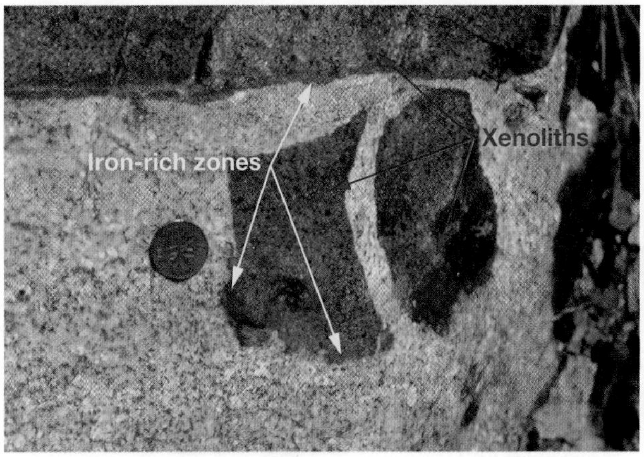

Iron-rich zones
Xenoliths

(c) In the figure above (right), a thin brown zone of weathering highlights iron-rich zones that separate the host granite from the mafic xenoliths. Based on the magmatic processes described above, suggest an origin for these zones.

____

____

____

5.5 Igneous Rocks and Plate Tectonics

Igneous rock types are not distributed equally or randomly on Earth, because melting occurs in different ways in different tectonic settings. As you read the next sections, remember that assimilation, magma mixing, and magmatic differentiation can produce exceptions to just about every generalization; so keep in mind the phrase "are generally found." With that warning, look at **FIGURE 5.9**, which illustrates the tectonic settings of igneous rocks, and **TABLE 5.3**, which lists the igneous rocks that are *generally* found in those settings. In Exercises 5.10, 5.11, and 5.12, you will apply what you've learned about the origin of igneous rocks to suggest tectonic settings for specimens in your rock sets.

FIGURE 5.9 Tectonic settings for major igneous rock types.

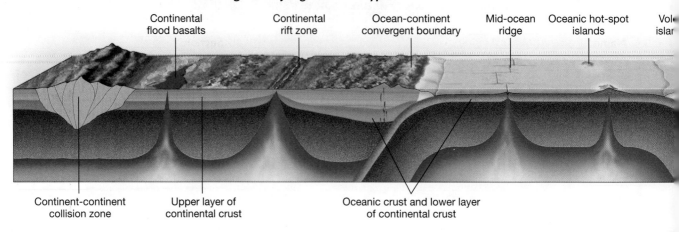

FIGURE 5.9 Tectonic settings for major igneous rock types.

TABLE 5.3 Tectonic associations of igneous rocks.

Type of Plate Boundary	Geologic Setting	Rock Types
Convergent	Volcanic island arcs (ocean-ocean convergence)	Basalt, minor andesite, very minor rhyolite (and intrusive equivalents)
	Andean-type mountains (ocean-continent convergence)	Andesite, rhyolite, minor basalt (and intrusive equivalents)
	Continental collision zones (continent-continent convergence)	Granite, rhyolite, peridotite in ophiolites
Divergent	Mid-ocean ridge	Basalt—a special kind called MORB (mid-ocean ridge basalt)
	Continental rift	Rhyolite and basalt
Igneous rocks not associated with plate boundaries	Oceanic hot-spot islands	Basalt
	Ocean floors	Basalt (MORB), peridotite (locally along faults)
	Continents	Granite, rhyolite, basalt, and gabbro; andesite and diorite by magma mixing

5.5.1 Tectonic Settings of Ultramafic Rocks (Peridotites)

Several lines of evidence indicate that the mantle consists of the ultramafic rock peridotite, making it the most abundant rock on Earth. But peridotite is not a common rock, because its magma is so much denser than the crust that it doesn't have the buoyancy to rise to the surface. It *is* found in divergent mid-ocean ridges and continental rifts where extreme crustal stretching has exposed the underlying mantle, and at convergent plate boundaries where mountain building has thrust slices of the upper mantle (called **ophiolites**) into the crust.

5.5.2 Tectonic Settings of Mafic Igneous Rocks (Basalt and Gabbro)

Basalt and gabbro are the most abundant igneous rocks in the lithosphere and occur in all tectonic settings—oceanic crust (formed at mid-ocean ridges), continental rifts, oceanic and continental volcanic arcs, and hot spots. These rocks are the result of partial melting in the asthenosphere, but the cause of melting is different in each setting.

5.5.3 Tectonic Settings of Intermediate Rocks (Andesite and Diorite)

In the mid-twentieth century, geologists were puzzled by the fact that nearly all andesites occur next to trenches in continental arcs and in some oceanic island arcs. Today we understand that most intermediate magmas are produced at subduction zones by the mixing of magmas produced during subduction with rocks of the overlying plate.

5.5.4 Tectonic Settings of Felsic Rocks (Granite and Rhyolite)

Granite and rhyolite are most abundant on the continents—in continental volcanic arcs, continent-continent collision zones, rifts, and where plumes rise beneath continents. They form largely by partial melting of the upper (granitic) layer of continental lithosphere and to a lesser extent by differentiation of mafic magmas. Some rhyolite and granite form in subduction zones by differentiation of mafic and intermediate magmas and/or by assimilation. Only very small amounts of rhyolite form in oceanic hot-spot islands by extreme fractional crystallization.

EXERCISE 5.10   **Origin of Mafic Magmas in Different Tectonic Settings**

Name: _____   Section: _____

Course: _____   Date: _____

Explain which type of melting (addition of volatiles, decompression, heat transfer) is responsible for mafic magmas in each of the following tectonic settings and why the resulting magma is mafic.

(a) Mid-ocean ridges

_____

_____

_____

(b) Continental rifts

_____

_____

_____

(c) Oceanic and continental volcanic arcs

_____

_____

_____

(d) Hot spots

_____

_____

_____

EXERCISE 5.11   Origin of Intermediate Magmas in Subduction Zones

Name: _____   Section: _____

Course: _____   Date: _____

Melting above the subducted slab produces mafic magma, as described above, yet intermediate rocks (andesite, diorite) are common in many subduction zones. Considering the magmatic processes discussed in section 5.4, and the difference between oceanic and continental lithosphere, explain how this intermediate magma forms.

(a) A relatively small amount of intermediate magma occurs in oceanic volcanic arcs. Remembering that melting above the subducted slab produces mafic magma, explain the origin of the intermediate magma.

_____

_____

_____

(b) Much more intermediate magma erupts in continental arcs. Why?

_____

_____

_____

EXERCISE 5.12   Origin of Granite and Rhyolite in Continental Rifts

Name: _____   Section: _____

Course: _____   Date: _____

Continental rifts typically contain large amounts of basalt and rhyolite. Your answer to Exercise 5.10b explained the origin of the basalt, but fractional crystallization of mafic magma can produce only a very small amount of felsic magma. Explain the origin of large volumes of felsic magma in continental rifts.

_____

_____

_____

_____

EXERCISE 5.13   Interpreting Tectonic Setting of Igneous Rocks

Name: _____   Section: _____

Course: _____   Date: _____

Based on the information in Table 5.3 and your answers to Exercises 5.10, 5.11, and 5.12, add possible tectonic settings for the igneous rocks in your study set to your rock identification charts.

5.6 Volcanic Hazards

As we saw in earlier in the chapter, when magma reaches the surface, it spews onto the Earth in an event called a volcanic eruption. As it erupts, it sometimes flows across the land as a glowing **lava flow**, at temperatures ranging from 700°C to 1200°C, or it may be blasted by gases into the air and fall back to Earth pyroclastic debris—solid particles ranging from tiny grains of ash to blocks several feet across. Eruptions can be beneficial. The state of Hawaii is a series of volcanoes, and the lava and pyroclastic material that created (and are still creating) it are extremely fertile. But lava and pyroclastic material can also be extremely dangerous, burying cities and killing thousands of people.

To understand these hazards, we first need to know the types of volcanoes, the types of eruptions that produce them, and where they occur. Then we can consider the risk of volcanoes in parts of North America and how a large eruption might affect how and where we live.

5.6.1 Volcanic Landforms

When asked to describe eruptions, most people immediately think of the mountains we call volcanoes, built from within the Earth by lava and/or pyroclastic material. It is first important to note that not all eruptions build volcanoes; for example, highly fluid lavas flowed out from long fissures and today underlie broad flat areas like the Columbia Plateau in Washington, Oregon, and Idaho (**FIG. 5.10**).

However, most eruptions *do* build volcanoes, which fall into four types based on whether they are made of lava, pyroclastic material, or a unique combination of the two kinds of material (**FIG. 5.11**). Two are made almost entirely of lava. **Shield volcanoes** (Fig. 5.11a) are made mostly of highly fluid lava. The lava's fluidity produces broad structures with gentle slopes—the largest volcanoes on Earth (and Mars) are shield volcanoes. Because ketchup or maple syrup are similar to the consistency of lava, you can simulate the formation of a shield volcano by pouring them on a flat surface. **Lava domes** (Fig. 5.11b) are much smaller, often forming in the craters of other kinds of volcano. Lava in these volcanoes is extremely viscous—thicker and stickier than normal lava—so it can form steeper slopes than those of shield volcanoes. To simulate formation of a lava dome, squeeze a blob of epoxy or model-building plastic cement onto a flat surface. While eruptions of lava domes and shield volcanoes are the least violent of all volcanoes, lava from Hawaiian shield volcanoes has buried homes, forests, and fields, displacing people and blocking roads at a cost of millions of dollars.

FIGURE 5.10 The Columbia Plateau, a volcanic landform without volcanoes.

FIGURE 5.11 Four types of volcanoes.

(a) Shield volcano (fluid lava), Mauna Loa, Hawaii.

(b) Lava dome (viscous lava) in the crater of Kelud Volcano, Indonesia.

(c) Cinder cone (pyroclastic material), Sunset Crater, Arizona.

(d) Stratovolcano (pyroclastic material plus lava), (Mt. Mayon, Philippines.

Cinder cones (Fig. 5.11c) are made mostly from relatively fine-grained pyroclastic material and, unlike other volcanoes, form from *above*, when ash erupted from the volcano's vent falls to the ground, similar to sugar being poured into a pile. Eruptions that build cinder cones are relatively peaceful, and the resulting volcanoes are much smaller than shield volcanoes and stratovolcanoes, with moderate slopes.

Stratovolcanoes (Fig. 5.11d) are made of large amounts of both lava and pyroclastic material. They have the steepest slopes of all volcanoes because the ash is so hot when it accumulates that the particles are welded together, and the pile is not flattened by the pull of gravity. Stratovolcanoes are commonly found in subduction zones, and their explosive eruptions are caused by large amounts of steam produced by flux melting associated with the subducted slab.

5.6.2 Impact of Eruptions

While many volcanic eruptions are peaceful, they of course can be quite dangerous. Stratovolcanoes often generate pyroclastic flows—avalanches of hot ash and gas—and **lahars,** which are mudslides made of volcanic ash, water, and debris picked up along the way that can obliterate anything in their path. Volcanic gases include poisonous acids, and explosive eruptions in the oceans can produce massive, destructive events called tsunamis, which will be discussed in a later chapter. Famously, falling ash from Mt. Vesuvius buried the Roman town of Pompeii and its citizens in A.D. 79, and when Mount St. Helens in Washington State erupted in 1980, it killed fifty-seven people, spread ash that caused people to wear surgical masks to avoid the caustic, glassy airborne particles, and destroyed close to fifty bridges. Very recently, ash from Iceland's Eyjafjallajökull in 2010 disrupted European air traffic for a week at a cost of more than $200 million. In Exercise 5.14, we will consider some of the practical aspects of living in an area with volcanic activity and then examine the damage that might occur if a disastrous eruption occurred today.

EXERCISE 5.14 **Living with Volcanoes**

Name: _____ Section: _____
Course: _____ Date: _____

Let's consider the map below. It shows current and potential volcanic activity for the United States.

(a) What area of the United States has the most volcanic activity? Why is this so?

_____

_____

_____

(b) What volcanic hazards do residents of the Pacific Northwest have to be concerned about that residents of Hawaii do not? Explain.

_____

_____

_____

(continued)

Name: _____   Section: _____

Course: _____   Date: _____

(c) Is there a risk of volcanoes occurring on the Atlantic Coast? Why or why not?

_____

_____

_____

? **What Do You Think** Every time Old Faithful spews steam into the air in Yellowstone National Park, it reminds us that there is hot magma beneath the park capable of producing an eruption. This activity in Yellowstone is not the first volcanism in the Yellowstone area. An enormous explosive eruption 640,000 years ago blasted ash over much of the central and south western United States, blanketing the area shown in the figure below.

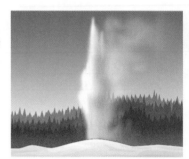

Suppose an eruption of the same size occurred tomorrow in Yellowstone National Park and blasted a layer of ash across the country in an area similar to that of the eruption 640,000 years ago. The map below projects this area onto a standard map showing major cities and highways.

Extent of ash deposits from the Yellowstone eruptive center 640,000 years ago.

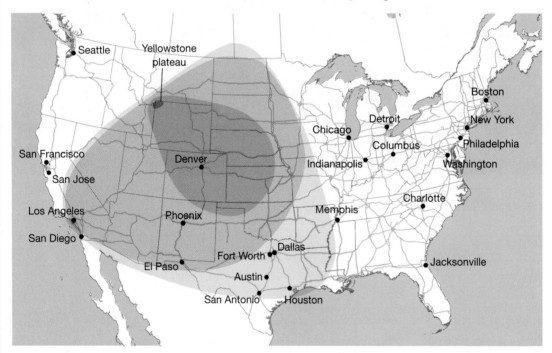

(d) What do you think would be the effect on this area of the country? Which cities and states would suffer the most and why? Consider effects on the economy, health, food supply, and transportation network in this area.

_____

_____

(continued)

Name: _____   Section: _____
Course: _____   Date: _____

_____

_____

_____

_____

(e) Similarly, though not covered by the explosion, what do you think would be the effect on the rest of the country that would not be directly covered by ash?

_____

_____

_____

_____

_____

_____

_____

_____

5.7 Visiting Localities Where Igneous Rocks Are Forming or Have Formed in the Past

Software such as *Google Earth*™ and *NASA World Wind* make it possible to tour the world to see current igneous activity at these tectonic settings and view ancient igneous rocks revealed by erosion. For example, use either software to visit

■ **Yellowstone National Park in Wyoming, a continental hot spot.** About 2.1 million years ago, an incredibly powerful explosive volcanic eruption shook the northwestern corner of Wyoming, spreading immense quantities of ash across the landscape. The sudden removal of so much magma caused the ground to subside, producing a large basin called a **caldera**. This was just the first of three similar explosive events; the most recent, only 640,000 years ago, formed the huge bowl-shaped depression in which Yellowstone National Park is located. Magma close to the surface is responsible for heating the groundwater, which gushes to the surface as hot springs and geysers like Old Faithful.

■ **The Palisades in New Jersey, a divergent plate margin.** About 180 million years ago, a supercontinent began to split apart to form the modern Atlantic Ocean. Rift valleys, like those in east Africa, formed along the east coast of North America from Nova Scotia to South Carolina. Part of the asthenosphere melted, and basaltic magma moved upward toward those valleys. Some flowed as lava, but the Palisades along the Hudson River intruded nearly horizontal sedimentary rocks beneath the surface.

- **Hawaii, an oceanic hot spot.** The "Big" island of Hawaii is the youngest of a long chain of hot-spot volcanic islands that helps us track the motion of the Pacific Plate. The island consists of five huge volcanoes, one of which (Kilauea) has been erupting nearly continually since 1983.

- **the Central Rift Valley in Iceland.** Iceland is a unique place where a hot spot lies beneath a segment of an ocean ridge. The Central Rift Valley is the divergent boundary—Iceland is growing wider as the valley sides move apart, and the lava here and at nearby volcanoes comes from magma rising into the space caused by divergence.

- **the Cascade Mountains of Washington, Oregon, and northern California, a subduction-related continental volcanic arc.** The volcanoes of the Cascade Mountains are a continental arc that developed as a small plate subducted beneath North America. Prominent peaks such as Mt. Rainier, Mt. Hood, and Mt. Baker overlook densely populated areas, much as Mt. Vesuvius stood over Pompeii (and still towers over Naples) in Italy. Thankfully, the most recent eruption in the Cascades arc was of Mount St. Helens, which is in a sparsely populated part of Washington.

- **the Sierra Nevada mountains in California.** Large volumes of the continental crust under eastern and east-central California melted between 210 and 85 million years ago, because of heat and hydrothermal fluids generated from a now-defunct subduction zone. The result was granitic magma, which cooled to form several large batholiths. Tens of millions of years of erosion have exposed the resulting granitic rocks in the Sierra Nevada mountains as seen in the Yosemite, Kings Canyon, and Sequoia national parks.

Name _____

IGNEOUS ROCKS STUDY SHEET

Sample number	Texture	Minerals present (approximate %)	Name of rock	Cooling history; source of magma; tectonic setting

Name _____

IGNEOUS ROCKS STUDY SHEET

Sample number	Texture	Minerals present (approximate %)	Name of rock	Cooling history; source of magma; tectonic setting

6

Using Sedimentary Rocks to Interpret Earth History

The beds of Grand Canyon National Park in Arizona are one of the most extensive sequences of sedimentary rocks in the world.

- Understand the processes by which sedimentary rocks form

- Understand how geologists classify sedimentary rocks based on their origin

- Become familiar with sedimentary rock textures and mineral assemblages

- Use texture to identify the agent of erosion involved in formation of sedimentary rocks

- Use sedimentary rock textures and minerals to interpret ancient geologic, geographic, and climatologic environments

MATERIALS
NEEDED

- Set of sedimentary rocks

- Magnifying glass or hand lens, microscope, and thin sections of sedimentary rocks

- Mineral-testing supplies (streak plate, glass plate, etc.)

- Glass beaker, concentrated Ca (OH)$_2$ solution, straws

6.1 Introduction

Like igneous rocks, sedimentary rocks form from previously existing rocks—but by very different processes. If you've ever seen shells on a beach, gravel along a riverbank, mud in a swamp, or sand in a desert, you've seen **sediment**. Sediment consists of loose mineral grains such as **clasts**, (fragments or grains produced by weathering of a rock—boulders, pebbles, sand, silt), shells and shell fragments, plant debris, and/or mineral crystals precipitated from bodies of water. **Sedimentary rock** forms at or near the Earth's surface by the cementation and compaction of accumulated layers of these different kinds of sediment. Sedimentary rocks preserve a record of past environments and ancient life and thus tell the story of Earth history. For example, the fact that a type of sedimentary rock called limestone occurs throughout North America means that, in the past, a warm, shallow sea covered much of the continent. Geologists have learned to read the record in the rocks by comparing features in sedimentary rocks to those found today in environments where distinct types of sediment form, move, or accumulate. Remember, the present is the key to the past. In this chapter, you will learn to read this record, too.

6.2 Sediment Formation and Evolution

6.2.1 The Origin of Sediment

The material from which sedimentary rocks form ultimately comes from **weathering**—the chemical and/or physical breakdown of preexisting rock. Weathering produces both **clasts** and **dissolved ions**, which are charged atoms or molecules in a water solution. Once formed, clasts may be transported by water, wind, or ice to another location, where they are deposited and accumulate. Dissolved ions, meanwhile, enter streams and groundwater. Some of these ions precipitate from groundwater in the spaces between clasts forming a **cement** that holds the clasts together. Others get carried to lakes or seas, where organisms extract them to form shells. Finally, some dissolved ions precipitate directly from water to form layers of new sedimentary minerals. In some environments, sediment may also include organic material, the carbon-containing compounds that remain when plants, animals, and microorganisms die.

6.2.2 Weathering and Its Influence on Sediment Composition

The minerals that occur in sediment at a location depend both on the composition of the source rock(s) *and* on the nature of weathering the minerals were exposed to. Here, we look at the weathering process and how it influences sediment composition.

Physical weathering, like hitting a rock with a hammer, breaks the rock into fragments (clasts) *but does not change the minerals* that make up the rock. Initially, clasts are simply small rock pieces and may contain many grains or crystals that retain their original sizes and shapes. But when physical weathering progresses, the clasts break into smaller pieces, perhaps consisting only of a single mineral. While moving along in wind and water, clasts grind and crash against each other and become progressively smaller. As a result, the size and shape of transported clasts do not tell us whether the source rock itself was coarse or fine or whether grains in the source rock interlocked or were cemented together (**FIG. 6.1**). But because physical weathering does not change the composition of clasts, we can get a sense of the

FIGURE 6.1 Physical weathering of coarse-grained granite.

(a) A rock breaks along the dashed lines

(b) The rock fragment preserves grain sizes and relationships.

(c) The fragment breaks into both smaller rock fragments and mineral grains.

composition of the source rock by looking at these clasts. For example, sediment derived from physical weathering of granite would contain clasts of quartz, K-feldspar, plagioclase, and biotite, whereas sediment derived from physical weathering of basalt would contain clasts of plagioclase, pyroxene, amphibole, and olivine.

Chemical weathering is the process during which rock chemically reacts with air, water, and acidic solutions. In other words, chemical weathering involves **chemical reactions**, the breaking and forming of chemical bonds. These reactions can destroy some of the original minerals and can produce new minerals. Minerals that are easily weathered are called *unstable* or *nonresistant*, whereas those that can survive weathering or are produced by weathering are called *stable* or *resistant*. Chemical

FIGURE 6.2 Chemical weathering of felsic and mafic igneous rocks.

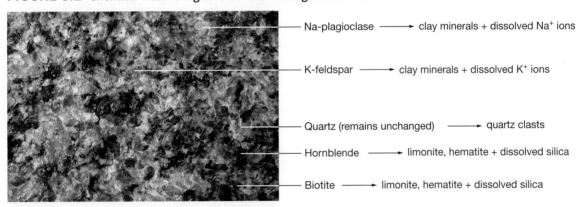

Na-plagioclase ⟶ clay minerals + dissolved Na⁺ ions

K-feldspar ⟶ clay minerals + dissolved K⁺ ions

Quartz (remains unchanged) ⟶ quartz clasts

Hornblende ⟶ limonite, hematite + dissolved silica

Biotite ⟶ limonite, hematite + dissolved silica

(a) Granite weathers to clay minerals, limonite, hematite, and quartz plus dissolved potassium, sodium, and silica.

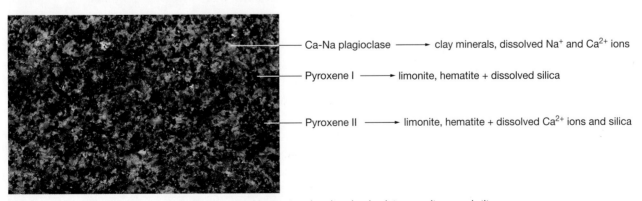

Ca-Na plagioclase ⟶ clay minerals, dissolved Na⁺ and Ca²⁺ ions

Pyroxene I ⟶ limonite, hematite + dissolved silica

Pyroxene II ⟶ limonite, hematite + dissolved Ca²⁺ ions and silica

(b) Gabbro weathers to clay minerals, limonite, and hematite plus dissolved calcium, sodium, and silica.

weathering has hardly any effect on quartz because quartz is stable, but it transforms feldspar, which is an unstable mineral, into clay and ions (K^+, Ca^{2+}, Na^+). Similarly, chemical weathering converts ferromagnesian minerals such as olivine, pyroxene, amphibole, and biotite into hematite and limonite (iron oxide minerals) along with ions of silicon and oxygen.

Relatively few minerals are stable at the Earth's surface, so *chemical weathering generally reduces the number of minerals in sediment* over time. As a result, weathering of very different source rocks can yield surprisingly similar sedimentary mineral assemblages (**FIG. 6.2**). Weathering of granite, for example, produces quartz, clay minerals, and iron oxides. Gabbro also weathers into clay minerals and iron oxides. Note that only the proportions of minerals (more iron oxides in the weathered gabbro) and the occurrence of quartz (none in weathered gabbro) help us distinguish sediments derived by chemical weathering of granite from those derived from gabbro.

6.2.3 Mineralogical Maturity

Not all sediments have undergone the same amount of weathering. Sediments that contain minerals which are susceptible to chemical weathering (unstable minerals) are said to be **mineralogically immature**. While sediments that contain minerals which are resistant to chemical weathering (stable minerals) are known as **mineralogically mature**. The effects of weathering are examined further in Exercise 6.1.

EXERCISE 6.1 **Looking at Weathering Products**

Name: _____   Section: _____
Course: _____   Date: _____

During chemical weathering, water, oxygen, and carbon dioxide combine with minerals in previously existing rocks to destroy some minerals and create new ones.

(a) How does the mineral assemblage in chemically weathered sediment change with the *amount* of chemical weathering?

_____

_____

(b) What weathering history can you interpret from the following mineral assemblages found in *clastic* sedimentary rocks? For example: What was the source rock? Was it weathered for a short time or a long time? Think broadly and remember that there may be more than one possible interpretation.
  (i) All quartz grains
  (ii) Nearly equal amounts of quartz, K-feldspar, and Na-plagioclase with a small amount of hematite
  (iii) All fine-grained clay minerals with some hematite and limonite
  (iv) A mixture of quartz grains and clay minerals
  (v) Rock fragments composed of Ca-plagioclase and pyroxene

_____

_____

_____

_____

_____

(continued)

Name: _____   **Section:** _____
Course: _____   **Date:** _____

(c) Rank the following environments in the order in which you would expect sediments to become mineralogically mature, from slowest to fastest. (*Hint*: Think about the temperature and the amount of rainfall available.)

    desert _____   tropical rain forest _____   temperate climate _____

Explain your reasoning.

_____

_____

_____

(d) Which of the following mineral assemblages (as seen in Exercise 6.1b) are mineralogically mature? Which are mineralogically immature?
  (i) All quartz grains _____
  (ii) Nearly equal amounts of quartz, K-feldspar, and Na-plagioclase with a small amount of hematite _____
  (iii) All fine-grained clay minerals with some hematite and limonite _____
  (iv) A mixture of quartz grains and clay minerals _____
  (v) Rock fragments composed of Ca-plagioclase and pyroxene _____

6.3 The Basic Classes of Sedimentary Rocks

Geologists distinguish among many kinds of sedimentary rocks, each with a name based on the nature of the material that the rock contains and the process by which the rock forms. To organize this information, we sort various rocks into classes, though not all geologists use the same classification. A simple scheme groups rocks into four classes—clastic, chemical, biochemical, and organic—based on the nature of the particles that make up rocks. Note that this classification scheme is not based on composition (the mineral present). In this section, we will first examine these classes and then we will conclude by introducing an overlapping classification scheme that focuses on composition.

6.3.1 Clastic Sedimentary Rock

Clastic sedimentary rocks are formed from clasts (mineral grains or rock fragments) derived from previously existing rocks. These rocks have a characteristic clastic texture in which discrete grains are held together by a chemical cement or by a very fine-grained clastic matrix. Of the most common clastic sedimentary rocks, the majority are derived from silicate rocks and thus contain clasts composed of silicate minerals. (To emphasize this, geologists sometimes refer to them as *siliciclastic rocks.*) Formation of clastic sedimentary rock involves the following processes (**FIG. 6.3**):

1. **Weathering:** The process of weathering reduces solid bedrock into a pile of loose (separate mineral) grains or loose clasts, also known as *detritus*.
2. **Erosion:** This occurs as moving water (streams and waves), moving air (wind), and moving ice (glaciers) pluck and/or pick up the clasts.

FIGURE 6.3 The five steps in clastic sedimentary rock formation.

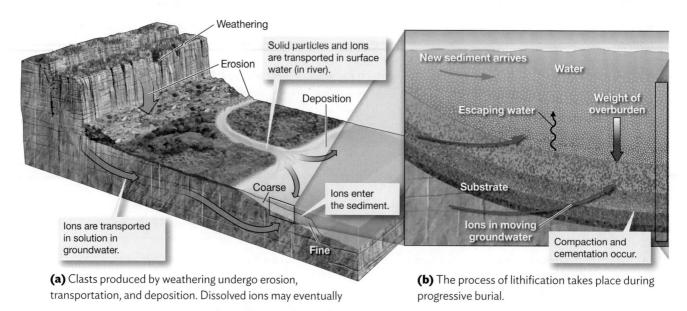

(a) Clasts produced by weathering undergo erosion, transportation, and deposition. Dissolved ions may eventually

(b) The process of lithification takes place during progressive burial.

3. **Transportation:** Moving water, wind, or ice carry clasts away from their source.
4. **Deposition:** When moving water or wind slows or when the ice melts, clasts settle out and accumulate. The places where this occurs are called **depositional environments** (such as the land surface, the sea floor, or a riverbed).
5. **Lithification:** This process transforms loose sediment into solid rock through compaction and cementation. Over time, accumulations of clasts are buried. When this happens, the weight of overlying sediment squeezes out air and/or water, thereby fitting the clasts more tightly together. This process is called **compaction**. As ion-rich groundwater passes through the compacted sediment, minerals precipitate and bind, or "glue," the clasts together. This process is called **cementation**, and the mineral glue between clasts is called **cement**. Cement in sedimentary rock typically consists of either calcite or quartz and may contain minor amounts of hematite or pyrite.

A final process is called **diagenesis**, where the application of pressure and the circulation of fluids over time may gradually change characteristics of sediments and sedimentary rock (e.g., grain size and composition of cement or the nature of grain boundaries), even at temperatures below those required for metamorphism. Any chemical, physical, or biological process that alters the rock after it is formed is called diagenesis.

As we see later in this chapter, the names of *clastic* sedimentary rocks are based primarily on the size of the clasts they contain. Geologists refer to the Wentworth scale (**TABLE. 6.1**) to distinguish among grain size. From coarsest to finest, clastic sedimentary rocks and their grain sizes include:

- **Conglomerate/breccia/diamictite** consists of pebbles and/or cobbles. Clasts in conglomerate are rounded (have no sharp corners), whereas those in breccia are angular (have sharp corners). Diamictite has large clasts in mud. Modifiers are sometimes added when one kind of clast dominates (e.g., quartz conglomerate or limestone breccia).
- **Sandstone/arkose/lithic sandstone/wacke** The name **sandstone** applies to any clastic sedimentary rock composed predominantly of sand-sized grains.

TABLE 6.1 Classification of clastic sedimentary rocks

Clast size terminology (Wentworth scale)	Clast size	Appearance	Size in mm	Clast character	Rock name (alternative name)
Coarse to very coarse	Boulder			Rounded pebbles, cobbles, and boulders	Conglomerate
Medium to coarse	Cobble		256	Angular pebbles, cobbles, and boulders	Breccia
Medium to coarse	Pebble		64	Pebbles, cobbles, and boulders in muddy matrix	Diamictite
Medium	Sand		2	Sand-sized grains • quartz grains only • quartz and feldspar sand • sand-sized rock fragments • quartz sand and sand-sized rock fragments in a clay-rich matrix	Sandstone • quartz sandstone (quartz arenite) • arkose • lithic sandstone • wacke (informally called graywacke)
Fine	Silt		0.06	Silt-sized clasts	Siltstone
Very fine	Clay		0.004	Clay and/or very fine silt	Shale, if it breaks into platy sheets Mudstone, if it doesn't break into platy sheets

Four types of sandstone are distinguished based on the composition of the grains they contain:

– **Quartz sandstone** (or quartz arenite) consists of $> 95\%$ quartz grains. It is the most common type because quartz is the most common resistant mineral in chemical weathering and the most mineralogically mature.

– **Arkose** contains quartz with $> 25\%$ feldspar grains. It is less mineralogically mature than quartz sandstone and forms close to a source of sediment in which feldspar has not been removed by chemical weathering.

– **Lithic sandstone** contains clasts of quartz, feldspar, and $> 5\%$ fragments of previously existing rocks (*lithic* fragments).

– **Wacke** (informally graywacke) is a sandstone containing quartz, feldspar, and lithic clasts surrounded by a matrix of mud.

■ **Siltstone** consists of silt-sized clasts.

■ **Shale/mudstone** consist mostly of clay-sized particles, invisible even with a hand lens.

6.3.2 Chemical Sedimentary Rock

This class consists of rocks formed from mineral crystals precipitated directly from a water solution. Chemical sedimentary rock forms when water solutions of dissolved ions become oversaturated and excess ions bond together to form solid mineral grains, which often display a crystalline texture as they grow and interlock with each other. These crystals either settle out of the solution or grow outward from the walls of the container holding the solution. Exercise 6.2 allows you to simulate the chemical precipitation process in the lab to see how the textures of these rocks develop.

Groundwater, oceans, and saline lakes all contain significant quantities of dissolved ions and can serve as a source of chemical sedimentary rock. Precipitation to form chemical sedimentary rocks happens in many environments, including (1) hot springs, where warm groundwater seeps out at the Earth's surface and cools; (2) cave walls, where groundwater seeps out, evaporates, and releases CO_2; (3) the floors of saline lakes or restricted seas, where salt water evaporates; (4) within sedimentary rocks, when reactions with groundwater result in the replacement of the original minerals with new minerals; and (5) on the deep-sea floor, where the shells of plankton dissolve to form a gel-like layer that then crystallizes. The composition of a chemical sedimentary rock depends on the composition of

---

EXERCISE 6.2   **Simulating Chemical Sedimentary Textures**

Name: _____   Section: _____

Course: _____   Date: _____

(a) Place a beaker with seawater (or homemade salt water) on a hot plate and heat it gently until the water evaporates. Partially fill a second beaker with a clear, concentrated solution of calcium hydroxide [$Ca(OH)_2$]. Using a straw, blow *gently* into the solution until you notice a change.

   Describe what happened in each demonstration, and sketch the resulting textures.

   (i)   Evaporated seawater description   Sketch

     _____

     _____

     _____

     _____

     _____

     _____

   (ii) $Ca(OH)_2$ solution description   Sketch

     _____

     _____

     _____

     _____

     _____

     _____

(continued)

Name: _____   Section: _____
Course: _____   Date: _____

(b) Compare the texture in (i) with that of a granite, a rock formed by cooling of a melt. Describe and explain the similarities in texture.

_____

_____

_____

_____

the solution it was derived from—some chemical sedimentary rocks consist of salts (e.g., halite, gypsum), whereas others consist of silica or carbonate. In some chemical sedimentary rocks, the grains are large enough to see (**FIG. 6.4a**). But in others, the grains are so small that the rock looks somewhat like porcelain (**FIG. 6.4b**). Such rocks are called **cryptocrystalline rocks**, from the Latin *crypta-*, meaning hidden.

Geologists distinguish among different types of chemical sedimentary rocks based primarily on composition:

- **Evaporites** Chemical sedimentary rocks in thick deposits composed of crystals formed when salt water evaporates. Rock salt and rock gypsum are two common examples.
- **Travertine (chemical limestones)** Rocks composed of crystalline calcium carbonate (CaCO$_3$) formed by chemical precipitation from groundwater that has seeped out of the ground surface either in hot- or cold-water springs or from the walls of caves.

FIGURE 6.4 Textures of chemical sedimentary rocks.

(a) Coarse, interlocking halite grains in rock salt.

(b) Cryptocrystalline silica grains (chert). Note the almost glassy appearance and conchoidal fracturing.

- **Dolostone** Carbonate rocks differing from limestone in that they contain a significant amount of dolomite, a mineral with equal amounts of calcium and magnesium. Most dolostone forms from limestone during diagenesis when groundwater adds magnesium to calcite ($CaCO_3$), forming the mineral dolomite [$CaMg(CO_3)_2$].
- **Chert** Sedimentary rocks composed of very fine-grained silica. Some chert, called replacement chert, forms when cryptocrystalline quartz replaces calcite in limestone. Another kind (biochemical chert) is formed during diagenesis when microscopic shells made of quartz dissolve partially, form an ultrafine-grained gel, and then solidify. *Note:* Biochemical and chemical forms of chert cannot be distinguished in hand samples.

6.3.3 Biochemical and Organic Sedimentary Rocks

While alive, organisms extract chemicals from their environment and use these to produce new molecules for building bodies and shells. For example, some invertebrates (such as clams and oysters), along with certain types of algae and plankton, extract dissolved ions from water to produce shells composed of either carbonate or silica. Plants extract CO_2 from the air to produce the cellulose of leaves and wood. And all organisms produce *organic chemicals*, meaning chemicals containing rings or chains of carbon atoms bonded to hydrogen, nitrogen, oxygen, and other elements. When organisms die, the materials that compose their cells and/or shells can accumulate in depositional settings, just like sediment. Carbonate or silicate shells are quite durable—after all, they're composed of fairly hard minerals such as calcite or quartz. Organic chemicals are less durable, and they commonly decompose and oxidize. But in special depositional settings where there is relatively little oxygen and the organisms get buried rapidly, organic chemicals can be preserved in rocks.

Rock composed primarily of the remains of once-living organisms are classified as **biochemical sedimentary rocks** if they consist of the hard shells of those organisms or **organic sedimentary rocks** if they contain significant quantities of the organic material from the soft, carbon-rich parts of those organisms. Continuing diagenesis tends to modify both biochemical and organic rocks significantly after initial lithification. For example, rocks containing abundant calcite may undergo pervasive recrystallization, so that the original grains are replaced by new, larger interlocking crystals. Diagenesis may also dissolve some grains and precipitate new cements in pores, and it may drive the transformation of calcite ($CaCO_3$) into dolomite [$CaMg(CO_3)_2$]. Organic chemicals, over time, lose certain elements and transform into either pure carbon or hydrocarbon (carbon + hydrogen). The texture of biochemical and organic rocks, like other sedimentary rocks, reflects the nature of the depositional environment. In some environments, shells are transported by currents or waves and break into fragments that behave like sedimentary clasts.

Geologists distinguish among many types of biochemical and organic sedimentary rocks based on composition and texture. Common examples are as follows.

Biochemical rocks

- Limestone is a general class of rocks formed from shells composed of calcite. Commonly, limestone tends to form as chunky blocks, light gray to dark-bluish gray in color. Geologists recognize distinct subcategories of biochemical limestone, based on texture.
    - **Fossiliferous limestone** contains abundant visible fossils in a matrix of fossil fragments and other grains.
    - **Micrite** is very fine-grained limestone formed from the lithification of carbonate mud. The mud may come from tiny spines of sponges or from the shells of algae or bacteria, or it may form after burial by diagenesis.

- **Chalk** is a soft, white limestone composed of the shells of plankton.
  - **Coquina** consists of a mass of shells that are only poorly cemented together and have undergone minimal diagenesis.
- **Biochemical chert** forms from a silica gel that forms when shells of silica-secreting plankton accumulate on the sea floor. We use the word *biogenic* to distinguish this chert from another occurrence, replacement chert, which forms by diagenesis in already lithified limestone.

Organic rocks

- **Coal** is composed primarily of carbon, derived from plant material (wood, leaves) that was buried and underwent diagenesis. Geologists distinguish three *ranks* of coal based on the proportion of carbon: lignite (50%); bituminous (85%); anthracite (95%).
- **Oil shale**, like regular shale, is composed mostly of clay; but unlike regular shale, oil shale contains a significant amount of kerogen (a waxy organic chemical derived from fats in the bodies of plankton and algae).

6.4 Sedimentary Rock Identification

You are now familiar with the nature and origin of sediment, the major classes of sedimentary rocks, and the names of the most common sedimentary rock types. **TABLE 6.2** summarizes information from the preceding sections in a simplified

TABLE 6.2 Classification of common sedimentary rocks.

(A) Clastic Sedimentary Rocks (silicate mineral grains or rock fragments held together by cement)			
Composition	**Texture or Clasts (grain size in mm)**	**Clues to Identify**	**Rock Name**
Usually silicate minerals	Boulders, Cobbles, Pebbles (> 2)	Clasts are bigger than peas; you can measure clasts with a ruler.	**Conglomerate** (if clasts are rounded); **Breccia** (if clasts are angular) **Diamictite** (larger clasts are surrounded by a muddy matrix)
Usually silicate minerals	Sand (1/16 to 2)	Grains are easily visible and identifiable, but too small to measure with a ruler.	**Sandstone** (a general term) - **Quartz sandstone** (= arenite; > 95% quartz) - **Arkose** (quartz plus > 25% feldspar) - **Lithic sandstone** (contains > 5% lithic clasts*) - **Graywacke** (informal name; sand grains ± rock chips, surrounded by a clay matrix)
Usually silicate minerals	Silt (1/256 to 1/16)	Grains are visible but too small to identify; feels gritty.	**Siltstone**
Usually silicate minerals	Clay (<1/256)	Individual grains are not visible; composed dominantly of clay.	**Mudstone** (breaks into blocky pieces); **Shale** (breaks into thin plates)

* Lithic clasts are sand-sized rock fragments

TABLE 6.2 Classification of common sedimentary rocks (*continued*).

(B) Chemical Sedimentary Rocks (composed of grains that precipitated from a water solution)

Composition	Texture or Clasts (grain size in mm)	Clues to Identify	Rock Name
Halite	Crystal (generally > 2)	Clear to gray, with visible interlocking crystals; tastes salty.	**Rock salt**
Gypsum	Crystal (generally > 2)	Clear to whitish-pale gray or pinkish; soft, can be scratched with a fingernail.	**Rock gypsum**
Calcite	Grains appear like tiny balls	Very fine-grained; HCl test yields a vigorous fizz.	**Oolitic limestone**
Quartz	Grains not visible	Won't be scratched by a nail or knife; grains are too small to see; tends to be porcelainous; fractures conchoidally.	**Chert** - **Jasper** (reddish) - **Flint** (black)

(C) Biochemical Sedimentary Rocks (composed of minerals originally extracted by organisms to form shells)

Composition	Texture or Clasts (grain size in mm)	Clues to Identify	Rock Name
Calcite	Visible shells or a few shells in very fine grains (generally 1/256 to 2); some crystal appearance	Commonly grayish, but may be white, yellow, or pink; vigorously reacts with acid; some examples may consist of shell fragments cemented together and have a clastic texture; some have recrystallized to form a crystalline texture.	**Limestone** (general name) - **Fossiliferous limestone** - **Micrite** (very fine-grained; grains aren't visible)
Calcite	Visible shells (>2)	A weakly cemented mixture of shells.	**Coquina** (a type of limestone)
Calcite	Grains not visible	Whitish; can be used to write on slate. Vigorously reacts with HCl acid	**Chalk** (a type of limestone)
Dolomite	Grains not visible	Grayish to rusty tan; scratched-off powder has a moderate reaction with HCl acid.	**Dolostone**

(D) Organic (containing organic chemicals derived from the bodies of organisms)

Composition	Texture or Clasts (grain size in mm)	Clues to Identify	Rock Name
Clay and kerogen	Grains not visible	Dark gray to black; may have an oily smell; may burn; can't see grains.	Oil shale
Carbon (± clay and quartz)	Grains not visible	Black; may have a subtle bedding; typically breaks into blocks; may contain plant fossils.	Coal - **Lignite coal** (50% carbon; fairly soft) - **Bituminous coal** (85% carbon; medium hard; dull) - **Anthracite coal** (95% carbon; quite hard; shiny)

classification chart for sedimentary rocks based on texture and mineralogy. With this information, careful observations, and sometimes using the same tests you used for mineral identification (see Chapter 3, section 3.4), you will be able to identify the most common sedimentary rock types.

6.4.1 Steps in Identifying Sedimentary Rocks

Geologists approach sedimentary rock identification systematically, following the steps outlined below. As with mineral identification, we offer two different visual reference tools to help as you follow these steps: (1) Table 6.2 presents different classes of sedimentary rocks, with their basic characteristics, clues for identification, and their names, and (2) **FIG. 6.5** presents the same information in a flowchart. Identical colors identify the different classes of sedimentary rock in both schemes.

STEP 1 **Examine the rock's texture**. A simple textural observation will be easy for sedimentary rocks with grains coarse enough to see with the naked eye or a hand lens. Does the specimen consist of cemented-together grains, clasts, intergrown crystals, or fossils? If your specimen has grains too small to be seen even with a hand lens, go directly to step 2. Otherwise follow steps (a), (b), or

FIGURE 6.5 Flow chart for identifying sedimentary rocks.

STEP 1: Examine the rock's texture (if individual grains can be seen).

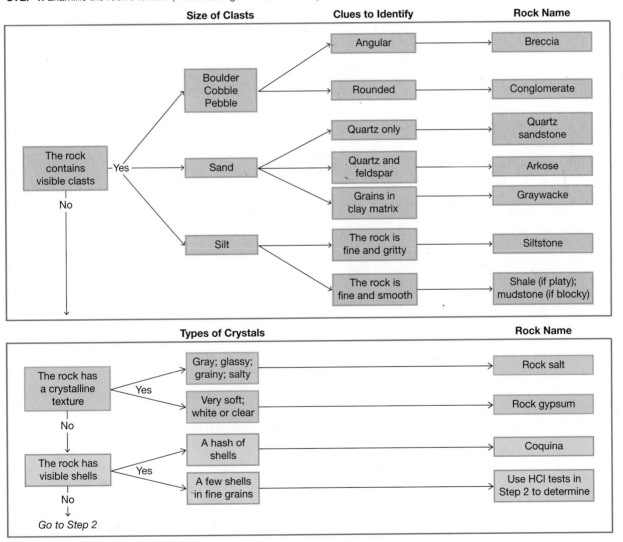

FIGURE 6.5 Flow chart for identifying sedimentary rocks. (*cont.*)

STEP 2: If the grains are too small to see, use physical (a) and HCl (b) tests to determine rock type.

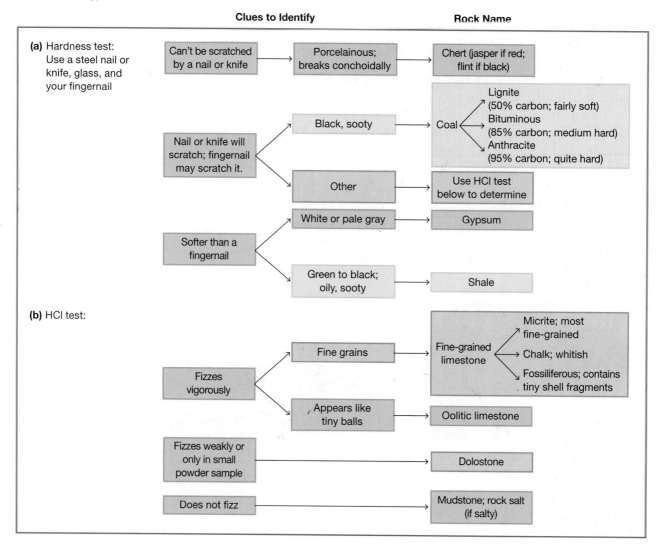

(c) depending on the texture of your specimen. Use Table 6.2 or Figure 6.5 to help make your determination.

(a) If the rock is clastic, determine grain *size* and *roundness*. First use the grain size chart in Table 6.1 to find the correct name for the grain size. Then, use other clues as identified in the Table 6.2 or Figure 6.5 to find the rock name. Remember that larger, rounded grains define a conglomerate and angular grains a breccia. If the clasts are smaller (at the size of sand or silt), you are identifying either a form of sandstone, siltstone, shale, or mudstone. Use the specifics in Table 6.2 or Figure 6.5 to choose the correct one.

(b) If the rock is crystalline, locate the crystalline rocks in Table 6.2 or Figure 6.5, and use the tests and observation skills you have learned for hardness, cleavage, luster, HCl reaction, and, if necessary, taste to identify the minerals. Then apply the appropriate name from the table or flowchart.

(c) If the rock is made entirely of cemented fossils and fossil fragments, it is *coquina*. If a few fossils are set in a fine-grained matrix, use HCl to determine if it is a fossiliferous limestone (it fizzes) or mudstone (it doesn't).

STEP 2 What to do when grains are too small to be seen. If the grains are too small to see, you will need to use the physical tests you have learned in Chapter 3.

(a) Use a steel nail (or knife) and fingernail to determine the rock's hardness.

    (i) If it is harder than the nail, it is chert—made entirely of silica—and has a Mohs hardness of about 7.

    (ii) If it is harder than the fingernail but softer than the steel nail (and it's not black and sooty—meaning it's a form of coal), go to step 2b to determine its chemical composition.

    (iii) If it is softer than the fingernail and white or pale gray, it is rock gypsum.

    (iv) If it is softer than the fingernail, black, and sooty, it is coal.

(b) Carefully add a very small drop of HCl to find out if the rock is a limestone or dolostone.

    (i) If the rock fizzes vigorously, it is a very fine-grained limestone. It could be chemical (micrite) or biochemical (chalk). However, if the texture appears like tiny balls it is oolitic limestone.

    (ii) If the rock fizzes weakly, scratch it again and put a drop of acid on the powder. If it reacts more strongly, it is a very fine-grained dolostone (dolomicrite).

    (iii) If the rock doesn't fizz, go to step 2c.

(c) Moisten your finger, rub it against the rock, and taste. If it tastes salty, the specimen is rock salt.

Exercise 6.3 will help you practice identifying rock samples by using these steps. Remember you can look to either the table or the chart for help. Both present the same information, so use whichever style you are more comfortable with.

EXERCISE 6.3　**Identifying Sedimentary Rock Samples**

Name: _____　Section: _____
Course: _____　Date: _____

Examine the rock samples in your set that you classified as sedimentary rocks. Read Section 6.4.1 closely to remind yourself of the steps. Fill in the rock study sheets at the end of this chapter to identify each sample. Keep these samples and your study sheets until you have finished the chapter. At that point, you will be able to add an interpretation of the histories of the rocks.

6.5 Interpreting Clastic Sedimentary Textures

If you are given a clastic sedimentary rock, identifying it is only part of the task. The texture of the rock is a rich source of information about the geologic history that eventually led to the deposition of the sediment from which the rock formed. Specific aspects of texture—grain size, sorting, and grain shape—provide clues to the amount of transport and the amount of weathering that the sediment has undergone between the time it eroded from its source to the time it became buried at the site of deposition.

6.5.1 Grain Size and Sorting

Clasts in sedimentary rocks range from the size of a house to specks so small they can't be seen without an electron microscope. As noted earlier, geologists use familiar words like *sand* and *pebble* to define clast size (see illustration in Table 6.1). The size of grains in a rock reflects the kinetic energy of the agent that transports it—an agent with more kinetic energy can move bigger clasts.

The kinetic energy of a transporting agent depends on both its mass and velocity. Air has much less mass than running water, so at the same velocity a stream can move larger particles than air can. Similarly, a small stream flowing at the same velocity as a large river has less kinetic energy (less mass) and therefore cannot move clasts as large as the river can. Another important factor for glaciers—the only *solid* transporting agent—is that ice is less dense than running water (remember, ice cubes float). A given volume of water therefore weighs more than the same volume of ice, and running water moves far faster than a glacier can advance. Yet, glaciers transport blocks of material far larger than those that even the largest river can move. There are two reasons for this: first, the mass of the ice is enormous; second, its solid nature provides a permanent strength absent in running water. Even though glacial ice moves slowly, it can carry debris weighing tons.

The **sorting** of a clastic sedimentary rock is a measure of the uniformity of grain size (**FIG. 6.6**). It shows the degree to which clasts have been separated by flowing currents, and it can also help identify the transporting agent. For example, aeolian (wind-deposited) sediment is very well sorted because wind can pick up only clasts of a narrow size range. A fast-moving, turbulent stream can carry clasts ranging from mud- to boulder-sized. But as a stream slows, sediment of different sizes progressively settles out. The coarsest grains (boulders and cobbles) drop out first, then the mid-size grains (pebbles and sand), and only when the water is moving very slowly can silt and mud settle out. Thus streams do sort clasts, but not as completely as the wind. Glacial ice is solid, so it can carry clasts of all sizes. As a result, glacially transported deposits are not sorted.

6.5.2 Grain Shape

The shape of clasts is another clue to the agent and distance of transportation. Clasts carried by water or wind collide frequently with one another as they move. As transport progresses, collisions knock off sharp corners and edges, eventually rounding the clasts. First the grains become **subrounded**, with smoothed edges and corners. Eventually, when they are almost spherical, they are called **rounded**. Thus, the farther streams and wind carry clasts, in general, the more rounded the grains become. The clasts also become smaller as a result of these collisions. **FIGURE 6.7**

FIGURE 6.6 Sorting in sedimentary rocks.

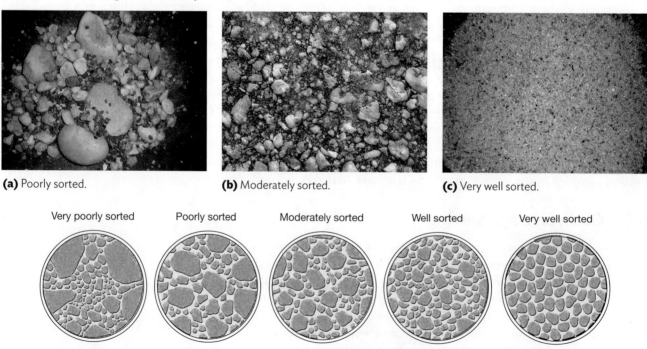

(a) Poorly sorted.

(b) Moderately sorted.

(c) Very well sorted.

Very poorly sorted   Poorly sorted   Moderately sorted   Well sorted   Very well sorted

(d) In a poorly sorted sediment, there is a great variety of different clast sizes, whereas in a well-sorted sediment, all the clasts are the same size.

FIGURE 6.7 Degrees of grain roundness in clastic sedimentary rocks.

Angular   Subangular   Subrounded   Rounded

(a) Angular clasts

(b) Rounded clasts

shows different degrees of roundness in two sedimentary rocks. In contrast, grains that have not moved far from their source tend to have sharp edges and corners because they have not had the opportunity to collide as much and are called **angular**.

EXERCISE 6.4   **What Could Move the Clasts?**

Name: _____   Section: _____

Course: _____   Date: _____

Some agents of transport can move sediment that ranges widely in grain size. Match the letter of the transport agent from the left-hand column with the **maximum grain size** it can transport listed in the right-hand column.

(a) house-sized block   ____ turbulent stream during a flood
(b) boulder   ____ very slow-moving stream
(c) cobble   ____ ocean waves or desert winds
(d) sand   ____ glacial ice
(e) silt   ____ quiet water
(f) mud   ____ fast-moving stream

EXERCISE 6.5   **Interpreting Sorting**

Name: _____   Section: _____

Course: _____   Date: _____

Look at the three sets of clasts in Figure 6.6. Suggest possible transporting agents that could be responsible for each.

(a) Poorly sorted: _____

(b) Moderately sorted: _____

(c) Very well sorted: _____

Name: _____   Section: _____

Course: _____   Date: _____

Look back at the definitions of breccia and conglomerate. Based on these definitions, which of these two rock types contains clasts that have been transported a longer distance? Explain.

_____

_____

_____

Angular clasts also occur in sediments deposited by glaciers, because these clasts are frozen into position and thus can't collide with one another.

6.5.3 Sediment "Maturity"

We discussed mineralogical maturity earlier in this chapter. We can expand this concept into the broader idea of **sediment maturity**—the degree to which a sediment has evolved from a crushed-up version of the original rock into a sediment that has lost its easily-weathered minerals and become well sorted and rounded. The changes occur as the sediment is transported and include the loss of easily-weathered minerals by chemical reaction and progressive rounding and sorting.

Specifically, if the resulting sediment accumulates close to the source, it can contain a variety of grain sizes and remain poorly sorted. Also, because chemical weathering has not progressed to completion, relatively unstable minerals (e.g., feldspar, mica, and amphibole) remained mixed with stable minerals (e.g., quartz). Finally, since grains have not traveled far, they may retain an angular shape. Geologists refer to a sediment with these characteristics as *immature*. If, however, the sediment is carried a long distance by a river and/or is washed by wave action along a shore, and if the sediment has time to undergo substantial chemical weathering so that unstable minerals transform into clay and wash away, it will be quite different—it will be well sorted, it will consist almost entirely of stable minerals, and grains will be well rounded. Geologists refer to sediments with this character as *mature* (**FIGURE 6.8**).

FIGURE 6.8 Mature sediment. As sediments are transported progressively farther, weatherable sediments such as feldspar break down and convert to clay, which washes away, so the proportion of sediment consisting of resistant minerals such as quartz increases. Further, the physical bouncing and grinding that accompanies the transport of sediment progressively rounds the quartz grains and sorts them.

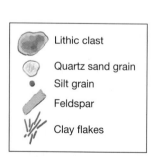

- Lithic clast
- Quartz sand grain
- Silt grain
- Feldspar
- Clay flakes

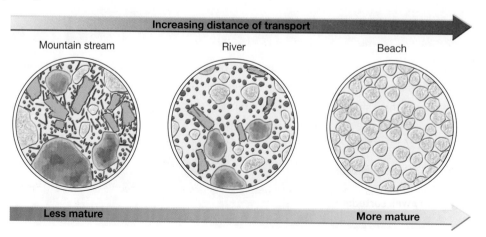

Increasing distance of transport

Mountain stream   River   Beach

Less mature   More mature

EXERCISE 6.7 | Recognizing Sediment Deposited by Streams, Wind, and Glaciers

Name: _____ Section: _____

Course: _____ Date: _____

Fill in the following table to summarize the characteristics of sediment deposited by the different continental agents of erosion.

Textural feature	Agent of transportation		
	Streams	Wind	Glaciers
Grain size			
Sorting			
Grain shape			

6.5.4 Cements in Clastic Rocks

The most common cements found in clastic sedimentary rocks are quartz or calcite, for the ions making these minerals can occur in relatively high concentrations in the groundwater from which cements precipitate. Which mineral occurs as the dominant component of the cement depends on the source of the groundwater and on the chemical environment in which precipitation takes place.

In some cases, the cement that holds clasts together provides a clue to the depositional environment. For example, the formation of a hematite cement requires oxygen, so sedimentary rocks in which the cement contains hematite, and thus has a reddish color, come from environments where water contained dissolved oxygen. In contrast, sedimentary rocks containing pyrite in the cement formed in environments low in oxygen, because if oxygen had been present, the pyrite would have dissolved. Reddish sedimentary rocks, or *redbeds*, generally indicate deposition in terrestrial environments (e.g., rivers, alluvial fans). You can practice this concept in Exercise 6.8.

EXERCISE 6.8 | Identifying Cements

Name: _____ Section: _____

Course: _____ Date: _____

How can you tell what minerals occur in a cement? Simply remember the basic physical properties of the minerals (e.g., hardness, the ability to react with acid, color). With this in mind, answer the following questions concerning sandstone, a type of sedimentary rock consisting of cemented-together grains of quartz sand.

(a) The cement (but not the grains) in a sedimentary rock reacts when in contact with dilute HCl. The cement consists of
_____.

(b) The rock, overall, has a reddish color. The cement contains _____.

(c) The cement is very strong, and when scratched with a steel needle it has the same hardness as the grains. The cement consists of _____.

6.6 Sedimentary Structures: Clues to Ancient Environments

6.6.1 Beds and Stratification

Gravity causes all sediment to settle to the floor of the basin in which it was deposited. Over time, layers of sediment called **beds** accumulate. Beds range from a millimeter to several meters thick, depending on the process involved. Each bed represents a single depositional event, and the different colors, grain sizes, and types of sediment from each event distinguish one bed from another (**FIG. 6.9**).

Many beds are fairly homogeneous, with uniform color, mineralogy, and texture, and have smooth surfaces. But some contain internal variations or have distinct features on their surfaces. These **sedimentary structures** provide important information about the rock's history. In the next section, we look at a few examples of sedimentary structures.

6.6.2 Sedimentary Structures

Graded Beds Graded beds are layers in which the grain size decreases progressively from the bottom to the top. These form during a submarine avalanche when poorly sorted sediment speeds down a submarine slope (**FIG. 6.10**). When the avalanche, called a *turbidity current*, slows, the coarsest grains settle first while the finer materials remain in suspension longer. With time, the finer sediments settle, and the finest ones settle last. To make a graded bed in the lab, put water, sand, silt, and gravel in a jar or graduated cylinder. Shake to mix thoroughly and watch as the grains settle.

Ripple Marks, Dunes, and Cross Beds *Ripple marks* are regularly spaced ridges on the surface of a sedimentary bed at right angles to the direction of the current flow. They form when sand grains are deposited by air or water currents. Small ripple

FIGURE 6.9 Sedimentary rock beds.

(a) Horizontal beds of sandstone and siltstone in the Painted Desert of Arizona. White, dark red, and light red beds contain different amounts of hematite cement.

(b) Horizontal beds of sandstones, siltstones, and limestones in the Grand Canyon in Arizona. Beds are distinguished by differential resistance to weathering as well as color.

FIGURE 6.10 Graded bedding.

marks are shown in **FIGURE 6.11**. Some ripple marks have systematically oriented steep and gentle sides (Fig. 6.11a). These **asymmetric ripple marks** are produced by a current that flowed from the gentle side toward the steep side. **Symmetric ripple marks** (Fig. 6.11b) have steep slopes on both sides and form from oscillating currents. Larger mounds of sediments that resemble asymmetric ripple marks in shape are called *dunes*. The largest dunes occur in deserts, where they build from windblown sand; however, dunes can also form underwater.

In some beds of sediment, subtle curving surfaces, delineated by coarser and/or denser grains, lie at an angle to the main bedding surfaces. These inclined surfaces are called **cross beds**. They form when sediment moves up the windward (or up-current) side and then slips down the lee side as dunes and ripples build (**FIG. 6.12**).

Mud Cracks and Bedding-Plane Impressions Mud cracks are arrays of gashes in the surface of a bed formed when mud dries out and shrinks. Typically, mud cracks are arrayed in a honeycomb-like pattern. Sand deposited over the mud layer fills the cracks, so when the sediment lithifies, the shape of the cracks remains visible (**FIG. 6.13**).

FIGURE 6.11 Ripple marks.

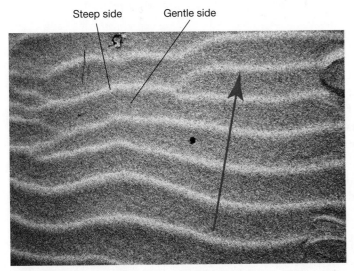

(a) Asymmetric ripple marks in modern sand. The arrow indicates the direction of the current that formed the ripples.

(b) Symmetric ripple marks in ancient sandstone. Note the equally steep slopes on both sides of the ripples.

FIGURE 6.12 Cross bedding, a type of sedimentary structure within a bed.

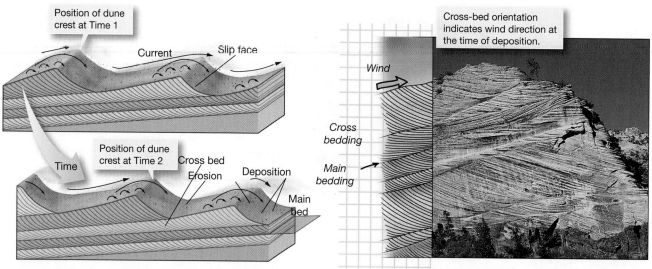

(a) Cross beds form as sand blows up the windward side of a dune or ripple and then accumulates on the slip face. With time, the dune crest moves.

(b) A cliff face in Zion National Park, Utah, displays large cross beds formed between 200 million and 180 million years ago, when the region was a desert with large sand dunes.

FIGURE 6.13 Formation of mud cracks.

(a) Mud cracks in modern sediment. The circular impressions were made by raindrops hitting the mud before it dried.

(b) Mud cracks preserved in 180 million-year-old mudstone.

Moving objects can leave impressions on unconsolidated sediment before the bed hardens into rock. The imprints of these objects can, in some instances, provide fascinating glimpses of animals living in the sediment or of processes acting on it.

Imprints may be made by raindrops falling on a muddy floodplain (Fig. 6.13a), logs dragged by currents along the bottom of a stream, worms crawling along the sea floor in search of food (**FIG. 6.14a**), and the footprints of animals walking through a forest (**FIG. 6.14b**). Indentations also may be scoured into a bed surface by the turbulence of the fluid flowing over it.

6.7 Fossils: Remnants of Past Life

Fossils are the remains or traces of animals and plants that are preserved in rocks—in effect, they are sedimentary structures. Generally, fossils reveal the shape of the hard

FIGURE 6.14 Impressions on bedding planes.

(a) Worm burrows on siltstone in Ireland (lens cap for scale).

(b) Footprints of different-sized dinosaurs.

Gaining Insight into Depositional Environments of Sedimentary Rocks

Name: _____   Section: _____
Course: _____   Date: _____

Sedimentary structures, fossils, cements, and other features of sedimentary rocks provide insight into the environment in which the sediment was deposited. Match each of the features in the left-hand column to an aspect of the depositional environment listed in the right-hand column by placing the corresponding letter in the proper blank.

Rock and/or feature of the rock

(a) poorly sorted arkose

(b) contains fossils of intact coral

(c) red mudstone containing dinosaur footprints

(d) black shale containing some pyrite crystals

(e) sandstone containing symmetrical ripples

(f) cross-bedded sandstone

(g) contains > 50% carbon and fossil leaves

(h) contains large, rounded pebbles and cobbles

(i) contains fossils of feathers

(j) very angular grains

Aspect of the depositional environment

_____ deposited in a desert dune or in a current

_____ formed from sediments accumulated in a swamp

_____ deposited in very quiet (stagnant) water

_____ deposited in an anoxic marine setting

_____ deposited by a swiftly moving stream

_____ formed from warm-water, shallow marine reef

_____ has not undergone a lot of transport

_____ immature, deposited close to the source

_____ formed from terrestrial mud (riverbank deposits)

_____ deposited on a beach

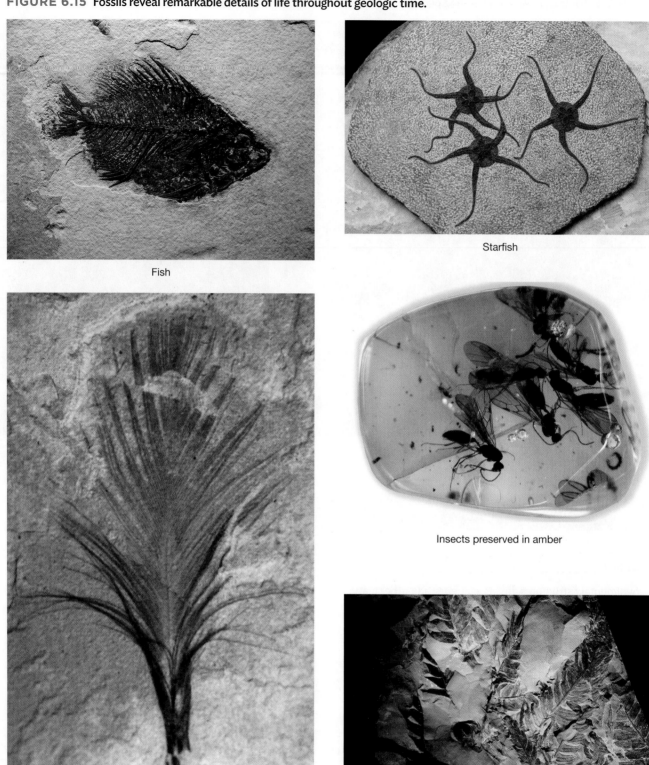

Fish

Starfish

Feather

Insects preserved in amber

Leaves

(a) Relatively fragile organisms.

FIGURE 6.15 Fossils reveal remarkable details of life throughout geologic time (*continued*).

Brachiopod

Tree trunks

Apatosaurus (formerly called Brontosaurus)

(b) Relatively robust organisms.

parts (shells or skeleton) of an organism. Soft-bodied organisms are preserved only in special cases, where deposition happens before the organism decays (**FIG. 6.15**). Fossils are the subject of a whole subdiscipline of geology called paleontology, for the study of fossils reveals the history of life on Earth and, as discussed elsewhere in this book, provides clues to the age of strata. We mention them briefly here because most fossils are found in sedimentary rocks, due to the environments in which sedimentary accumulation occurs are the same as those in which animals and plants lived.

Because different kinds of organisms live in different environments, the types of fossils in a sedimentary rock provide clues to the depositional environment in which the sediment composing the rock accumulated. Applying the *principle of uniformitarianism,* we can deduce that sediments containing fossil corals probably formed in warm, clear seawater; sediments with fossils of tropical plants were probably deposited in equatorial latitudes; sediments with fossils of dinosaurs generally accumulated on land rather than in the deep ocean; and sediments containing fossil starfish accumulated in the oceans, not on land.

FIGURE 6.16 Examples of depositional environments.

Glacial: unsorted sediment containing angular clasts—commonly, such sediment may contain large clasts suspended in a much finer matrix

Mountain stream

Alluvial fan: conglomerate and arkose

Delta

Swamp: coal

Shallow marine

Deep marine: shale, chert, micrite, or chalk

Lagoon

Reef

Fluvial (river): channels of sandstone or conglomerate, in some places surrounded by layers of mudstone, and hematite containing cement

Lagoon

Reef

Laminated mud from a lake bed

Desert dune: well-sorted sandstone with large cross beds

Beach: well sorted, cross-bedded sandstone

6.8 Applying Your Knowledge to Stratigraphy

Ultimately, geologists use observational data about sedimentary rocks to interpret the depositional setting (conditions of deposition) in which the sediment forming the rock accumulated. Examples of depositional settings are listed in **FIGURE 6.16**, along with a simplified description of characteristics found there. Successions of beds, or strata, are like pages in a book that record the succession of depositional environments. Further study in geology will add additional detail to this image, but even with your basic knowledge, you can interpret outcrops in the field. You will do this in Exercise 6.10.

EXERCISE 6.10 **Interpreting Outcrops**

Name: _____   Section: _____
Course: _____   Date: _____

(a) Look at the sedimentary rocks in the figures below. Brief "field notes" are provided with each photograph. In each case, identify the rock and any sedimentary structures present, and interpret the depositional setting in which the rock formed.

Interpreting sedimentary outcrops.

Entrance to Zion National Park in Utah. Large cliff face containing thick beds of medium-grained white, gray, and beige sandstones displaying excellent cross bedding.

_____

_____

_____

_____

Sandstone with preserved dinosaur trackway and asymmetric ripple marks. Note different orientation of ripple marks in the foreground.

_____

_____

_____

(continued)

Name: _____ Section: _____

Course: _____ Date: _____

Excavation along a lumber road in eastern Maine. Very poorly sorted sediment with large clasts of granite, sandstone, and shale in a matrix of gravel, sand, and large amounts of clay. Based on this information and the photo to the left, what is the most likely depositional environment and what is the most likely agent of erosion responsible.

_____

_____

_____

_____

(b) Using *Google Earth*™, fly to the following coordinates. Describe what you see.

- Lat 24°44'40.42" S, Long 15°30'5.34" E _____

- Lat 23°27'03.08" N, Long 75°38'15.68" W _____

- Lat 67°3'38.31" N, Long 65°33'2.88" W _____

- Lat 25°7'48.94" N, Long 80°59'3.18" W _____

- Lat 30°10'2.73" N, Long 94°48'58.24" W _____

- Lat 5°24'54.48" N, Long 6°31'26.25" E _____

(c) What do regions underlain by sedimentary rocks look like? Fly to the following coordinates (Lat 47°48'02.31" N, Long 112°45'30.97" W) using *Google Earth*™ and see an example of a region where sedimentary beds are well exposed. Note that the bedding appears as a series of parallel ridges if the layers are tilted. That's because softer beds (shale) weather faster than harder beds (sandstone).

?What Do You Think You are the project manager for a company and you're in charge of completing large projects on time, with millions of dollars on the line in costs and profits. While bulldozing land for a new housing development, one of your crews uncovered a large dinosaur trackway, like the one shown on page 161. It is clearly an important find because nothing like this has ever been reported in the area. Reporting the trackway to the State, however, could delay the project, lowering the anticipated profits. But, not reporting it would mean losing this priceless artifact forever.

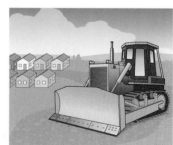

What do you think? Is it possible to preserve the trackway without affecting the timeline, cost, and profit of your project? Or should you just continue with the excavation? What would you recommend and why? (Answer on a separate sheet of paper.)

SEDIMENTARY ROCKS STUDY SHEET

Sample	Texture (grain size, shape, sorting)	Components			Minerals/rock fragments present (approximate %)	Name of sedimentary rock	Rock history (transporting agent, environment, etc.)
		Clastic	Chemical	Biogenic			

SEDIMENTARY ROCKS STUDY SHEET

Name _____

| Sample | Texture (grain size, shape, sorting) | Components | | | Minerals/rock fragments present (approximate %) | Name of sedimentary rock | Rock history (transporting agent, environment, etc.) |
		Clastic	Chemical	Biogenic			

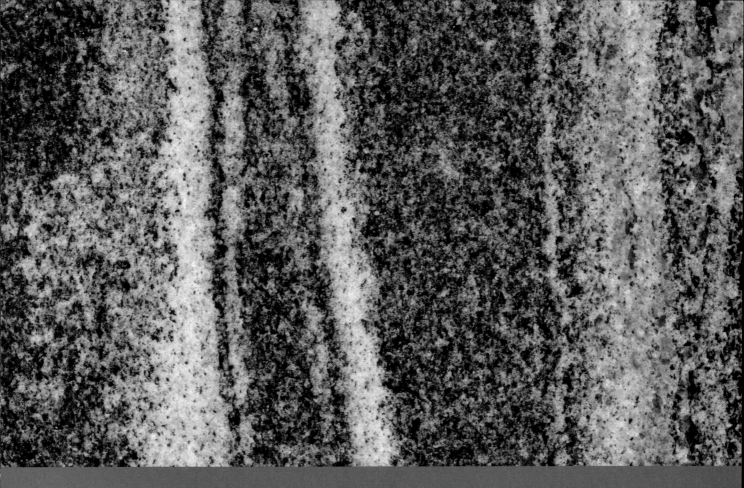

7

Using Metamorphic Rocks to Interpret Earth History

This polished specimen of gneiss is a layered and foliated metamorphic rock.

- Understand the agents that bring about metamorphism
- Understand the changes that occur during formation of metamorphic rocks
- Become familiar with metamorphic textures and mineral assemblages
- Interpret the history of metamorphic rocks from their textures and mineral assemblages

MATERIALS
NEEDED

- A set of metamorphic rocks
- Magnifying glass or hand lens and, ideally, a microscope and thin sections of metamorphic rocks
- Standard supplies for identifying minerals (streak plate, glass plate, etc.)
- Play-Doh, small rods (plastic coffee stirrers cut into short pieces), pennies

7.1 Introduction

Biologists use the term *metamorphosis* (from the Greek *meta*, meaning change, and *morph*, meaning form) to describe what happens when a caterpillar turns into a butterfly. Geologists use the similar term **metamorphism** for certain processes of change that a rock undergoes when exposed to physical and chemical conditions significantly different from those under which it first formed. Logically, the end product of metamorphism is a **metamorphic rock**. Note that we used the word *certain* in the definition of metamorphism. That's because not all changes that a rock can undergo are considered to be metamorphism. For example, metamorphism does not include the processes of weathering, diagenesis, melting, or solidification from a magma. By definition, metamorphism occurs at higher temperatures than those of diagenesis (i.e., greater than about 250°C) and only under conditions in which the rock remains solid (i.e., metamorphic rocks are not igneous). Because metamorphism takes place without either fragmentation or melting, geologists describe metamorphism as a "solid-state" process.

Metamorphism can cause many kinds of changes in a rock or, in some cases, just one type of change. For example, during metamorphism one or more of the following may take place as the parent rock (called the *protolith*) is converted to a metamorphic rock:

- The **texture** of the protolith may change, sometimes dramatically, by:
  - *Recrystallization*: when small grains of a mineral coalesce to form larger grains or large grains separate into smaller ones.
  - *Pressure solution*: if water is present, some minerals dissolve partially and the dissolved ions re-precipitate, changing relationships between grains.
  - *Alignment of grains*: in some instances, randomly oriented protolith minerals are aligned parallel to one another. The result is foliation and lineation—two metamorphic textures described Section 7.4.2.

- The **mineralogy** of the protolith may change. The original minerals in the protolith may be changed into new **metamorphic minerals** that are more stable under the new temperature and pressure conditions. This may occur by:
  - *Phase change*: one polymorph (remember Chapter 3) changes to another more stable at the new conditions. There is no change in the chemical composition. Example: andalusite (Al_2SiO_5) converts to sillimanite (Al_2SiO_5) with increasing heat.
  - *Neocrystallization*: a process of chemical reactions between protolith minerals during which ions from two or more minerals combine to create new minerals.

  Example: Muscovite + quartz → sillimanite + potassic feldspar + water
  $$KAl_2(AlSi_3O_{10})(OH)_2 + SiO_2 \rightarrow Al_2SiO_5 + KAlSi_3O_8 + H_2O$$

Sometimes a rock changes only slightly during metamorphism, so that most of its original characteristics are still recognizable. In many cases, however, the changes are so drastic that a metamorphic rock can look as different from its protolith as a butterfly is from a caterpillar (**FIG. 7.1**).

Most metamorphism is caused by tectonic and igneous processes that we cannot observe because they take place deep within the Earth. For example, what happens to rocks as the opposite sides of the San Andreas Fault grind past one another? Or to rocks along the west coast of South America as part of the Nazca Plate is subducted beneath the continent? Or to rocks deep below Mt. Everest as India collides with Asia? How did rocks in California change when they were intruded by granitic magma that today forms the Sierra Nevada batholith?

FIGURE 7.1 Examples of changes from protolith (left) to metamorphic rock (right).

(a) Changes in texture (grain size and shape).

    i. Very fine-grained calcite in micrite recrystallizes to coarse-grained calcite in marble.

    ii. Coarse quartz and feldspar grains in granite change to very fine grains in ultramylonite.

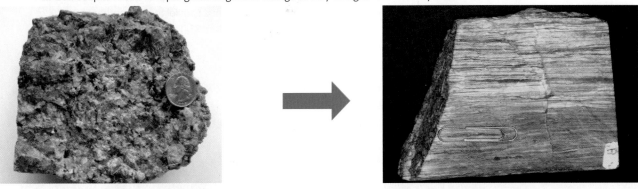

    iii. Grain shape changes: round clasts in conglomerate are flattened in metamorphosed conglomerate.

(b) New minerals form—protolith minerals react to form new minerals more stable at the metamorphic conditions.

    i. Clay minerals in mudstone react with quartz to form biotite and garnet in gneiss.

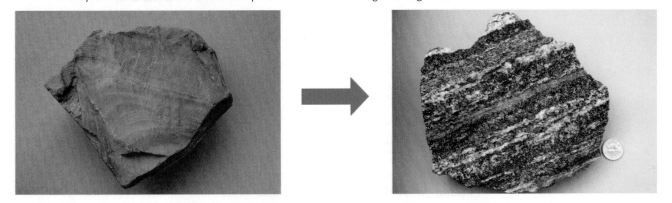

ii. Dolomite and quartz in micrite react to produce actinolite (green) in marble.

(c) Both texture changes and new minerals form.

i. Foliation: muscovite flakes formed from clay minerals in mudstone are aligned parallel to one another to produce foliation in schist.

ii. Lineation: rod-shaped amphiboles produced from pyroxene in diorite are aligned parallel to one another to produce lineation in a metamorphosed diorite.

Metamorphic rocks provide answers to these and many other questions about Earth history and Earth processes. Because they form at depths below which sedimentary rocks form and above which melting occurs, metamorphic rocks yield important clues to conditions and processes that igneous and sedimentary rocks cannot.

This chapter explains how you can read the Earth history recorded in metamorphic rocks. We begin by examining the causes of metamorphism and then examine characteristic features of metamorphic rocks, kinds of metamorphic rocks, and the information we can glean from a sample about the conditions under which it was metamorphosed. Finally, we introduce the concept of the intensity or **grade** of metamorphism and discuss how to interpret metamorphic grade from a specimen.

7.2 Agents of Metamorphism

Geologists refer to the heat, pressure, and fluids that cause metamorphism as agents of metamorphism. Let's consider how each of these may trigger change in a rock. Keep in mind, however, that even though we address each agent of metamorphism separately, all of them may act simultaneously in many geologic environments.

7.2.1 The Effect of Heat

A rock heats up when it either becomes buried to great depth, as can happen either during mountain building—when part of the crust is pushed up and over another part—or when magma intrudes nearby. As the temperature rises, atoms vibrate

FIGURE 7.2 Growth of metamorphic minerals.

(a) Garnet crystals.

(b) Andalusite crystals.

more rapidly, and chemical bonds holding the atoms into the lattice of minerals begin to stretch and break. The freed atoms migrate slowly through the solid rock by a process called **diffusion**, much as atoms of ink spread when dropped into a glass of water. Eventually, the wandering atoms bond with other atoms to produce metamorphic minerals that are more stable under the higher temperatures. Put another way, heat drives chemical reactions that can produce a new assemblage of metamorphic minerals.

If the protolith has a very simple composition (e.g., quartz only, or calcite only), metamorphic reactions may produce new crystals of the same minerals. For example, metamorphism of a pure quartz sandstone (a sedimentary rock composed of SiO_2 grains) produces a metamorphic rock composed of new grains of quartz. But if the protolith contains a variety of elements, metamorphism can produce completely different minerals. For example, metamorphism of shale (composed of clay, which can contain Ca, Fe, Al, Mg, and K) produces a metamorphic rock composed of mica, quartz, and garnet. Diffusion is much slower in solid rock than in magmas or aqueous solutions, but over thousands to millions of years, it can completely change a rock. In some cases, new metamorphic minerals can grow to impressive sizes (**FIG. 7.2**).

7.2.2 The Effect of Pressure

Two kinds of pressure are important in metamorphism (**FIG. 7.3**). **Lithostatic pressure** results from the burial of rocks within the Earth and, like hydrostatic pressure in the oceans or atmospheric pressure in air, it is equal in all directions. Lithostatic pressure can become so large at great depths beneath mountain belts or in association with subduction that it forces atoms closer together than the original mineral's structure can allow. As a consequence, a new mineral with a more compact atomic arrangement can form that is more stable under the high-pressure conditions. Such a change is called a **phase change**. Diamond, for example, is a metamorphic mineral formed at very high pressures from graphite. Diamond and graphite are both made of pure carbon, but the diamond structure is much more compact than that of graphite and therefore more stable at high pressure. Minerals like diamond and graphite that have identical compositions but different internal structures are called **polymorphs**.

When subjected to lithostatic pressure, the shape of mineral grains in a rock doesn't necessarily change, because the rock is being pushed *equally* from all directions (Fig. 7.3a). During mountain-building processes and along plate boundaries, bodies of rock undergo more intense squeezing in one direction than in others. To picture this vise-like squeezing, imagine what happens when you press a ball of

FIGURE 7.3 Effects of lithostatic pressure and differential stress.

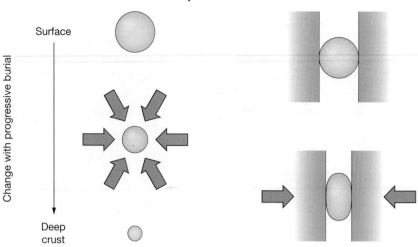

(a) Lithostatic pressure. Pressure acts equally from all directions; grain remains spherical but gets smaller.

(b) Differential stress. Greater pressure in one direction than others; spherical grain is flattened.

clay or dough between a book and a table—it flattens into a pancake. In addition, the crust may move, producing shear, during which one part of a rock body moves sideways relative to another—to picture shear, imagine what happens when you smear a deck of cards across a table. Simplistically, we can refer to the type of pressure in such rock due to vise-like squeezing and/or shear as **directed pressure**, or **differential stress**. Under conditions of elevated temperature, differential stress can cause grains to change shape, flattening them in one direction and elongating or stretching them in others (Fig. 7.3b).

7.2.3 The Effect of Temperature and Pressure Combined

It's important to keep in mind that pressure changes in most metamorphic environments are accompanied by temperature changes, because as the depth of burial increases, *both* temperature and pressure rise. But the temperature change that occurs with increasing depth at one location is not necessarily the same as that which occurs at another. For example, at a given depth, the temperature near an intruding pluton of magma may be higher than in a region at a great distance from the magma. The particular minerals that form during metamorphism depend on both pressure and temperature. We can represent this fact on a graph, called a **phase diagram**, that has pressure and depth on the vertical axis and temperature on the horizontal axis. **FIGURE 7.4** shows the phase diagram for the aluminum silicate minerals—three different minerals that exist in nature but have the same chemical formula. Note that at points Y and Z, the pressure is the same but the temperature is different—at a lower temperature, kyanite forms, whereas at a higher temperature, sillimanite forms. Point X is at the same temperature as Y, but at a lower pressure; under these conditions, andalusite forms at low pressure while kyanite forms at high pressure.

FIGURE 7.4 Phase diagram for aluminum silicate (Al_2SiO_5).

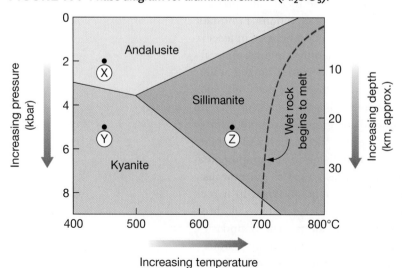

7.2.4 Metasomatism: The Effect of Hydrothermal Fluids

In many environments, rocks contain high-temperature liquids, gases, and "supercritical fluids" (fluids that have characteristics of both a gas and liquid) composed of H_2O and/or CO_2 in which ions of other elements have been dissolved. Collectively, these fluids are called **hydrothermal fluids** (from *hydro*, meaning water, and *thermal*, meaning hot). They may be released from magma when the magma solidifies, may be produced by metamorphic reactions, or may be formed where groundwater has percolated deep into the crust. Hydrothermal fluids can be "chemically active" in that they can dissolve ions and transport them throughout the crust and can provide ions that bond to other ions and produce new, metamorphic minerals. Thus, reactions with hydrothermal fluids tend to change the chemical composition of rocks—the fluids can be thought of as buses that pick up and drop off ions as they pass through a rock. In some cases, the fluids pick up certain chemicals and flush them entirely out of the rock. When the chemical composition of a metamorphic rock changes significantly due to reaction with hydrothermal fluids, we say that the rock has undergone **metasomatism**.

Exercise 7.1 will help you understand how metamorphic agents bring about changes in metamorphic rocks.

EXERCISE 7.1   **Effects of Metamorphic Agents**

Name: _____   Section: _____
Course: _____   Date: _____

(a) Andalusite, sillimanite, and kyanite are polymorphs of Al_2SiO_5 with identical chemical compositions but different crystal structures. Based on the phase diagram in Figure 7.4, and keeping in mind how pressure tends to affect the compactness of crystals, which of these minerals has the lowest specific gravity? Explain.

_____

_____

_____

_____

(b) During metamorphism at a given location, the overall amount of the element calcium in a metamorphic rock changes substantially. Which metamorphic agent is likely to have caused this metamorphism? Why?

_____

_____

_____

_____

(c) Would you expect to find aligned, flattened grains in a rock that has undergone heating but has not been subjected to differential stress during metamorphism? Why?

_____

_____

_____

_____

7.3 Types and Environments of Metamorphism

Metamorphism may be caused by a single agent or by a combination of two or three, depending on the tectonic and geologic settings in which agents are applied. Geologists recognize six types of metamorphism, each characterized by a unique agent or combination of agents and associated with a different geologic or tectonic environment. Each type produces distinctive metamorphic rocks, enabling geologists to interpret ancient tectonic settings from the rock types exposed on the surface.

Heat plays a role in all types of metamorphism, helping to break bonds in minerals. It is a dominant agent in some types, but plays a relatively minor role in others where lithostatic pressure or differential stress dominate. The types of metamorphism are as follows:

- **Contact metamorphism** (also called thermal metamorphism) occurs where rocks are subjected to elevated heat without a change in pressure and without the application of differential stress. This happens where an igneous intrusion comes in contact with a rock (**FIG. 7.5a, b**).

- **Regional metamorphism** (also called dynamothermal metamorphism) occurs where rocks in a large region of crust are subjected to increases in temperature and pressure *and* are subjected to differential stress (squeezing and shearing). Such metamorphism typically happens during mountain-building processes (Fig. 7.5a,b).

- **Burial metamorphism** happens when rocks are simply buried very deeply by overlying sediment (**FIG. 7.5c**).

- **Dynamic metamorphism** occurs where rock undergoes differential stress in response to shear along a fault zone but does not undergo a change in temperature and/or lithostatic pressure (**FIG. 7.5d**).

- **Shock** (or impact) **metamorphism** occurs where a meteorite hits the Earth and its enormous kinetic energy is converted instantaneously to heat and differential stress. Impact metamorphism is rare on Earth (**FIG. 7.5e**).

- **Metasomatism** occurs where hydrothermal fluids are present, especially as

FIGURE 7.5 Environments of metamorphism.

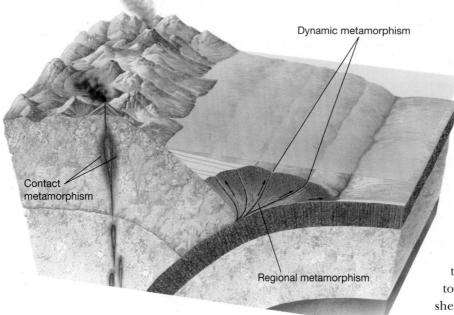

(a) Convergent plate boundaries (e.g., Japan).

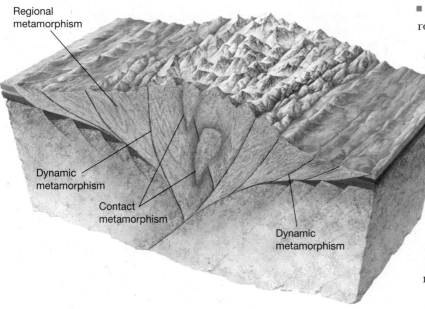

(b) Continent-continent collision zone (e.g., Appalachian Mountains).

FIGURE 7.5 Environments of metamorphism (*continued*).

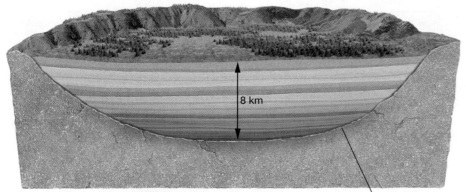

(c) Deep sedimentary basin (e.g., Rio Grande Rift).

8 km

Burial metamorphism

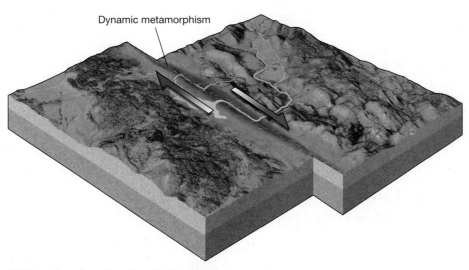

Dynamic metamorphism

(d) Transform fault (e.g., San Andreas Fault).

Impact metamorphism

(e) Meteorite crater (e.g., Barringer Crater, Arizona).

superheated water from metamorphic reactions adds dissolved ions to or removes ions from rock, changing its composition.

TABLE 7.1 summarizes the types of metamorphism and notes how the agents of metamorphism present in each type tend to operate.

TABLE 7.1 Types, environments, and agents of metamorphism.

Agent(s) of metamorphism*	How the agents of metamorphism are applied	Type of metamorphism	Geologic environment
Heat (pressure not a major cause)	Heat given off from cooling magma affects rocks adjacent to an intrusive body or lava flow.	Contact or thermal	Contact between pluton or lava flow and host rock
Lithostatic pressure and a little heat	Gravity causes increased lithostatic pressure as rocks are buried deeper.	Burial	Lower parts of deep sedimentary basins
Differential stress and a little heat	Two large blocks of rock grind past one another in a fault zone. The moving blocks generate the differential stress; heat comes from depth in the Earth and a very minor amount from friction.	Dynamic or fault zone	Fault zones, including transform faults
Hydrothermal fluids	Superheated water from magma or metamorphic reactions adds dissolved ions to or removes ions from rock, changing its composition.	Metasomatism	Continental collision zones; pluton-host rock contacts; mid-ocean ridges
Heat, lithostatic pressure, differential stress	Lithosphere plates collide in subduction zone or continent-continent collision. Collision creates intense differential stress; heat and lithostatic pressure come from depth in the Earth.	Regional	Convergent plate boundaries: subduction zones and continent-continent collisions
Heat, differential stress	Meteorite survives passage through the atmosphere and collides with Earth.	Impact	Meteorite impact crater

* Hydrothermal fluids may play a role in any type of metamorphism.

7.4 Metamorphic Rock Classification and Identification

Classification of metamorphic rocks, like that of their igneous and sedimentary relatives, is based on mineral content and texture. Both kinds of information are needed to identify a rock successfully. In this section, we will first review mineral content or composition, then texture, and then see how we can use both to identify metamorphic rocks.

7.4.1 Compositional Classes of Metamorphic Rocks

Geologists recognize four major classes of metamorphic rocks based on their mineral (and therefore chemical) composition:

■ **Aluminous** (or **pelitic**) metamorphic rocks contain a relatively large proportion of aluminum, silicon, and oxygen, with lesser amounts of potassium, iron, and magnesium. These rocks were derived from shale or mudstone, which are made up mostly of clay minerals rich in aluminum. Common metamorphic minerals of aluminous rocks include muscovite, biotite, chlorite, garnet, andalusite, sillimanite, and kyanite.

■ **Quartzo-feldspathic** metamorphic rocks contain large amounts of silicon, oxygen, potassium, and sodium, less aluminum than aluminous rocks, and small amounts of calcium, iron, and magnesium. Possible protoliths include felsic igneous rocks like granite and rhyolite or sedimentary rocks like arkose. Typical minerals, as the name suggests, are quartz, potassic feldspar, and plagioclase feldspar.

■ **Calcareous** metamorphic rocks are composed mostly of calcium, carbon, and oxygen, with some magnesium. These are derived from carbonate sedimentary rocks—limestones and dolostones. If the protolith contained significant amounts of clay and quartz, the additional aluminum, silicon, and oxygen permit crystallization of "calc-silicate" metamorphic minerals like garnet, the amphibole actinolite, or the pyroxene diopside.

■ **Mafic** metamorphic rocks contain calcium, iron, magnesium, aluminum, silicon, and oxygen. As the name suggests, their protoliths were basalts and gabbros. Typical mafic metamorphic minerals include epidote, amphiboles, pyroxenes, and garnet.

A fifth class—**organic**—is needed for the relatively uncommon instances in which an organic sedimentary rock is metamorphosed. The most common of these is anthracite coal.

Exercise 7.2 will show how minerals in metamorphic rocks can help identify the composition of the rock's protolith.

EXERCISE 7.2 **Metamorphic Minerals as a Key to Protolith Composition**

Name: _____   Section: _____
Course: _____   Date: _____

(a) Identify the minerals in the four metamorphic rocks provided by your instructor, and record the data in the following table. Add the chemical compositions for these minerals from Appendix 3.2.

Major minerals and their compositions	Specimen 1	Specimen 2	Specimen 3	Specimen 4

(b) Based on the minerals and their chemical compositions, what were the most abundant elements in the protoliths of these specimens?

Specimen 1	Specimen 2	Specimen 3	Specimen 4

(c) Is it likely that these metamorphic rocks came from the same parent rock? Explain.

_____

_____

_____

_____

7.4.2 Textural Classes of Metamorphic Rocks

All aspects of a rock's texture—grain size, grain shape, grain orientation—may change during metamorphism. The textural criterion used in naming metamorphic rocks is whether or not their minerals display **preferred orientation**. Unlike igneous and sedimentary rocks, grain *size* is not important in classification even though it may change drastically. It *is*, however, helpful in interpreting a rock's metamorphic history (see section 7.5).

Preferred Orientation You saw in Figure 7.3 that differential stress can flatten grains. It also causes original grains from the protolith and newly crystallized metamorphic minerals to show two kinds of **preferred orientation** that geologists refer to as **foliation** and **lineation**. A rock is said to display **foliation** if it contains a stack of parallel features, either alternating bands of different minerals [see the chapter opening photograph and the right side of Fig. 7.1b (i)] or alignment of platy minerals as shown in Figure 7.1c (i). A rock has a **mineral lineation** if rod-shaped minerals are aligned parallel to one another [as in Fig. 7.1c (ii)]. **FIGURE 7.6** shows the relationships among aligned grains in rocks displaying foliation and lineation.

FIGURE 7.6 Mineral alignment: foliation and lineation.

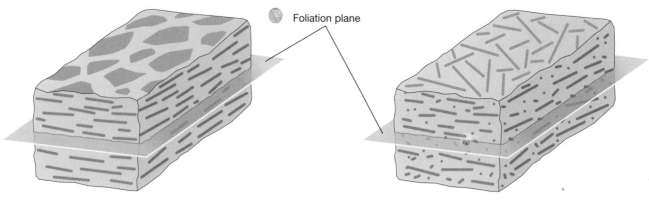

Foliation plane

(a) Mineral foliation.
Stack of platy minerals (e.g., mica) aligned parallel to one another.

Rod-shaped minerals are not parallel to one another, but their long dimensions all lie in parallel planes (foliation planes).

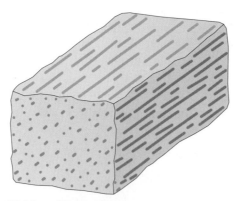

(b) Mineral lineation.
The long dimensions of rod-shaped minerals (e.g., amphibole) are parallel to one another, producing a streaky appearance on some surfaces and a dotted pattern on others.

Exercise 7.3 shows how differential stress produces preferred mineral orientation in metamorphic rocks. *Note:* Foliations are given different names depending on the intensity of metamorphism (see section 7.4.3), but all that is needed to name a rock is the presence or absence of any kind of foliation or lineation.

EXERCISE 7.3	Visualizing Preferred Orientation Due to Reorientation

Name: _____   Section: _____

Course: _____   Date: _____

We can gain insight into how preferred orientation develops with a simple laboratory experiment.

(a) Randomly insert pennies (flat grains) and small sticks (elongate grains) into a ball of Play-Doh. Flatten the dough into a pancake by pressing on it with a book (i.e., apply differential stress). What happens to the alignment of the objects? Are the sticks parallel to each other in the plane of the pancake?

_____

_____

(b) Clean off the pennies and sticks, and reinsert them into another ball of Play-Doh. This time, roll the dough into a cigar shape. What happens to the alignment of the objects now? Are the sticks parallel to each other? How are they oriented relative to the length of the cigar?

_____

_____

Absence of Preferred Orientation Minerals in some metamorphic rocks are oriented randomly and therefore have neither foliation nor lineation. Such rocks are said to have a **granoblastic** texture (**FIG. 7.7**). Preferred orientation may be absent because (a) there was no differential stress during metamorphism, or (b) there were no platy or rod-shaped minerals to align, or (c) both (a) and (b).

7.4.3 Identifying Metamorphic Rocks

TABLE 7.2 is a classification scheme for common metamorphic rocks based on observable mineralogy and texture. Each column corresponds to one of the five major compositional classes and each row to textural variations. Different names are given to some textural varieties depending on whether they result from low-, moderate-, or high-grade metamorphism. For example, slate, phyllite, and schist are all foliated aluminous rocks, but they represent different metamorphic grades.

To name a metamorphic rock, begin by identifying its minerals. This will place the rock in the correct compositional type and column in Table 7.2. Move downward through the table until you come to the row with the appropriate texture. In some cases there is only one choice; in others there may be a few possibilities. The following descriptions will help you choose.

Slate: A low-grade metamorphosed shale or mudstone composed of strongly aligned flakes of clay. The foliation is revealed by slate's tendency to split into thin plates. Typically, slate is gray, black, red, or green.

Phyllite: As grade increases, clay is replaced by tiny flakes of muscovite and/or chlorite. Grains are larger than those in slate but may still be too small to be seen without magnification. Phyllite has a silky sheen caused by reflection of light from the aligned platy minerals.

FIGURE 7.7 Granoblastic texture (random mineral alignment).

(a) Interlocking calcite grains in coarse-grained marble.

(b) Photomicrograph of interlocking calcite grains in a marble.

(c) Interlocking quartz grains in quartzite.

(d) Photomicrograph of quartz grains fused together in quartzite.

Schist: This is a medium- to high-grade rock with a strong foliation defined by large mica flakes that can be identified with the naked eye. Porphyroblasts (large crystals) of other metamorphic minerals (e.g., garnet) may occur in schist. We distinguish among different types of schist based on composition. Aluminous schist contains mica and chlorite; mafic schist contains chlorite, talc, and hornblende; and calcareous schist contains Ca-mica and Ca-amphiboles.

Gneiss: A high-grade metamorphic rock of any composition in which minerals have segregated into light and dark layers. The foliation is defined by compositional banding.

Mylonite: Dynamically metamorphosed rock in which very tiny grains formed. The foliation is defined by flattened and stretched grains. (A *protomylonite* is only slightly sheared, so less than 50% of the rock is very fine-grained. Some grains from the protolith may still be present. An *ultramylonite* is intensely sheared, so more than 90% is extremely fine grained.)

Nonfoliated aluminous rocks: These rocks are typically dense, hard, very fine-grained, dark gray or brown. Some contain porphyroblasts of metamorphic index minerals (like the andalusite in Fig. 7.2b). Many rocks that fall into this category are found next to igneous plutons and are clearly the result of contact metamorphism. These are properly called **hornfels** (plural, **hornfelses**). Others, however, could form during burial metamorphism. Without the field data, you can't tell. For an introductory course in geology where we want to keep names to a minimum, there are essentially two choices: hornfels or "something similar to hornfels."

Name: _____   **Section:** _____

Course: _____   **Date:** _____

Using Table 7.2 and what you read in this section, determine the mineralogy and texture of the metamorphic rocks in your study set. Record the information in the rock study sheets at the end of the chapter, and name each specimen. Don't worry about the columns for type of metamorphism and metamorphic grade. We will look next at how to interpret the metamorphic history of these rocks.

Greenschist: Low-grade mafic metamorphic rock composed of chlorite and plagioclase feldspar with epidote and/or actinolite. Foliation is defined by alignment of chlorite and actinolite.

Greenstone: A nonfoliated (granoblastic) equivalent of greenschist.

Serpentinite: Typically fine-grained; various shades of light to medium green. May contain fibrous minerals ("asbestos").

Soapstone: Very soft (can be scratched with a fingernail) and greasy; composed almost entirely of talc. Color ranges widely: gray, white, green.

Hornblende schist: A higher-grade version of greenschist in which foliated and often aligned hornblende crystals and plagioclase replace the lower-grade chlorite-epidote-actinolite assemblage. This rock is more commonly called *amphibolite*. Of note, some amphibolites have no lineation.

Quartzite: A monomineralic, granoblastic, metamorphic rock composed of quartz.

Marble: A calcareous metamorphic rock composed mostly of calcite or dolomite. Marble may also contain some of the calc-silicate index minerals. Marble is commonly granoblastic.

Anthracite coal: A black, shiny, hard rock made of carbon that has no visible mineral grains and has relatively low specific gravity. It may exhibit conchoidal fracture.

TABLE 7.2   Simplified classification scheme for metamorphic rocks.

Texture		Composition				
		Aluminous muscovite, chlorite, biotite, garnet, staurolite, kyanite, sillimanite	**Calcareous** calcite, dolomite, amphibole (actinolite), garnet, pyroxene	**Mafic** talc, serpentine, chlorite, biotite, amphibole, pyroxene, garnet + plagioclase	**Quartzo-feldspathic** dominantly quartz, potassic feldspar, plagioclase feldspar	**Organic**
Foliated	Fine grained	Slate* Phyllite*	Rarely observed	Greenschist Hornblende Schist Amphibolite	Rarely observed	
	Medium and coarse grained	Schist*				
	Foliated and layered	Gneiss*	Calc-silicate gneiss	Mafic gneiss	Gneiss*	
	Foliated, layered, smeared	Mylonite	Mylonite	Mylonite	Mylonite	
	Nonfoliated (granoblastic)	Hornfels or other nonfoliated rock (see text)	Marble	Serpentinite Soapstone (talc) Greenstone	Quartzite Metasandstone Metaconglomerate	Anthracite coal

Increasing metamorphic grade (arrow pointing down in Aluminous column)

* Add more detail with dominant minerals (e.g., muscovite-biotite phyllite; garnet-staurolite schist).

7.5 What Can We Learn from a Metamorphic Rock?

Metamorphic rocks may preserve evidence of both their metamorphic history and that of their protoliths. There's a lot to learn when we study a metamorphic rock. First, we would like to know the *nature of the protolith*. For example, did the metamorphic rock we see today originate as a granite, sandstone, basalt, or shale? Second, we would like to know the *agents and type of metamorphism* responsible for causing the change. Did metamorphism result from a change in temperature, pressure, or both? Was the rock subjected to differential stress during deformation or not? Third, we would like to know the *"intensity" of metamorphic conditions*. In this context, the informal term *intensity* refers to the highest temperature and pressure at which the metamorphism took place and/or the overall amount of shearing and squeezing during deformation. Finally, we hope to define the *tectonic setting* in which metamorphism occurred. For example, did metamorphism occur in a collisional mountain belt, along a fault zone, or next to a hot pluton? This, however, cannot be done without information about field relationships and will not be discussed further.

7.5.1 Identifying the Protolith

During metamorphism, ions in the protolith minerals commonly recombine with one another to form new minerals that are more stable under the new temperature and pressure conditions. The original protolith minerals may disappear completely, but because the new mineral assemblage recycles the available protolith ions, it can tell us a lot about the protolith. But chemistry alone cannot tell exactly what kind of rock it was. For example, a quartzo-feldspathic rock could have been igneous or sedimentary. If igneous, it might have been either intrusive (granite) or extrusive (rhyolite). A calcareous rock can be identified as a limestone or dolostone, but not as a specific kind (fossiliferous, micrite, oolitic, coquina, etc.). Other factors not available to you in the classroom can help in these cases, as shown in Exercise 7.5. It will show how the minerals in a metamorphic rock can be used to determine the composition of that rock's protolith.

7.5.2 Identifying the Type of Metamorphism

It is not always possible to determine the type of metamorphism that created a laboratory specimen because field evidence is often critical to that decision, and that information is lost if you didn't collect the specimens yourself. For example, a metamorphic rock with random grain orientation adjacent to an igneous intrusion almost certainly resulted from contact metamorphism. But if you look at a piece of that rock in the laboratory, you have no idea whether it was the result of contact or burial metamorphism. Or, if it is a calcareous or quartzo-feldspathic rock, you don't know whether it was subjected to regional metamorphism.

Some conclusions, however, can be made based on what you have learned about the agents of metamorphism and metamorphic textures. *The key is the presence or absence of preferred mineral orientation*—foliation or lineation—which is produced by differential stress. Differential stress is an important agent of regional and dynamic metamorphism (see Table 7.1). If it is present in a metamorphic rock, one of these types of metamorphism was responsible.

The absence of foliation or lineation (i.e., a granoblastic texture) is trickier to interpret and depends on whether or not platy (e.g., mica) or rod-shaped (e.g., amphibole) grains that could have been foliated or lineated are present. If these minerals are present and the rock is granoblastic, then differential stress was not involved, and contact

Name: _____   Section: _____

Course: _____   Date: _____

(a) In Exercise 7.2, your instructor provided you with four rocks and you identified their minerals and chemical composition. Compare the mineral content and most abundant ions from the four rocks you were given with the four main compositional classes of metamorphic rocks. What protoliths produced the four specimens you studied?

	Specimen 1	Specimen 2	Specimen 3	Specimen 4
Protolith				

(b) How would your interpretation of protoliths be affected if you learned that the area had experienced significant metasomatism? Explain.

_____

_____

_____

_____

or burial metamorphism is the likely origin. If they are not present, but rather equant grains that cannot be aligned, then any type of metamorphism is possible.

Exercise 7.6 will help clarify which types of metamorphism produce preferred mineral orientation and which do not.

7.5.3 Determining the Intensity of Metamorphism

Geologists can determine the temperatures and pressures at which metamorphism occurred but only with the use of sophisticated (and very expensive) instruments. In an introductory geology class without those instruments, you can still describe the **metamorphic grade**—a broad, informal approximation of how much a rock has changed. This generally relates to how much heat and pressure the rock experienced during metamorphism. We estimate that metamorphic *low-grade* rock has changed little, still preserving much of the character of its protolith; a *moderate-grade* rock retains fewer aspects of the protolith; and *high-grade* rock has changed so much that no traces of its protolith's minerals or texture are preserved. As an example, consider a rock that starts out as a sequence of alternating layers of sandstone and shale. At low grade, the sandstone may have recrystallized and the shale may have developed a subtle foliation, but the bedding is still visible and the original minerals are still present. At high grade, the rock may contain a totally different assemblage of minerals with a very strong foliation and no hint at all of the original bedding.

Textural Evidence for Metamorphic Grade Three textural features help estimate metamorphic grade: grain size, relationships between grains, and degree of preferred orientation. In general, the longer a rock remains at its peak metamorphic

Name: _____   Section: _____

Course: _____   Date: _____

(a) Summarize the relationship between mineral alignment and type of metamorphism by checking the appropriate box(es) in each row in the following table.

Texture	Type of metamorphism				
	Contact	Regional	Dynamic	Burial	Can't tell from texture alone
Foliation					
Lineation					
Granoblastic with micas					
Granoblastic without micas					
Compositionally layered but not foliated					
Compositionally layered and foliated					

(b) Examine the metamorphic rocks in your study set. Which are foliated? Lineated? Granoblastic? Write your answers in the appropriate column of the study sheets at the end of the chapter. Suggest, wherever possible, the type of metamorphism that produced each sample.

temperature, the greater the amount of solid-state diffusion that can occur and the larger its grains can grow. Thus, in most cases, **grain size** increases with increasing metamorphic grade. This is illustrated in **FIGURE 7.8**: compare the original fine-grained micrite protolith (Fig. 7.8a, b) with its metamorphosed equivalent (Fig. 7.8c, d). Details of fine-grained rocks are best examined with a microscope, as seen in photomicrographs in Figure 7.8b and d.

Some metamorphic minerals called **porphyroblasts** grow much larger than others in the rock. Garnet porphyroblasts in **FIGURE 7.9** resemble phenocrysts in an igneous rock, but they have nothing to do with the rate of cooling. Instead, they result from solid-state diffusion and concentration of ions. In general, the larger the porphyroblast, the higher the metamorphic grade and the longer the rock stayed at its peak metamorphic temperature.

Relationships between grains also change, as shown in photomicrographs of a quartzose sandstone protolith and its metamorphic product, quartzite. During metamorphism, the rounded quartz clasts and fine-grained cement that hold them together (**FIG. 7.10a**) recrystallize, and the cement grains grow larger. The final result is an angular, interlocking granoblastic texture that has obliterated the original clast/cement relationships (**FIG. 7.10b**). At low metamorphic grade, some of the clast outlines are preserved, but with progressively higher grades they eventually disappear.

Foliation may develop during the earliest stages of metamorphism in which differential stress plays a role, particularly in aluminous rocks (**FIG. 7.11**). Grain size increases and foliation becomes more permanent with the increasing of the metamorphic grade. Initially fine-grained aligned micas in slate (Fig. 7.11a) grow with increasing metamorphic grade, becoming large enough to produce a

FIGURE 7.8 Fine-grained parent rock of a common metamorphic rock.

(a) Hand specimen of micritic limestone (metamorphic equivalent shown in Fig. 7.8c).

(b) Photomicrograph of micritic limestone (metamorphic equivalent shown in Fig 7.8d).

(c) Hand specimen of coarse-grained marble.

(d) Photomicrograph of marble.

visible sheen in phyllite (Fig. 7.11b) and finally a coarse-grained shiny surface in schist (Fig. 7.11c).

At high metamorphic grades, **gneissic banding** may develop— a compositional banding defined by alternating light- and dark-colored layers. If mica is present, it is usually foliated parallel to the banding. But at very high grades, the micas may be consumed in making new minerals, and the gneiss will have a granoblastic texture.

Dynamic metamorphism is an exception to the principle that grain size increases with increasing metamorphic grade (see Fig. 7.8), but the progressively more intense foliation is the same. At temperatures high enough that rocks do not fracture and break up when sheared, dynamic metamorphism can transform a coarse-grained rock into a very fine-grained rock with strong foliation and lineation. The product is called **mylonite.** Such transformation occurs, simplistically, because the stress causes large crystals of quartz to subdivide into very

FIGURE 7.9 Porphyroblastic texture (red garnet porphyroblasts in a metamorphosed mafic igneous rock).

FIGURE 7.10 Photomicrographs showing recrystallization of a clastic sedimentary texture.

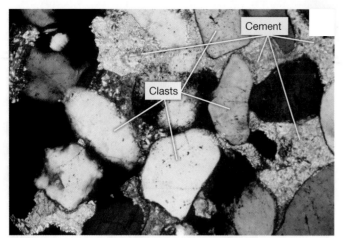

Cement

Clasts

(a) Sandstone protolith: rounded clasts distinguished clearly from interstitial cement that holds them together.

(b) Metamorphosed sandstone (quartzite): cement and clasts recrystallized, forming network of interlocking quartz grains. Original clasts and cement are indistinguishable.

FIGURE 7.11 Change in foliation in aluminous rocks with increasing metamorphic grade.

(a) Low grade (slaty texture).

(b) Low-moderate grade (phyllitic texture).

(c) High grade (schistose texture).

(d) Very high grade (gneissic texture).

small grains that fit together like a mass of soap bubbles. **FIGURE 7.12** shows how progressively intense dynamic metamorphism affected a coarse-grained granite from the Norumbega fault system in east-central Maine (Fig. 7.12a).

The process of transforming a rock into mylonite takes place gradually, so we may see intermediate stages preserved in outcrops. [In fact, geologists sometimes distinguish between *protomylonite*, rock that has just started transformation (Fig. 7.12b); *mylonite*, in which the transformation has affected most of the rock (Fig. 7.12c); and *ultramylonite*, in which the transformation is complete (Fig. 7.12d).] Blocky feldspar grains tend to be aligned during the process and grow tails of fine-grained mica, whereas some quartz grains smear into very thin ribbons (Fig. 7.12b). In some cases, relatively large feldspar grains remain after the surrounding rock has changed into a mass of very fine mica and quartz; these leftover grains of feldspar are called **porphyroclasts** (Fig. 7.12c).

Mineralogic Evidence for Metamorphic Grade We have seen that metamorphic minerals form from protolith minerals as temperature and pressure increase.

FIGURE 7.12 Effects of dynamic metamorphism on coarse-grained granite.

(a) Undeformed granite showing original coarse-grained igneous texture.

(b) Protomylonite: potassic feldspars (pink grains) are stretched and aligned parallel to the white line; quartz grains (gray) have coalesced to form continuous ribbons.

(c) Mylonite: grain size of the original granite has been reduced to form a dark, very fine-grained matrix with remnant feldspars smeared into ovoid grains.

(d) Ultramylonite : the intensely metamorphosed granite now consists entirely of fine grains and has strong foliation and lineation.

Some minerals (e.g., chlorite) are stable only at relatively low grade (low temperature), others (e.g., biotite) only at moderate grade, and still others (e.g., sillimanite) only at high grade. Minerals that therefore indicate the conditions of metamorphism are called **metamorphic index minerals** (**TABLE 7.3**). Note that some minerals like quartz and feldspar, which are stable over a wide range of temperature and pressure conditions, are *not* helpful at all in determining metamorphic grade.

Table 7.3 also shows that each compositional class of metamorphic rock has its own set of index minerals due to its unique chemical composition. Some minerals, like chlorite, biotite, and garnet, can occur in more than one compositional class because the appropriate ions can be present in more than one kind of protolith. The complete *assemblage* of minerals in a rock, however, makes it clear what the parent rock was. For example, biotite can be found in low- to moderate-grade aluminous and mafic metamorphic rocks. If biotite occurs in an assemblage with muscovite and quartz, the protolith must have been aluminous; but if it is found with plagioclase feldspar and actinolite, the protolith must have been mafic.

Variations in Metamorphic Grade The pattern of variation in metamorphic grade depends on the type of metamorphism (**FIG. 7.13**) and can be predicted by using a little common sense. For example, metamorphic grade in a wall rock due to thermal metamorphism is greatest at the contact between a pluton and its wall rock and decreases from the pluton (Fig. 7.13a).

Regional metamorphism occurs in a collisional mountain belt. In such a setting, the crust squeezes together horizontally and thickens vertically. Rocks that were originally near the surface of one continental margin may end up at great depth when thrust beneath the edge of the other continent. The rocks that are carried to

TABLE 7.3   Mineral assemblages as indicators of metamorphic grade.

Parent rock type	Increasing metamorphic grade ⟶				
	Low grade		Medium grade		High grade
Quartzo-feldspathic (quartz, potassic feldspar, plagioclase, ± micas)	Impossible to determine based on mineralogy alone				
Aluminous (muscovite, quartz, chlorite, garnet, staurolite, etc., ± quartz, feldspars)	Muscovite Chlorite	Muscovite Biotite	Muscovite Biotite Fe-Mg garnet Staurolite	Muscovite Biotite Fe-Mg garnet Kyanite or sillimanite	Sillimanite K-feldspar Fe-Mg garnet
Calcareous (calcite, dolomite, ± calc-silicate minerals such as those listed to the right)	Calcite Talc Olivine	Calcite Tremolite	Calcite Diopside Ca-Al garnet		Diopside Wollastonite Ca-Al garnet
Mafic (plagioclase + ferromagnesian minerals such as those listed to the right)	Chlorite Na-plagioclase Epidote	Actinolite Na-Ca plagioclase Biotite	Hornblende Ca-Na plagioclase Fe-Mg garnet		Pyroxene Ca-plagioclase Fe-Mg garnet

FIGURE 7.13 Geographic distribution of metamorphic intensity.

(a) Contact metamorphism: intensity decreases outward from the intrusive pluton or lava flow.

(b) Regional metamorphism: intensity increases with depth and closeness to suture between colliding plates.

(c) Burial metamorphism: intensity increases with depth

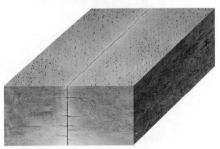

(d) Dynamic metamorphism: intensity decreases outward from the fault and increases with depth.

the greatest depth not only reach the highest temperature and the highest grade, but they also tend to be squeezed and sheared the most (Fig. 7.13b). Because burial metamorphism is due principally to lithostatic pressure associated with the thickness of overlying rocks, and to temperature due to the geothermal gradient, the burial metamorphic grade increases with depth (Fig. 7.13c). Dynamic metamorphic intensity is greatest where fault blocks are in contact, and it decreases outward in the two fault blocks (Fig. 7.13d).

You will put all of this information together in Exercise 7.7 to interpret the histories of the metamorphic rocks in your study set.

EXERCISE 7.7   **Interpreting Metamorphic Rock History**

Name: _____   Section: _____

Course: _____   Date: _____

(a) You have already identified the metamorphic rocks in your study set. Now, based on the information presented earlier, fill in the rest of the rock study sheet at the end of the chapter, saying as much as you can for each sample about possible types of metamorphism and approximate metamorphic grade.

(b) The map on the next page shows the distribution of rocks of different metamorphic grades. The local geologist has mapped a concentric pattern in which the highest metamorphic grade is in the center of the area and surrounded by rocks of progressively lower grade. All metamorphic rocks have a granoblastic texture; the coarsest grains are found in the central region. Although the granoblastic texture suggests contact metamorphism, the geologist has found no igneous rock anywhere in the region. Suggest an explanation for this metamorphic pattern.

(continued)

Mapped distribution of rocks with varying metamorphic grades.

0   5   10   15 km

| Unmetamorphosed | Low grade | Medium grade | High grade |

7.6 Applying Your Knowledge of Metamorphic Rocks to Geologic Problems

With your new skills, you can interpret the history of metamorphic rocks from textural and mineralogic data and information from the field. Now you're ready for a little detective work and geologic reasoning in Exercise 7.8.

Name: _____   Section: _____

Course: _____   Date: _____

? **What Do You Think**   Geologists in Africa have found economically valuable concentrations of chromite (the principal ore of chromium) that formed by gravity settling in thick mafic sills. In the same area of Africa, there are also mafic lava flows that look very much like the sills but which contain none of the valuable minerals. You have been hired as a geologic consultant by an investment firm that is considering purchasing the mineral rights for the mountain that includes the sills. The firm specifically needs to know whether the mafic igneous rock (in the photograph below) is a sill (i.e., intrusive) and therefore worth the cost of exploration or a lava-flow (i.e., extrusive) and therefore not worth a penny. What *metamorphic* evidence would you look for that would distinguish the two possibilities? On a separate sheet of paper, write a brief report to the firm explaining your reasoning and what you would expect to find in either case.

METAMORPHIC ROCKS STUDY SHEET

Name _____

Sample #	Minerals present (compositional group)	Texture (grain orientation, size, shape)	Type of metamorphism	Metamorphic grade (low, medium, or high grade)	Rock name

METAMORPHIC ROCKS STUDY SHEET

Name _____

Sample #	Minerals present (compositional group)	Texture (grain orientation, size, shape)	Type of metamorphism	Metamorphic grade (low, medium, or high grade)	Rock name

METAMORPHIC ROCKS STUDY SHEET

Name _____

Sample #	Minerals present (compositional group)	Texture (grain orientation, size, shape)	Type of metamorphism	Metamorphic grade (low, medium, or high grade)	Rock name

8

Studying Earth's Landforms: Maps and Other Tools

Horseshoe Bend in Arizona, showing the effectiveness of stream erosion.

- Become familiar with different ways to portray landforms and landscapes
- Learn the strengths and weaknesses of these representations of the Earth's surface
- Understand the essential elements of an accurate map (location, distance, direction, elevation), how they are measured, and how they are portrayed on maps

MATERIALS
NEEDED

- Clear plastic ruler with divisions in tenths of an inch and millimeters (in your toolkit at the back of this book)
- Circular protractor (in your toolkit at the back of this book)
- A globe and maps provided by your instructor that show major cities

8.1 Introduction

It is easier to study Earth's surface today than at any time in history. To examine surface features, we can now make detailed surface models from satellite elevation surveys and download images of any point on the planet with the click of a mouse. Geologists were quick to understand the scientific value of satellite imaging technology and adopted new methods as quickly as they were developed. Some of the images in this manual were not even available to researchers a decade ago. The study of the Earth's surface is almost as dynamic as the surface itself!

This chapter is an introduction to traditional maps and aerial photographs and some of the new methods used by geologists to view Earth's surface and understand how its landscapes form. Much of the new technology is available to you *free* for use on your computer. *Google Earth*™ and *NASA World Wind* provide free satellite images of the entire globe and can generate three-dimensional views of landforms. Archived topographic maps of most states are available online from the U.S. Geological Survey at http://ngmdb.usgs.gov/maps/TopoView/. And *Google Maps* is another exciting new tool for geologists; you can use it to zoom anywhere on Earth.

8.2 Ways to Portray Earth's Surface

The best way to study landforms is to fly over them for a bird's-eye view and then walk or drive over them to see them from a human perspective. That isn't practical for a college course, so we have to bring the landforms to you instead. To portray landforms, we will use a combination of traditional topographic maps, aerial photographs, satellite images, and digital elevation models (**FIG. 8.1**).

Figure 8.1 portrays an area of eastern Maine by four different methods. A **topographic map** (Fig. 8.1a) uses contour lines to show landforms (see Chapter 9). Topographic maps used to be drawn by surveyors who measured distances, directions, and elevations in the field. They are now made by computers from aerial photographs and radar data. **Aerial photographs** (Fig. 8.1b), including U.S. Geological Survey (USGS) Orthophotoquads, are photographs taken from a plane and pieced together to form a mosaic of an area. **Landsat images** (Fig. 8.1c) are made by a satellite that takes digital images of Earth's surface using visible light and other wavelengths of the electromagnetic spectrum. Scientists adjust the wavelengths to color the image artificially and emphasize specific features. For example, some infrared wavelengths help reveal the amount and type of vegetation. **Digital elevation models** (DEMs; Fig. 8.1d) are computer-generated, three-dimensional views of landforms made from radar satellite elevation data spaced at 10- or 30-meter (m) intervals on the Earth's surface. A new generation based on 1-m data is now being released that provides a more accurate model of the surface than anything available to the public 5 years ago. Exercise 8.1 asks you to make recommendations based on the images in Figure 8.1.

8.2.1 Map Projections

The portrayals of the area in Figure 8.1 are flat, two-dimensional pictures, but Earth is a nearly spherical three-dimensional body. Only a three-dimensional representation—a globe—can accurately show the *areas* and *shapes* of figures and the *directions* and *distances* between points. The process by which the three-dimensional Earth is converted to a two-dimensional map is called making a **projection**. There are many different projections, each of which distorts one or more

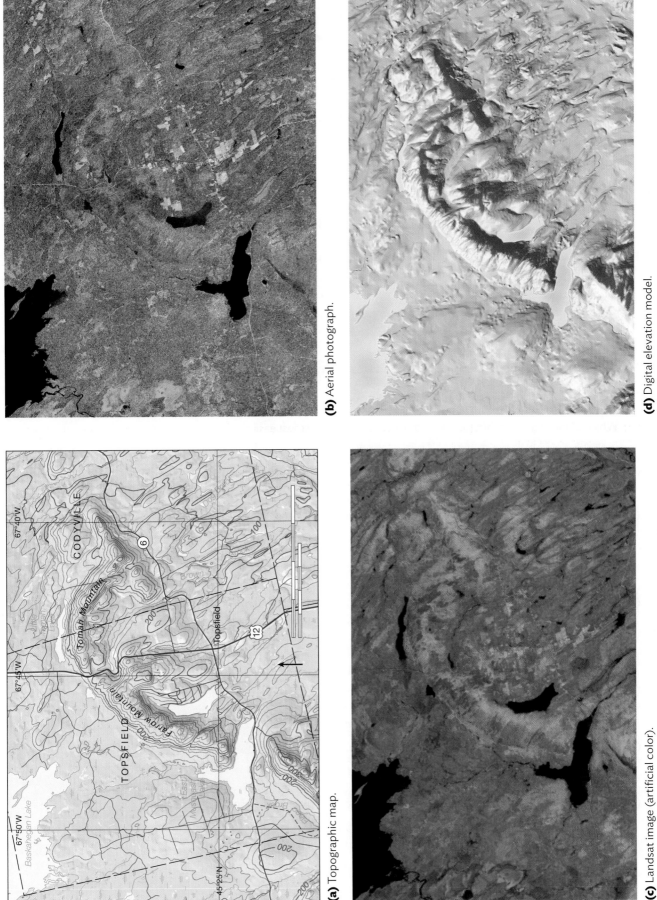

FIGURE 8.1 One area in eastern Maine represented by different imaging methods. [Scale bar and north arrow in (a) apply to all figures.]

(a) Topographic map.

(b) Aerial photograph.

(c) Landsat image (artificial color).

(d) Digital elevation model.

Name: _____   Section: _____
Course: _____   Date: _____

Each type of portrayal has strengths that make it useful for some purposes and weaknesses that prevent its use in others. This exercise examines the strengths and weaknesses of the four landscape representations in Figure 8.1.

(a) Examine the images in Figure 8.1 and rank them (on a scale of 1 to 4) in the table below by how well they show the map elements indicated (1 is most effective, 4 is least effective; ties are allowed).

	Topographic map	Aerial photograph	Landsat image	DEM
Location				
Direction				
Elevation				
Changes in slope				
Distance				
Names of features				

(b) Which of the images enables you to recognize the topography most easily? Why?

_____

_____

(c) Which is least helpful in trying to visualize the hills, valleys, and lakes? Why?

_____

_____

(d) Erosional agents often produce a "topographic grain," an alignment of elongate hills, ridges, and valleys. Which images show the topographic grain in this area most clearly? Once you've seen it on those images, can you recognize it on the others?

_____

(e) Which images show highways most clearly?

_____

(f) Which images show unpaved lumber roads most clearly?

_____

(g) Which image do you think is the oldest? The most recent? Explain your reasoning.

_____

_____

(continued)

Name: _____   Section: _____
Course: _____   Date: _____

? **What Do You Think** You work at a state university forestry department and are also a scout leader for young teenagers. As part of your "day job," you have been asked by the state forestry commissioner to discuss recent changes in forest cover with her staff. In addition, you were asked to organize an overnight wilderness hike by your scouting group. Which type of image would you want to use for each of the two activities? In the space below, indicate your choices and why you made them.

(a) Staff presentation: _____

_____

_____

(b) Wilderness hike: _____

_____

_____

of the four elements italicized above and each of which is useful for some purposes but unusable for others. Three common map projections are shown in **FIGURE 8.2**, and **TABLE 8.1** indicates what they distort and for what purposes they are best used.

FIGURE 8.2 **Three common map projections and their different views of the world.** (The equator is indicated in each projection by a red line.)

(a) Orthographic.

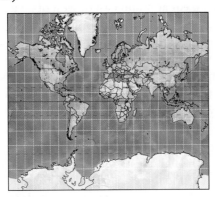

(b) Mercator.

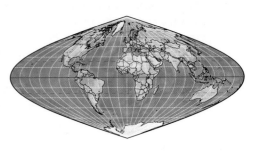

(c) Uninterrupted sinusoidal.

8.3 Map Elements

All accurate depictions of Earth's surface must contain certain basic elements: *location*, a way to show precisely where the area is; a way to measure the *distance* between features; and an accurate portrayal of *directions* between features. It is also important to know *elevations* of hilltops and other features and the *steepness of slopes*.

TABLE 8.1 The strengths and weaknesses of three common map projections.

	Orthographic	Mercator	Uninterrupted sinusoidal
Strengths	Directions between points are preserved. East–west distances are accurate.	Accurate near the equator. *Directions preserved in most areas.*	Areas of continents are represented accurately.
Weaknesses	Shapes and areas are distorted, especially near the edge of the projection. Distances other than in east–west direction are distorted.	Severe shape and area distortions away from the equator. Useless for north and south polar areas.	Scale is constant only along a central north–south line and the equator and changes elsewhere. Shapes are distorted for features distant from these reference lines.
Comments	Perspective view is similar to view of a globe from a great distance. Note accurate representation of size of Greenland compared with the Mercator projection.	Note vastly distorted polar areas of Greenland (shown much more accurately on the other projections) and Antarctica.	Note that the area of Greenland (much smaller than Africa and South America) is portrayed accurately compared with the Mercator projection.
Uses	Often used to provide context for images taken of Earth from space.	Nautical navigation charts because point-to-point directions are accurate.	Commonly used for features elongated north–south (e.g., maps of Africa and South America).

8.3.1 Map Element 1: Location

Road maps and atlases use a simple grid system to locate cities and towns (e.g., Chicago is in grid square A8). This is not very precise because many other places may be in the same square, but it is good enough for most drivers. More sophisticated grids are used to locate features on Earth precisely. Maps published by the USGS use three grid systems: latitude/longitude, the Universal Transverse Mercator (UTM) grid, and, for most states, the Public Land Survey System. The UTM grid is least familiar to Americans but is used extensively in the rest of the world.

Latitude and Longitude The latitude/longitude grid is based on location north or south of the **equator** and east or west of an arbitrarily chosen north–south line (**FIG. 8.3**). A *parallel of latitude* connects all points that are the same angular distance north or south of the equator. The maximum value for latitude is 90° N or 90° S (the North and South Poles, respectively). A *meridian of longitude* connects all points that are the same angular distance east or west of the **prime meridian**, a line that passes through the Royal Observatory in Greenwich, England. The maximum value for longitude is 180° E or 180° W, the international date line. *Remember: You must indicate whether a point is north or south of the equator **and** east or west of the prime meridian.*

Latitude and longitude readings are typically reported in **degrees** (°), **minutes** ('), and **seconds** ("), and there are 60' in a degree and 60" in a minute (e.g., 40°37'44" N, 73°45'09" W). *For reference, 1 degree of latitude is equivalent to approximately 69 miles (111 km), 1 minute of latitude is about 1.1 mile (1.85 km), and 1 second of latitude is about 100 feet (31 m).* The same kind of comparison can be made for longitude only *at the equator*, because the meridians converge at the poles and the distance between degrees of longitude decreases gradually toward the poles (Fig. 8.3b). Handheld global positioning system (GPS) receivers and those used in cars and planes can locate points to within a second. In Exercise 8.2 you will practice working with latitude and longitude.

FIGURE 8.3 The latitude/longitude grid.

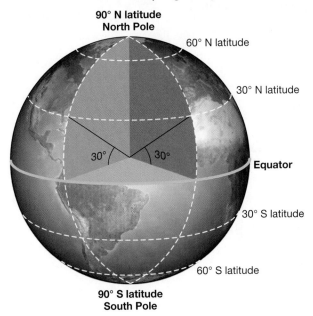

(a) **Latitude** is measured in degrees north or south of the equator.

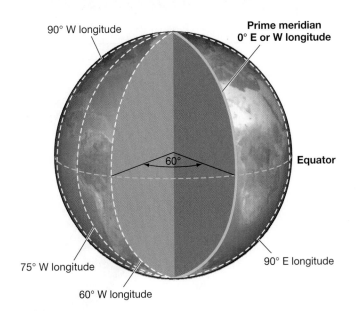

(b) **Longitude** is measured in degrees east or west of the prime meridian (Greenwich, England).

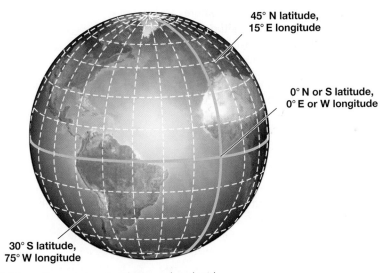

(c) Locating points using the completed grid.

EXERCISE 8.2   Locating Cities Using Latitude and Longitude

Name: _____   Section: _____
Course: _____   Date: _____

(a) For practice using the latitude/longitude system, describe the locations of the points (indicated by the stars) on the figure below. (*Remember* to indicate N or S for latitude, W or E for longitude.) *Note:* Latitude parallels and longitude meridians are spaced 15° apart.

i _____   ii _____   iii _____

(continued)

Name: _____   Section: _____
Course: _____   Date: _____

(b) With the aid of a globe or map, what geographic features are located at
  i. 45°00′00″ N latitude, 90°00′00″ W longitude? _____
  ii. 15°00′00″ N latitude, 30°00′00″ E longitude? _____
  iii. 30°00′00″ S latitude, 90°00′00″ W longitude? _____

(c) With the aid of a globe or map, determine the latitude and longitude of your geology laboratory as accurately as you can. How could you locate the laboratory more accurately?

_____

_____

(d) If you have access to a GPS receiver, locate the corners of your laboratory building. Draw a map below (or on a separate piece of paper) showing the location, orientation, and distances between the corners.

(e) Locate the latitude and longitude coordinates of the following U.S. and Canadian cities as accurately as possible.

Nome, Alaska _____   Seattle, Washington _____

Chicago, Illinois _____   Los Angeles, California _____

St. Louis, Missouri _____   Houston, Texas _____

New York, New York _____   Miami, Florida _____

St. Johns, Newfoundland _____   Ottawa, Ontario _____

Calgary, Alberta _____   Victoria, British Columbia _____

(continued)

Name: _____   Section: _____

Course: _____   Date: _____

(f) Which of the cities in (e) do you think is closest in latitude to each of the following world cities? Predict first, without looking at a map, globe, or *Google Earth*™, then check. Were you surprised by any?

City	Predicted best match	Latitude and longitude	Actual best match
Oslo, Norway			
Baghdad, Iraq			
London, England			
Paris, France			
Rome, Italy			
Beijing, China			
Tokyo, Japan			
Quito, Ecuador			
Cairo, Egypt			
Cape Town, South Africa			

Public Land Survey System   The Public Land Survey System was created in 1785 to provide accurate maps as America expanded from its thirteen original states. Much of the country is covered by this system, except for the original thirteen colonies, Kentucky, Maine, Tennessee, West Virginia, Alaska, Hawaii, Texas, and part of the southwestern states surveyed by Spanish colonists before the states joined the Union. Points can be located rapidly to within an eighth of a mile in this system (**FIG. 8.4**).

The grid is based on accurately surveyed north–south (**principal meridian**) and east–west (**base line**) lines for each survey region. Lines parallel to these at 6-mile intervals create grid squares 6 miles on a side, forming east-west rows called **townships** and north–south columns called **ranges** (Fig. 8.4). Townships are numbered north or south of the base line, and ranges are numbered east and west of the principal meridian. Each 6-mile square is divided into 36 **sections**, each 1 mile on a side, numbered as shown in Figure 8.4. Each section is divided into **quarter sections** ½ mile on a side, and each of these is further quartered, resulting in squares ¼ mile on a side. The location of the yellow star in the orange box in Figure 8.4 is described in the series of blow-ups:

- **T2S R3E** locates it somewhere within an area of 36 square miles (inside a 6 mi × 6 mi square).
- **Section 12, T2S R3E** locates it somewhere within an area of 1 square mile.
- **SW ¼ of Section 12, T2S R3E** locates it somewhere within an area of ¼ square mile.
- **SW ¼ of the SW ¼ of Section 12, T2S R3E** locates it within an area of ¹⁄₁₆ square mile.

In Exercise 8.3, you will practice using the Public Land Survey System.

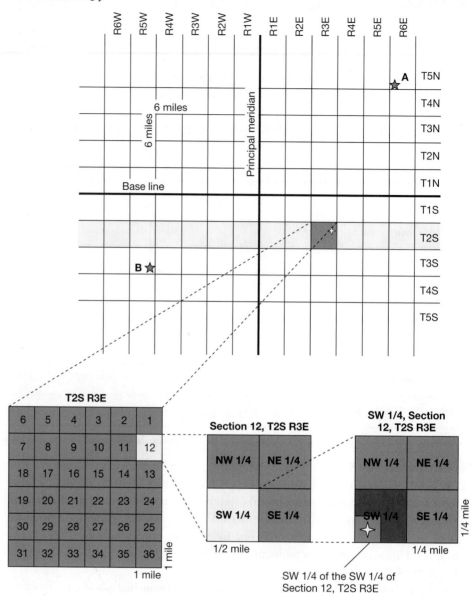

FIGURE 8.4 The Public Land Survey System grid. (Points A and B are for use with Exercise 8.3.)

Universal Transverse Mercator Grid The UTM grid divides the Earth into 1,200 segments, each containing 6° of longitude and 8° of latitude (**FIG. 8.5**). North–south segments are assigned letters (C through X); east–west segments are called **UTM zones** and are numbered 1 to 60 eastward from the international date line (180° W). Thus, UTM zone 1 extends from 180° to 174° W longitude, zone 2 from 174° to 168° W longitude, and so on. The forty-eight conterminous United States lie within UTM zones 10 through 19, roughly 125° to 67° W longitude. The UTM grid is based on a Mercator projection, in which the north and south polar regions are extremely distorted: Greenland and Antarctica appear much larger than their actual sizes. Because of this distortion, the grid does not extend beyond 80° N and 80° S latitudes.

To locate a point, begin with the grid box in which the feature is located. For example, the red box in Figure 8.5 is grid S22. UTM grid readings tell *in meters* how

Name: _____   Section: _____

Course: _____   Date: _____

(a) Determine the location of points A and B in Figure 8.4.

    A _____   B _____

(b) Locate the following points on Figure 8.4 by writing the number on the figure:
    1. NE ¼ of Section 36, T3N R4W
    2. SE ¼ of Section 18, T1N R1E
    3. NW ¼ of Section 3, T4S R5E

(c) Determine the location of points indicated by your instructor on topographic maps.

_____

_____

_____

far north of the equator (*northings*) and east of the central meridian for each zone (*eastings*) a point lies. The central meridian for each UTM zone is a line of longitude that runs through the center of the zone (**FIG. 8.6**) and is arbitrarily assigned an easting of 500,000 m so that no point has a negative easting. Points east of a central meridian thus have eastings greater than 500,000 m, and those west of the central meridian have eastings less than 500,000 m.

FIGURE 8.5 The UTM grid.

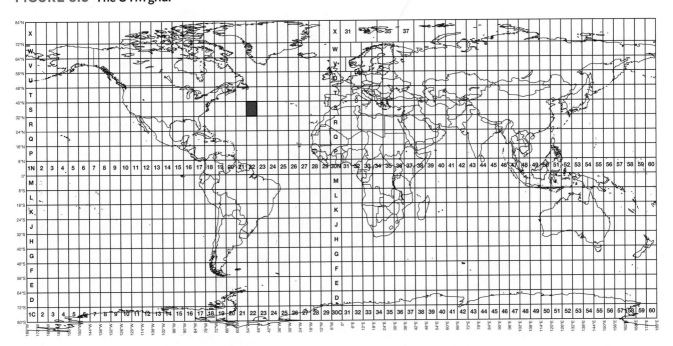

FIGURE 8.6 UTM zones for the forty-eight conterminous United States.

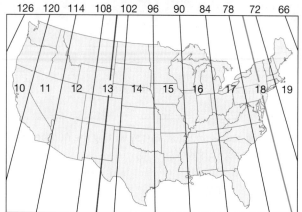

The red line is the central meridian (105° W) for UTM zone 13; the blue line is the central meridian for zone 18.

FIGURE 8.7 shows how the UTM grid appears on the most recent USGS topographic maps. Each grid square is exactly 1,000 m (1 km) on a side. Labels along the top and (as here) bottom of the map are **eastings (E)**—the distance in meters from the central meridian for the UTM zone (19 as shown in Fig. 8.6). Labels along the east and west sides of the map are **northings (N)**—the distance in meters north or south of the equator. These values are written out fully near the corner and elsewhere in

FIGURE 8.7 Southeast corner of the Greenfield quadrangle, Maine, showing UTM grid and marginal UTM grid values. (Map scale = 1:24,000. UTM tool is shown in position to locate the blue point.)

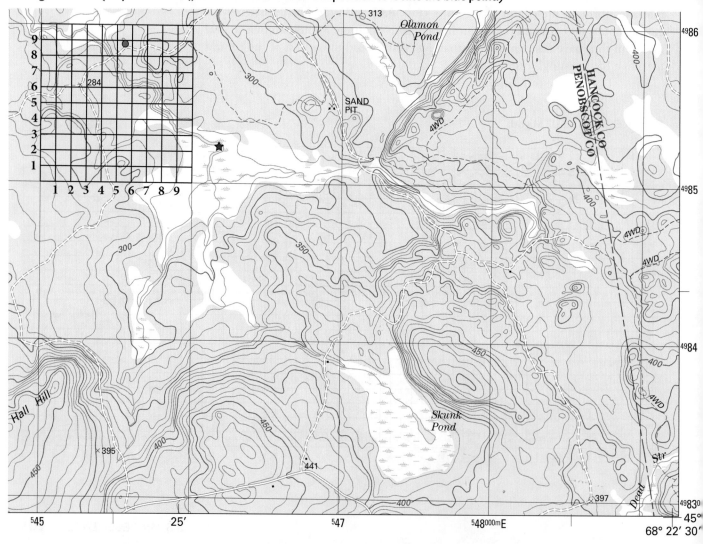

FIGURE 8.9 An area in eastern Maine shown at three common map scales. The red bar in each figure is 1 kilometer (km) long.

(a) Scale = 1:100,000.   **(b)** Scale = 1:62,500.

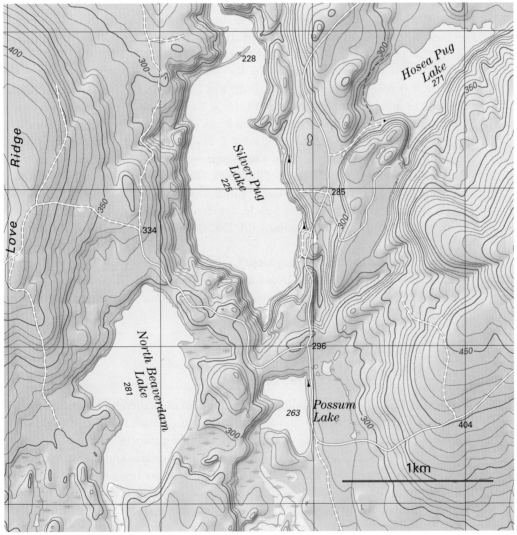

(c) Scale = 1:24,000.

scaled down so we can relate inches or centimeters on the map to real distances on the ground. **FIGURE 8.9** shows three maps of the same general area made at different scales. The more we scale down an area, the more detail we lose; the closer the map is to the real size, the more detail we can see.

The three map segments in Figure 8.9 show approximately the same area of land, but are different sizes because each is scaled down a different amount. Figure 8.9a has been scaled down more than four times as much as Figure 8.9c (1:100,000 versus 1:24,000) and therefore can cover the same area in much less space.

Different Ways to Describe Map Scale Map scale may be expressed verbally, proportionally, or graphically. A **verbal scale**, used on many road maps, uses words like "1 inch equals approximately 6.7 miles" to describe the scaling of map and real distances. A driver can estimate distances between cities, but not very accurately.

The most accurate way to describe scale is a **proportional scale**, one that tells exactly how much the ground has been scaled down to fit onto the sheet of paper. For example, a proportional scale of 1:100,000 (read "one to one hundred thousand") means that distances on the map are 1/100,000 of the distance on the ground. The proportions are the same for all units of measurement: 1 inch on such a map corresponds to 100,000 inches on the ground (1.58 miles) *and* 1 centimeter (cm) on the map corresponds to 100,000 cm on the ground (1 km). The larger the number in a proportional scale, the less space is needed to portray a given area on a map (compare Fig. 8.9a, b, and c). Note in Figure 8.9 that while it takes more space on a map to show an area at a scale of 1:24,000 than at 1:62,500 or 1:100,000, the 1:24,000 map shows much more detail.

The metric system is ideally suited for scales like 1:100,000,000 or 1:100,000 because it is based on multiples of 10. In the United States, we measure ground distance in miles but map distance in inches. Unfortunately, relationships among inches, feet, and miles are not as simple as in the metric system. There are 63,360 inches in a mile (12 in per foot $\times$ 5,280 ft per mile), so the proportional scale 1:63,360 means that 1 inch on a map represents exactly 1 mile on the ground. Old topographic maps use a scale of 1:62,500. For most purposes, we can interpret this scale to be approximately 1 inch = 1 mile, even though an inch on such a map would be about 70 feet short of a mile. Other common map scales are 1:24,000 (1 in. = 2,000 ft), 1:100,000 (see above), 1:250,000 (1 in. = 3.95 mi), and 1:1,000,000 (1 in. = 15.8 mi).

Map scale can also be shown **graphically**, using a **bar scale** (**FIG. 8.10**) to express the same relation as the verbal scale. Depending on how carefully you measure, a bar scale can be more accurate than a verbal scale, but not as accurate as a proportional scale.

8.4 Vertical Exaggeration: A Matter of Perspective

DEMs show the land surface in three dimensions and must therefore use an appropriate *vertical* scale to indicate how much taller one feature is than another. It would seem logical to use the same scale for vertical and horizontal distances, but we don't usually do so because mountains wouldn't look much like mountains and hills would barely be visible. Landforms are typically much wider than they are high, standing only a few hundred or thousand feet above or below their surroundings. At a scale of 1:62,500, 1 inch represents about a mile. If we used the same scale to make a three-dimensional model, a hilltop 400 feet above its surroundings would be less than one-tenth of an inch high. A mountain rising a mile above its base would be only 1 inch high.

We therefore exaggerate the vertical scale compared to the horizontal to show features from a human perspective. For a three-dimensional model of a 1:62,500

FIGURE 8.10 Scale bars used with three common proportional scales.

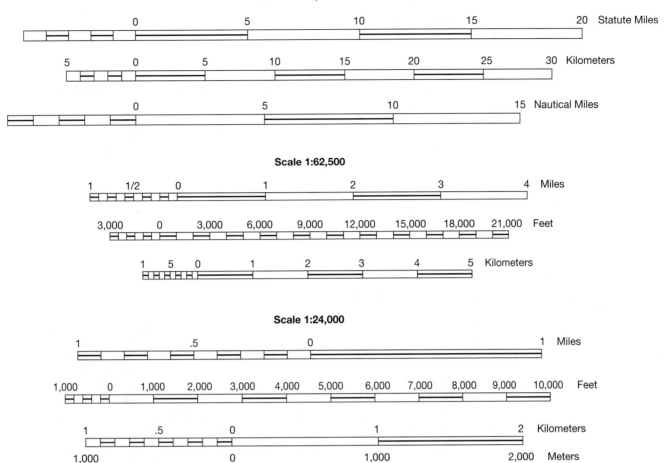

map, a vertical scale of 1:10,000 would exaggerate apparent elevations by a little more than six times (62,500/10,000 = 6.25). A mountain rising 1 mile above its surroundings would stand 6.25 inches high in the model; a 400-foot hill would be about half an inch tall, which is more realistic than the 0.1 inch if the 1:62,500 horizontal scale had been used vertically.

The degree to which the vertical scale has been exaggerated is, logically enough, called the **vertical exaggeration**. **FIGURE 8.11** shows the effects of vertical exaggeration on a DEM. With no vertical exaggeration, the prominent hill in the center of Figure 8.11a is barely noticeable. One of the authors of this manual has climbed that hill several times and guarantees that climbing it is far more difficult than Figure 8.11a would suggest. In contrast, Figure 8.11d exaggerates too much; the hill did not seem that steep, even with a pack loaded with rocks.

Is there such a thing as too much vertical exaggeration? The basic rule of thumb is not to make a mountain out of a molehill. Vertical exaggerations of two to five times generally preserve the basic proportions of landforms while presenting features clearly. We return to the concept of vertical exaggeration when we discuss drawing topographic profiles from topographic maps in Chapter 9.

FIGURE 8.11 DEMs of part of the area in Figure 8.1 showing the effects of vertical exaggeration (VE).

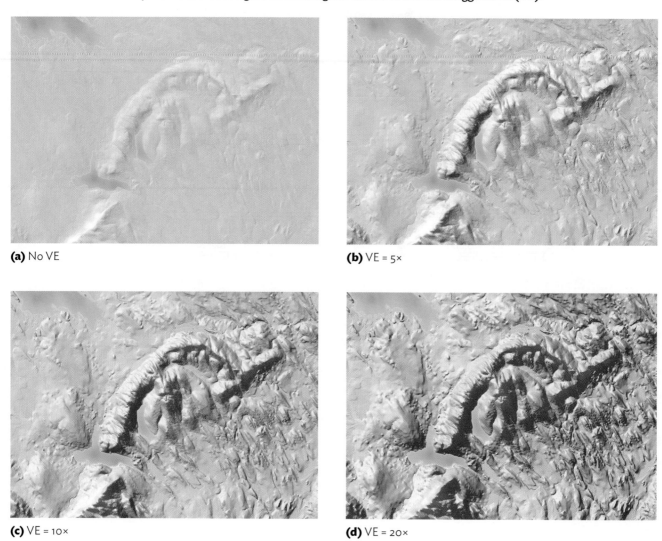

(a) No VE

(b) VE = 5×

(c) VE = 10×

(d) VE = 20×

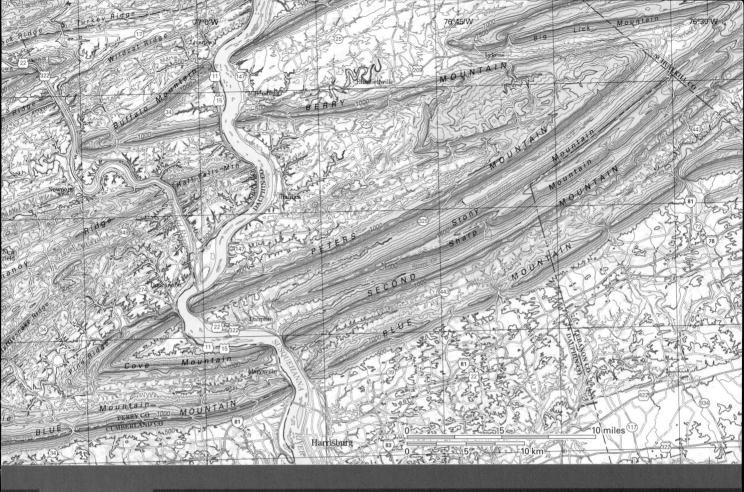

9

Working with Topographic Maps

A topographic map of the Appalachian Valley and Ridge province in Pennsylvania

LEARNING OBJECTIVES

- Understand how contour lines on topographic maps represent Earth's surface

- Practice constructing topographic profiles to illustrate and interpret landforms

- Use topographic maps to recognize potential natural disasters and solve economic problems

MATERIALS NEEDED

- Pencil, ruler, protractor, and tracing paper

- Topographic maps provided by your instructor

9.1 Introduction

In today's modern world, we have become very familiar with tools that help picture the world around us. With the click of a mouse, *Google Earth*™ and *NASA World Wind* provide satellite images of any point on the planet; digital elevation models (DEMs) are available to help to visualize topography; GPS satellites circling the Earth can help locate exactly where you are standing; and sophisticated Geographic Information System (GIS) software can locate points, give elevations, measure lengths of meandering streams, and construct detailed profiles of an area. But what would you do if you were in the field and did not have access to any sophisticated equipment or could not get a reliable signal? A geologist in the field would use a **topographic map**—a special type of map that uses contour lines to show landforms. Topographic maps cost almost nothing, weigh much less than a digital device, and withstand rain, swarming insects, and being dropped better than computers. A recent topographic map gives the names and elevations of lakes, streams, mountains, and roads, and it outlines fields and distinguishes swamps and forests as well as a satellite image does. With a little practice, you can learn more about landforms from topographic maps than from a satellite image or DEM. This chapter explains how topographic maps work and helps you develop map-reading skills for identifying landforms, planning hikes, solving environmental problems—and possibly, even saving your life.

9.2 Contour Lines

Like aerial photographs and satellite images, topographic maps show location, distance, and direction very accurately. Topographic maps show the shapes of landforms, elevations, and the steepness of slopes with a special kind of line called a **contour line**.

A contour line is a line on a map that connects points of the same value for whatever is being measured. Contour lines can show many types of features on a map, like population density and average income, as well as elevation. You are already familiar with one common type of contour line on weather maps, which shows a change in temperature in different areas (**FIG. 9.1**).

The contour lines in Figure 9.1b are *isotherms* or lines of equal temperature. Thus, the predicted temperature for every point on the 60° line is 60°F, every point on the 40° line 40°F, and so on. Each isotherm separates areas where the temperature is higher or lower than that along the line itself. Thus, all points in the area north of the 30° contour line have predicted high temperatures for the day lower than 30°F, and those on the south side are predicted to be higher than 30°F. The map has a **contour interval** of 10°, meaning that contour lines represent temperatures at 10° increments.

To get experience with contour lines before looking at topographic maps, the following two exercises ask you to create two basic contour maps. In Exercise 9.1, you will create a contour map showing areas of different tree heights in a small woodland area. Then, Exercise 9.2 asks you to create your own contour map of different temperature areas in the United States, like you see in Figure 9.1b and like you see on many news websites as shown in Figure 9.1c.

FIGURE 9.1 Contour map showing predicted high temperatures for the United States.

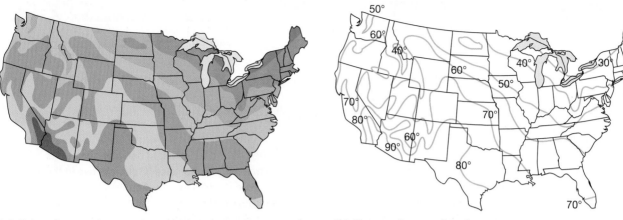

(a) Colored zones show predicted high temperature ranges in increments of 10°F.

(b) Contour lines outline the same zones.

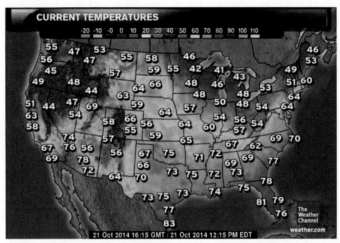

(c) You will see these types of maps frequently on the news. This map is from weather.com.

EXERCISE 9.1   **Making a Simple Contour Map**

Name: _____   Section: _____
Course: _____   Date: _____

Last year, foresters planted pine tree saplings in a grid pattern in a small area. This year, they noted that some trees were taller than others and want to understand the factors that caused those trees to grow faster. The map in this exercise shows the heights of saplings throughout the area.

To help the foresters, create a contour map that shows areas of different tree heights (and thus growth rates). Use the map with a contour interval of 10 cm on the next page, and sketch your contour lines in pencil. You may want to copy the next page to practice your initial attempts.

(continued)

Name: _____   Section: _____
Course: _____   Date: _____

Hints

- As you saw with the temperatures in Figure 9.1b, heights that are multiples of 10 cm will lie on a contour line.
- Heights that are not multiples of 10 cm will lie between two contour lines. For example, tree heights between 10 and 20 cm will lie between the 10-cm and 20-cm contour lines.
- To estimate where a contour line passes between two points, examine the differences in height between the points. For example, we have placed a red line connecting the 6-cm and 14-cm tree heights. Because 10 cm is between these two numbers, the 10-cm contour line must pass between them. And because 10 cm is exactly 4 cm less than 14 cm and 4 cm more than 6 cm, the 10-cm contour will pass midway between these two points.
- Similarly, the 20-cm contour will be between the 14-cm and 22-cm areas. But as 20 cm is only 2 cm less than 22 cm but 6 cm more than 14 cm, this contour line should be drawn close to the 22-cm data point.
- When you are ready, sketch the 10-cm contour line first, then the 20 cm, then the 30 cm, then the 40 cm. Remember, multiple contour lines can pass between two heights—contour lines for 20 cm and 30 cm will pass between the 15-cm and 32-cm measurements.
- Some contour lines will form completely closed areas but others may end at a side of the map.

Map showing the heights (in centimeters) of pine saplings in different areas after 1 year of growth.

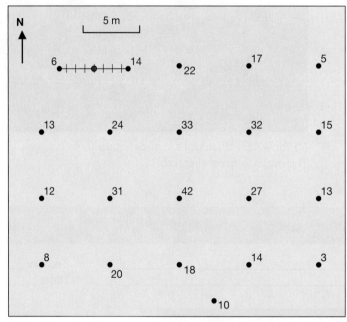

Contour interval = 10 cm

Name: _____   Section: _____
Course: _____   Date: _____

On a certain day in the middle of the summer, the U.S. Weather Bureau predicted maximum daily temperatures shown on the map that follows later in this exercise. Using the data included on the map, create a contoured map with a contour interval of 10°F. Sketch your contour lines in pencil on the map. You may want to copy this page to practice your initial attempts.

(continued)

Name: _____   Section: _____
Course: _____   Date: _____

Hints: As with Exercise 9.1, your main task is to estimate where contour lines will pass between temperatures that are not multiples of 10°.
- To get you started, we show how to locate the 80° isotherm between San Francisco and Los Angeles. San Francisco is 4° cooler than the 80° contour, whereas Los Angeles is 6° warmer, so the 80° contour is closer to San Francisco than Los Angeles as shown.
- Each contour line will be some multiple of 10° (70°, 80°, 90°, 100°), so that contour lines will pass through predicted temperatures that are multiples of 10°.
- Remember, more than one contour line could fall between cities with large differences in temperature. For example, the 80°, 90°, and 100° contour lines have to fit between San Francisco (76°) and Reno, Nevada (102°). Some lines can form completely closed areas while others may end at the sides of the map.

Predicted high temperatures (°F) for a day in the United States.

9.2.1 Contour Lines on Topographic Maps

Now that you understand the basic concept of contour lines and how to create them from map data, let's look at how they work on topographic maps.

A contour line on a **topographic** map connects points that are the same elevation above sea level (the reference for all elevations), and the complete set of lines accurately shows the three-dimensional shape of the surface. **FIGURE 9.2a** shows the shape of the island of Hawaii using a contour interval of 1,000 m. The first contour line is easy to draw for any island because the 0-m contour line is sea level and defines the outer shape of the island.

The elevation difference between contour lines is called the **contour interval.** Small contour intervals are used where there isn't much change in elevation, and large contour intervals are used where the difference in elevation is high. Typical contour intervals on topographic maps are 10, 20, 50, and 100 feet. A 1,000-m

FIGURE 9.2 Topographic maps of the island of Hawaii with different contour intervals.

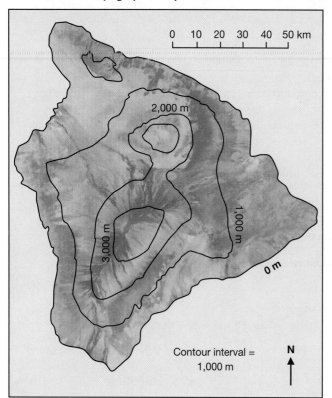

(a) 1,000 m contour interval

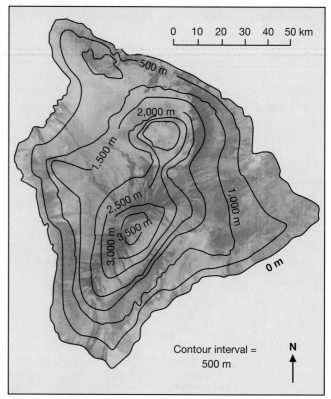

(b) 500 m contour interval

FIGURE 9.3 Useful information outside the borders of a topographic map.

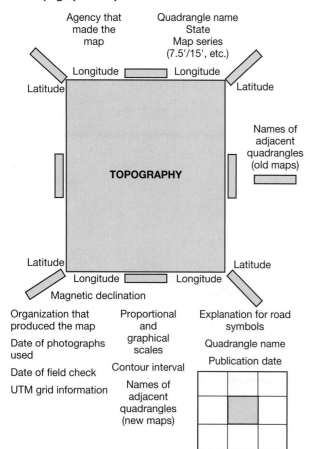

interval is used in Figure 9.2a because the island of Hawaii is so mountainous. If the contour interval had been 10 m, the map would have been unreadable with 100 contour lines for each contour line in the figure. The 1,000-m contour interval can give only a rough idea of the topography, but a smaller interval could give better details (**FIG. 9.2b**).

Geologists use the term **relief** to describe the range in elevation in an area. **High relief** indicates mountainous areas with great elevation differences between the highest and lowest places. **Low relief** refers to flat areas like plains and plateaus where there is very little elevation change over broad expanses.

9.3 Reading Topographic Maps

Because topographic maps come in many sizes and scales, cover areas of different sizes, and can use different contour intervals, the first thing you should do when looking at a map is to examine its borders for useful information like the contour interval, scale, and location. **FIGURE 9.3** is a guide for finding this information on the most recent topographic maps produced by the U.S. Geological Survey (USGS). Further, Appendix 9.1 shows the standard map symbols used to represent natural and man-made features on most maps. As an example, consider **FIGURE 9.4**, which shows a topographic map of an area in eastern Maine.

USGS maps use green to highlight forests, blue for bodies of water like streams, rivers, and lakes, and white for fields or marshes. If there were any urban areas in this part of Maine, they would be shown in pink. Human-built features like roads and buildings in and around the town of Topsfield are shown as white lines.

Take a moment to see how this map compares to a digital elevation model of the same area in **FIGURE 9.5**, a model created using sophisticated computing power. Note how instead of using contour lines to show hills, valleys, and ridges, the DEM uses shading and perspective to show elevations. In this next exercise, we will help you compare different areas of both maps to see how differences in contour lines communicate a variety of shapes in the landscape (hills, valleys, lakes, and flat areas), shapes that are easily seen on a DEM.

EXERCISE 9.3   **How Topographic Maps Show Landform Shapes**

Name: _____   Section: _____
Course: _____   Date: _____

For this exercise, you will use the topographic map in Figure 9.4 and the DEM in Figure 9.5, of the same location.

(a) **Slope:** Find a flat place on the DEM in Figure 9.5 and a place where the slope is steep. Now locate these places on the topographic map (Fig. 9.4). Compare the contour lines in the flat and steep places.
 (i) How does the spacing of the contour lines show the difference between gentle and steep slopes?

_____

_____

 (ii) Describe the slopes on both sides of Farrow Lake, the first lake west of Topsfield. Are the slopes equally steep on both sides of the lake or is one side steeper than the other? Explain your reasoning.

_____

_____

_____

 (iii) Describe the slopes of Farrow Mountain in words, and draw a sketch (using the graph paper provided at the end of this chapter) showing what it would look like to climb over the mountain from northwest to southeast.

_____

_____

_____

_____

_____

_____

(b) **Nested contour lines:** Colored circles in Figure 9.4 identify places where there are a series of concentric (nested) contour lines. Look at these features on the DEM (Figure 9.5).
 (i) What type of feature do nested contour lines indicate? _____
 (ii) Mark similar features with colored dots on the map.

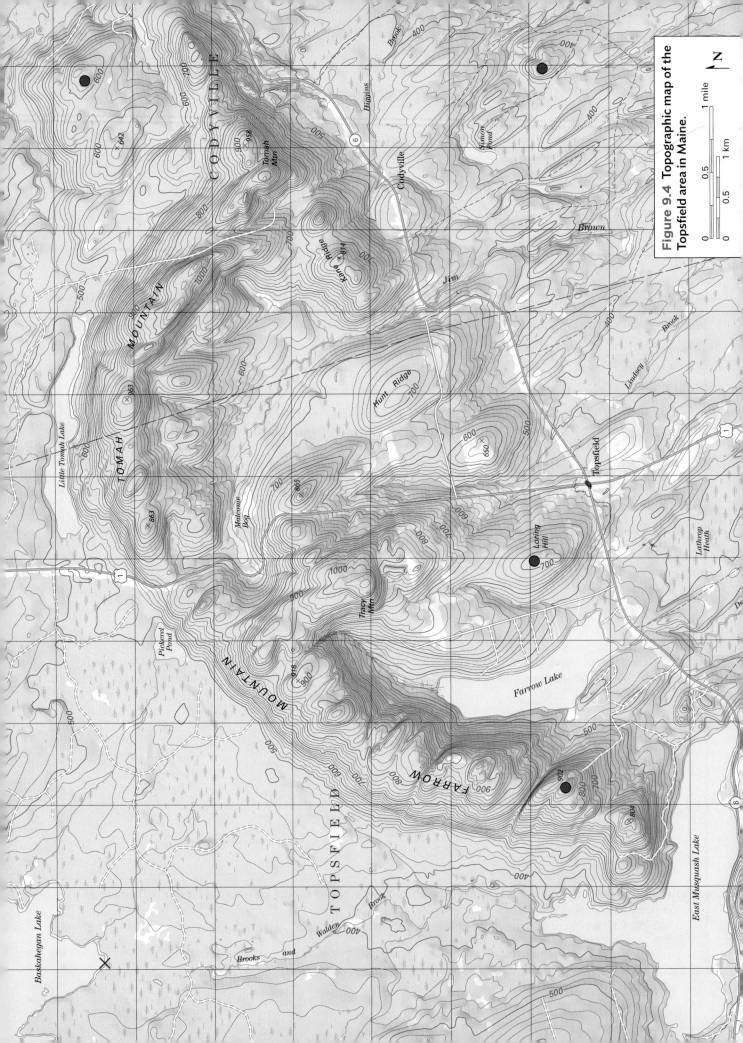

Figure 9.4 Topographic map of the Topsfield area in Maine.

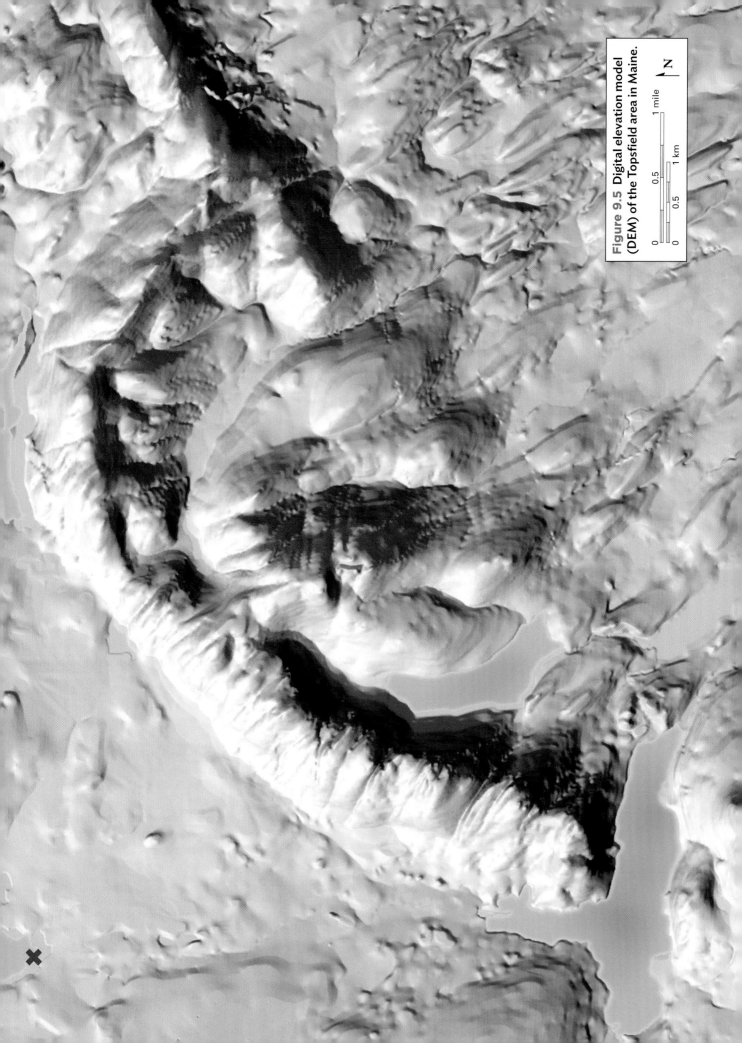

Figure 9.5 Digital elevation model (DEM) of the Topsfield area in Maine.

9.3.1 Contour Lines and Elevation

Maps provide several key features to help determine elevations. Two of the easiest to use are benchmarks and index contour lines.

A benchmark is an accurately surveyed point marked on the ground by a brass plaque cemented in place. It records the latitude, longitude, and elevation of that spot. On topographic maps, elevations of selected points (like hilltops, lake surfaces, and highway intersections) are indicated with a symbol and a number, such as $X_{1438'}$ or Δ_{561}. The latter symbol is the benchmark.

To find the elevation of other points on the map, however, we need to learn how to use the contour lines. First, check the contour interval shown at the bottom of every map for the difference in elevation between adjacent contour lines (see Fig. 9.3). Then notice that on maps like Figure 9.4, every fifth contour line is darker than those around it and has its elevation labeled. These are index contours, which represent elevations that are five times the contour interval (multiples of 5 × 20 in Fig. 9.4—300 feet, 400 feet, and so on). To determine the elevation of an unlabeled contour line, determine its position relative to an index contour. For example, a contour line immediately adjacent to the 500-foot index contour must be 480 or 520 feet—20 feet lower or higher than the index contour on a map where the contour interval is 20 feet.

What about the elevation of a point between two contour lines? All points on one side of a contour line are at higher elevations than the line itself, and those on the opposite side are at lower elevations—but how much higher or lower? This is just as easy as finding the temperature between two isotherms on a weather map. The contour interval in **FIGURE 9.6** is 50 feet, so a point between two adjacent contour lines must be less than 50 feet higher or lower than those lines. Either use a ruler or estimate as you did in Exercise 9.2: a point midway between the 500- and 550-foot contour lines would be estimated as 525 feet above sea level.

Depressions Concentric contour lines may indicate a hill (as in Exercise 9.3b) or a depression. To avoid confusion, it must be clear that contour lines outlining a depression indicate progressively *lower* elevations toward the center of the concentric lines rather than the *higher* elevations that would indicate a hill. To show this, small marks called hachures are added to the line, pointing toward the lower elevation (see the innermost two lines in **FIGURE 9.7**).

The feature in Figure 9.7 could be a volcano with a crater at its summit. Note that the nested contour lines on the flanks of the feature indicate increasing elevation just like those of Figure 9.4, but that the 300-foot contour line is repeated, with the

FIGURE 9.6 Reading elevations of hills and valleys from topographic maps.

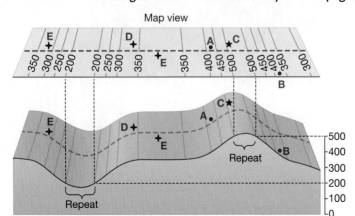

The elevation of Point A is between 400 and 450 feet. Because the slope of the hill is constant between the 400- and 450-foot contour lines and A is halfway between the contour lines, its elevation is 425 feet above sea level.

The elevation of Point B is between 350 and 400 feet. The slope of the hill is gentler close to the 350-foot line and steeper near the 400-foot line. We must therefore estimate elevation taking this into account. The approximate elevation of B is 355 feet above sea level.

The elevation of Point C is greater than 500 feet, but how much greater? If it was more than 550 feet above sea level, there would be another contour line (550 feet). Therefore, the most we can say about the elevation of C is that it is between 500 and 550 feet above sea level.

FIGURE 9.7 Hachured contour lines indicate depressions.

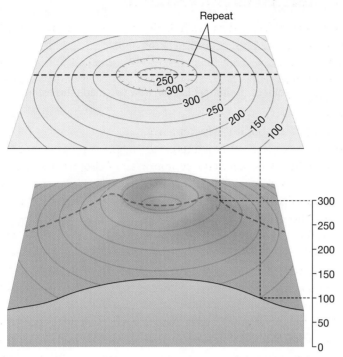

inner one hachured. This shows that the rim of the crater is higher than 300 feet but lower than 350 feet.

Using what you have learned and some of the figures you have studied, Exercise 9.4 helps you to practice determining different elevations on contour maps.

| EXERCISE 9.4 | Determining Elevations from Topographic Maps |

Name: _____   **Section:** _____
Course: _____   **Date:** _____

To determine the elevation of a point on a topographic map:

1. Determine the map contour interval.

2. Find the known elevation closest to the point. This may be a benchmark or a contour line. Remember that every fifth contour line is a heavier **index contour**. You may have to follow contour lines some distance before finding an elevation label.

3. Determine whether the point is higher or lower than the known elevation based on place names, such as "fire tower," "valley," and so forth.

4. Interpolate between the contour lines that bracket the point to get elevation.

Using this procedure and what you have learned in your reading, answer the following questions using figures found earlier in the chapter:

(a) In Figure 9.7, what is your estimate of the elevation of the highest point on the rim of the crater? Explain how you made this estimation.

_____

_____

_____

(continued)

Name: _____   Section: _____

Course: _____   Date: _____

(b) In Figure 9.7, what is your estimate of the lowest point of the crater? Explain how you made this estimation.

_____

_____

_____

(c) In Figure 9.5, a red "x" marks a spot on the shore of Baskahegan Lake, but a printed DEM doesn't give information about its elevation. See the topographic map (Fig. 9.4) to determine the elevation of that point as accurately as possible. _____ feet

(d) In Figure 9.4, what is the elevation of:

the highest point on Hunt Ridge? _____

the highest point on Farrow Mountain? _____

the intersection of U.S. Route 1 and Maine Route 6 in Topsfield? _____

Malcome Bog? _____

(e) What is the relief between Little Tomah Lake and the top of Tomah Mountain in Figure 9.4? _____

(f) What is the relief between the intersection of U.S. Route 1 and Maine Route 6 in Topsfield and East Musquash Lake in Figure 9.4? _____

9.3.2 Contour Lines and Streams: Which Way Is the Water Flowing?

Geologists study the flow of streams and drainage basins to prevent the downstream spread of pollutants and to determine where to collect water samples to find traces of valuable minerals. Streams flow downhill from high elevations to low elevations. You could determine the flow direction by looking for benchmarks along the stream or where different contour lines cross the stream, but there is an easier way as you will see in Exercise 9.5.

EXERCISE 9.5   **Understanding Stream Behavior from Topographic Maps**

Name: _____   Section: _____

Course: _____   Date: _____

(a) Refer to Figure 9.4, and look closely at the unnamed stream at the north end of Farrow Lake. Based on the elevations of the contour lines that cross the stream and the general nature of the topography, in which compass direction does this stream flow?

_____

(b) Apply the same reasoning to the stream at the east end of Malcome Bog. In which compass direction does this stream flow?

_____

(continued)

Name: _____   **Section:** _____
Course: _____   **Date:** _____

(c) Now look at the contour lines as they cross these two streams. Their distinctive V shape tells which way the stream is flowing. Suggest a "rule of V" that describes how the direction of stream flow is revealed by the contour lines that cross it.

_____

_____

_____

(d) Based on your rule of V, does the stream at the west side of Pickerel Pond flow into or away from the pond?

_____

(e) In what direction does Jim Brown Brook flow (in the southeast corner of the map)?

_____

9.4 Rules and Applications of Contour Maps

In the past few sections and exercises, you learned the most important points and techniques to get started with topographic maps. The basic "rules" for reading contour lines are mostly common sense, and you figured out the most important ones for yourself in Exercises 9.3 to 9.5. Now in Exercise 9.6, you will summarize what you have learned and then apply it. Exercise 9.7 asks you to create your own topographic map, and Exercise 9.8 shows how understanding these maps could help save your life.

EXERCISE 9.6   Rules of Contour Lines on Topographic Maps

Name: _____   **Section:** _____
Course: _____   **Date:** _____

In Exercise 9.3, you deduced for yourself the most important "rules" of contour lines, and you will use them in the next several chapters to study landforms produced by streams, glaciers, groundwater, wind, and shoreline currents. Complete the following sentences *using what you've just learned* to summarize the rules of contour lines.

(a) Two different contour lines cannot cross because _____.

(b) The spacing between contour lines on a map reveals the _____ of the ground surface. Closely spaced contour lines indicate _____ and widely spaced contour lines indicate _____.

(c) Concentric contour lines indicate a _____.

(d) Concentric *hachured* contour lines indicate a _____.

(e) Contour lines form a V when they cross a stream. The open part of the V faces the (upstream/downstream) direction.

Now that you understand the basic rules of contour lines, you can make your own contoured map from elevation data. The figure below shows elevation data for a coastal area. Using the figure below, construct a contour map showing the topography. Remember that nature has already drawn the first contour line: the coastline is, by definition, the 0-foot contour. You've already done something similar on the weather map, but topographic contours must follow the rules you have just learned. Use a contour interval of 20 feet. *Remember: Contour lines do not go straight across a stream—they form a V.*

Topographic contouring exercise.

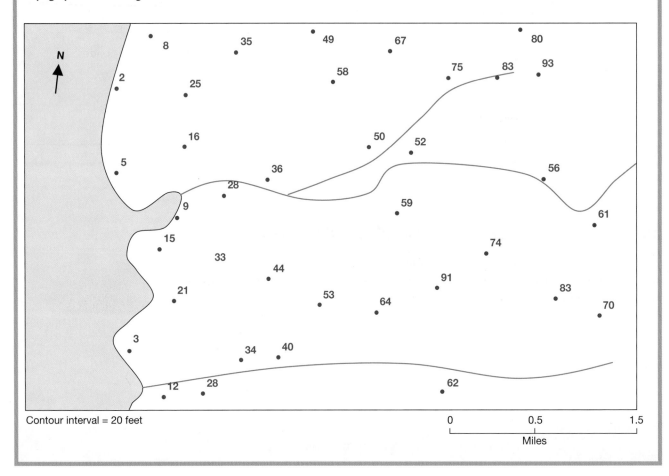

Contour interval = 20 feet

0   0.5   1.5
Miles

A small plane carrying you and a friend crashes near the northeast corner of an island at the spot indicated on the following map. The nearest human beings are a lighthouse keeper and his family living on the opposite side of the island, as shown on the map. Unfortunately, the plane's radio was destroyed during the crash, and your cell phones don't work. No one knows where you are, and the only way to save yourselves is to walk to the lighthouse.

(continued)

Name: _____   Section: _____
Course: _____   Date: _____

You and your friend were injured in the crash and can't climb hills higher than 75 feet or swim across rivers. You can only walk about 12 miles a day and carry only enough water to last 3 days. It gets worse: the rivers have crocodiles, there's a large area filled with quicksand, and there's a jungle filled with—yes—lions and tigers and bears. The good news is that you have a compass and a topographic map (on the next page) showing the hazards. You have figured out exactly where the plane crashed and know where you have to go. The map also shows a well where you can get drinking water—if you can get there in 3 days.

(a) With a pencil, protractor, and ruler, plan the shortest route from the crash site to the lighthouse, avoiding steep hills, rivers, quicksand, and hungry jungle carnivores. Record the direction (using the azimuth system) and distance of each leg of your trip in the table below.

(b) How many days will the trip take? _____

(c) Do you have to stop for water? If so, on what day do you to get to the well? _____

(d) How many days will it take to get from the well to the lighthouse? _____

(e) Where is the highest point on the island? _____ Give its elevation as precisely as you can. _____

(f) What is the elevation of the lowest point in Sinking Feeling Basin? _____

(g) Which is the steepest side of Deadman's Mountain? _____

Your route to safety on Survivor Island.

Leg #	Direction of leg (in azimuth degrees)	Length of leg (in miles)	Leg #	Direction of leg (in azimuth degrees)	Length of leg (in miles)
1			11		
2			12		
3			13		
4			14		
5			15		
6			16		
7			17		
8			18		
9			19		
10			20		
Total distance					

(h) Unfortunately, the plane didn't carry a life raft. If it had, what would have been the shortest route to sail or paddle from the crash site to the lighthouse? Give your answer with directions and distances for each leg on a separate piece of paper.

(continued)

Name: _____ Section: _____

Course: _____ Date: _____

Survivor Island.

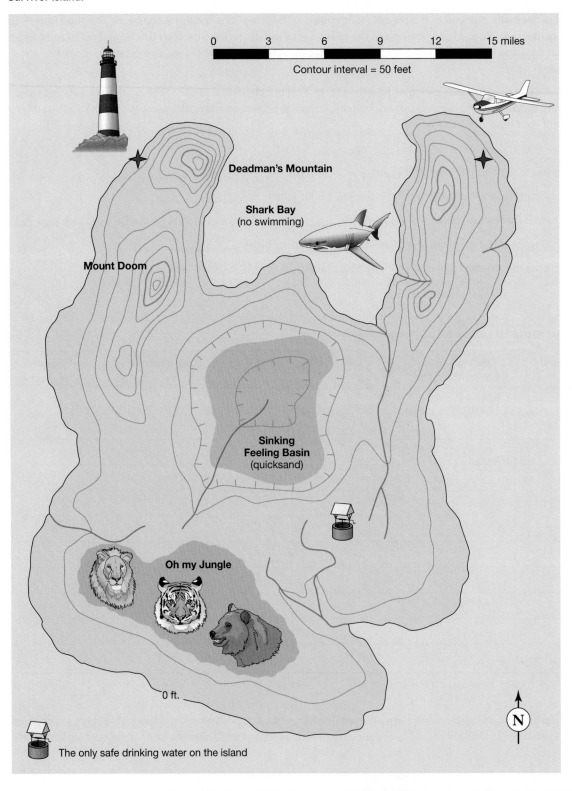

Contour interval = 50 feet

9.5 Topographic Profiles

Topographic maps show a view of an area from above; however, in many cases, seeing a cross section is extremely helpful. The next two sections of this chapter show how to create a simple cross section and how to understand vertical exaggeration. Then Exercise 9.9 examines a situation where this knowledge might help you get a job (or a raise).

9.5.1 Constructing a Topographic Profile

One benefit of using a topographic map is that it is easy to construct an accurate **topographic profile**, a cross-section view of the topography. Freeware, such as MICRODEM, or other GIS software, could draw the profile for you, but it is important to understand just what a profile can and cannot do. This is best learned by constructing profiles by hand—a simple process outlined below.

STEP 1: Place a strip of paper along the line of profile (A–B) shown in **FIGURE 9.8**, and label the starting and finishing points.

STEP 2: Draw a short line where the strip of paper crosses contour lines, streams, roads, and so on, and label each mark with its elevation (**FIG. 9.9**). For clarity, index contour labels are shown here.

STEP 3: Create a profile paper. Set up a vertical scale using the graph paper at the end of this chapter or blank paper with a series of evenly spaced horizontal lines corresponding to the elevations represented by contour lines along the traverse. Label each line so its elevation is recognizable (right side of **FIG. 9.10**).

STEP 4: Place the strip of paper with the labeled lines at the bottom of the vertical scale, and use a ruler to transfer the elevations to their correct positions on the profile paper. Place a dot where each contour line marker intersects the corresponding elevation line in the profile (Fig. 9.10).

STEP 5: Connect the dots, using what you know about contour lines to estimate the elevations at the tops of hills and bottoms of stream valleys along the profile (Fig. 9.10).

FIGURE 9.8 Map of the microwave tower project area.
Scale = 1:62,500; contour interval = 50 feet

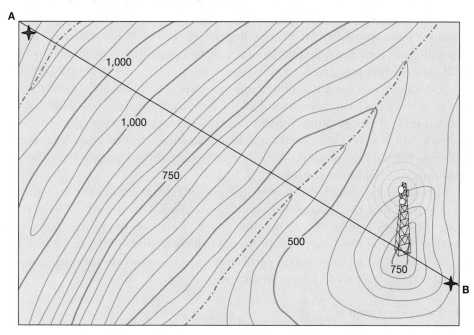

FIGURE 9.9 Constructing the profile.

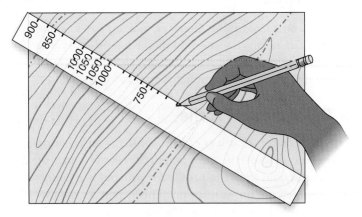

FIGURE 9.10 Completing the profile.

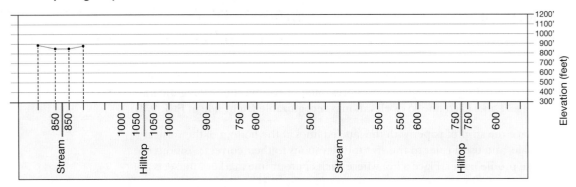

9.5.2 Vertical Exaggeration

Constructing a topographic profile is quick, even without a computer, and can yield important information that can't be obtained just by looking at a map. But care must be taken to make sure that the profile is a realistic view of the topography. It is possible to minimize a mountain so that it looks like an anthill or exaggerate an anthill to make it look like Mt. Everest, depending on the vertical scale you choose. A profile must have not only a horizontal scale to agree with the map but also a vertical scale to show topography. The elevation lines in Figure 9.10 actually define a vertical scale. When the horizontal (map) and vertical (profile) scales are the same, the profile is a perfect representation of topography. If the scales are different, the result is **vertical exaggeration**. Vertical exaggeration overemphasizes the vertical dimension with respect to the horizontal. This is illustrated by three profiles drawn at different vertical exaggerations across the same part of the Hanging Rock Canyon quadrangle in California (**FIG. 9.11**).

To find out how much the profile in Figure 9.10 exaggerates the topography, we need to know the vertical scale used in profiling. First, measure the vertical dimension with a ruler; this shows that 1 *vertical* inch on the profile represents nearly 900 feet of elevation, a proportional scale of 1:10,800 (900 ft × 12 in. per foot = 10,800 in.). To calculate vertical exaggeration, divide the map proportional scale by the profile proportional scale.

$$\textbf{Vertical exaggeration} = \frac{\text{Map scale}}{\text{Profile scale}} = \frac{62,500}{10,800} = 5.8\times$$

FIGURE 9.11 **The effect of vertical exaggeration on topographic profiles.**

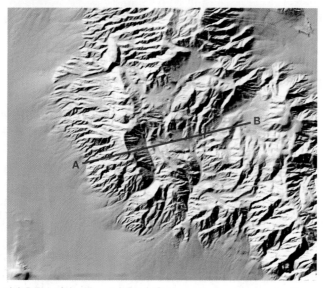

(a) DEM of the Hanging Rock Canyon quadrangle showing the line of profile.

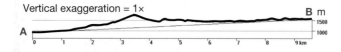

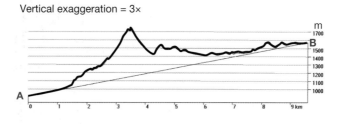

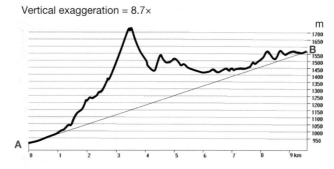

(b) Topographic profiles along the line A–B drawn with different vertical exaggerations. Note that the horizontal scale is the same in all three profiles; only the vertical scale chagnes.

Name: _____   Section: _____
Course: _____   Date: _____

A telecommunications company wants to place microwave relay towers in a new region to improve cell phone reception and plans to put a tower on the hilltop in the southeast corner of the map shown in Figure 9.8. Project managers are concerned that a prominent ridge might block the signal to areas to the northwest. It is not immediately obvious from the map if there will be a "dead zone." You have been hired as a consultant to answer this question. The best way to do so is to construct a topographic profile showing the hills and valleys.

(a) Construct a profile along the line indicated in Figure 9.8. The horizontal scale is set on Figure 9.8 but you must choose the vertical scale for the profile. *Do not use the same scale as the horizontal;* that is, one large box on the vertical scale will *not* represent the same distance as on the horizontal scale. (Use the graph paper at the end of the chapter to construct the profile.)

(b) What is the vertical exaggeration of your profile? This will depend on what you chose for the vertical scale in (a).

_____

(c) Now construct a profile (using the graph paper at the end of this chapter) along the same line with twice the vertical exaggeration.

? What Do You Think   Now that you've got the profile, what advice do you give the telecommunications company? Is there a direct line of sight so there won't be a dead zone for customers in that valley? If there isn't, how high would the tower have to be to guarantee service to everyone in the area? Write a brief summary of your recommendations.

_____

_____

_____

_____

_____

Examine the map again. Are there other potential problem areas? If so, sketch on the figure other lines of profile that would need to be evaluated more fully and submit this updated figure as your work.

Topographic Map Symbols

BATHYMETRIC FEATURES

Area exposed at mean low tide; sounding datum line***	
Channel***	=====
Sunken rock***	+

BOUNDARIES

National	
State or territorial	
County or equivalent	
Civil township or equivalent	
Incorporated city or equivalent	
Federally administered park, reservation, or monument (external)	
Federally administered park, reservation, or monument (internal)	
State forest, park, reservation, or monument and large county park	
Forest Service administrative area*	
Forest Service ranger district*	
National Forest System land status, Forest Service lands*	
National Forest System land status, non-Forest Service lands*	
Small park (county or city)	

BUILDINGS AND RELATED FEATURES

Building	
School; house of worship	
Athletic field	
Built-up area	
Forest headquarters*	
Ranger district office*	
Guard station or work center*	
Racetrack or raceway	
Airport, paved landing strip, runway, taxiway, or apron	
Unpaved landing strip	
Well (other than water), windmill, or wind generator	
Tanks	
Covered reservoir	
Gaging station	
Located or landmark object (feature as labeled)	
Boat ramp or boat access*	
Roadside park or rest area	
Picnic area	
Campground	
Winter recreation area*	
Cemetery	Cem

COASTAL FEATURES

Foreshore flat	Mud
Coral or rock reef	Reef
Rock, bare or awash; dangerous to navigation	
Group of rocks, bare or awash	
Exposed wreck	
Depth curve; sounding	18 23
Breakwater, pier, jetty, or wharf	
Seawall	
Oil or gas well; platform	

CONTOURS

Topographic

Index	6000
Approximate or indefinite	
Intermediate	
Approximate or indefinite	
Supplementary	
Depression	
Cut	
Fill	
Continental divide	

Bathymetric

Index***	
Intermediate***	
Index primary***	
Primary***	
Supplementary***	

CONTROL DATA AND MONUMENTS

Principal point**	3-20
U.S. mineral or location monument	USMM 438
River mileage marker	Mile 69

Boundary monument

Third-order or better elevation, with tablet	BM 9134 BM 277
Third-order or better elevation, recoverable mark, no tablet	5628
With number and elevation	67 4567

Horizontal control

Third-order or better, permanent mark	Neace Neace
With third-order or better elevation	BM 52 Pike BM393
With checked spot elevation	1012
Coincident with found section corner	Cactus Cactus
Unmonumented**	+

Topographic Map Symbols

CONTROL DATA AND MONUMENTS – *continued*

Vertical control

Third-order or better elevation, with tablet	BM × 5280
Third-order or better elevation, recoverable mark, no tablet	× 528
Bench mark coincident with found section corner	BM + 5280
Spot elevation	× 7523

GLACIERS AND PERMANENT SNOWFIELDS

Contours and limits	
Formlines	
Glacial advance	
Glacial retreat	

LAND SURVEYS

Public land survey system

Range or Township line	
Location approximate	
Location doubtful	
Protracted	
Protracted (AK 1:63,360-scale)	
Range or Township labels	R1E T2N R3W T4S
Section line	
Location approximate	
Location doubtful	
Protracted	
Protracted (AK 1:63,360-scale)	
Section numbers	1 - 36 1 - 36
Found section corner	+
Found closing corner	+
Witness corner	WC +
Meander corner	MC
Weak corner*	+

Other land surveys

Range or Township line	
Section line	
Land grant, mining claim, donation land claim, or tract	
Land grant, homestead, mineral, or other special survey monument	▯
Fence or field lines	

MARINE SHORELINES

Shoreline	
Apparent (edge of vegetation)***	
Indefinite or unsurveyed	

MINES AND CAVES

Quarry or open-pit mine	⤬
Gravel, sand, clay, or borrow pit	⤬
Mine tunnel or cave entrance	⤙
Mine shaft	▪
Prospect	X
Tailings	Tailings
Mine dump	
Former disposal site or mine	

PROJECTION AND GRIDS

Neatline	39°15' 90°37'30"
Graticule tick	55'
Graticule intersection	+
Datum shift tick	-+-

State plane coordinate systems

Primary zone tick	640 000 FEET
Secondary zone tick	247 500 METERS
Tertiary zone tick	260 000 FEET
Quaternary zone tick	98 500 METERS
Quintary zone tick	320 000 FEET

Universal transverse mercator grid

UTM grid (full grid)	273
UTM grid ticks*	269

RAILROADS AND RELATED FEATURES

Standard gauge railroad, single track	
Standard gauge railroad, multiple track	
Narrow gauge railroad, single track	
Narrow gauge railroad, multiple track	
Railroad siding	
Railroad in highway Railroad in road Railroad in light-duty road*	
Railroad underpass; overpass	
Railroad bridge; drawbridge	
Railroad tunnel	
Railroad yard	
Railroad turntable; roundhouse	

RIVERS, LAKES, AND CANALS

Perennial stream	
Perennial river	
Intermittent stream	
Intermittent river	
Disappearing stream	
Falls, small	
Falls, large	
Rapids, small	
Rapids, large	
Masonry dam	
Dam with lock	
Dam carrying road	

Topographic Map Symbols

RIVERS, LAKES, AND CANALS – *continued*

Perennial lake/pond	
Intermittent lake/pond	
Dry lake/pond	
Narrow wash	
Wide wash	
Canal, flume, or aqueduct with lock	
Elevated aqueduct, flume, or conduit	
Aqueduct tunnel	
Water well, geyser, fumarole, or mud pot	
Spring or seep	

ROADS AND RELATED FEATURES

Please note: Roads on Provisional-edition maps are not classified as primary, secondary, or light duty. These roads are all classified as improved roads and are symbolized the same as light-duty roads.

Primary highway	
Secondary highway	
Light-duty road Light-duty road, paved* Light-duty road, gravel* Light-duty road, dirt* Light-duty road, unspecified*	
Unimproved road Unimproved road*	
4WD road 4WD road*	
Trail	
Highway or road with median strip	
Highway or road under construction	
Highway or road underpass; overpass	
Highway or road bridge; drawbridge	
Highway or road tunnel	
Road block, berm, or barrier*	
Gate on road*	
Trailhead*	

SUBMERGED AREAS AND BOGS

Marsh or swamp	
Submerged marsh or swamp	
Wooded marsh or swamp	
Submerged wooded marsh or swamp	
Land subject to inundation	Max Pool 431

SURFACE FEATURES

Levee	
Sand or mud	
Disturbed surface	
Gravel beach or glacial moraine	
Tailings pond	

TRANSMISSION LINES AND PIPELINES

Power transmission line; pole; tower	
Telephone line	
Aboveground pipeline	
Underground pipeline	

VEGETATION

Woodland	
Shrubland	
Orchard	
Vineyard	
Mangrove	

* USGS–USDA Forest Service Single-Edition Quadrangle maps only.

In August 1993, the U.S. Geological Survey and the U.S. Department of Agriculture's Forest Service signed an Interagency Agreement to begin a single-edition joint mapping program. This agreement established the coordination for producing and maintaining single-edition primary series topographic maps for quadrangles containing National Forest System lands. The joint mapping program eliminates duplication of effort by the agencies and results in a more frequent revision cycle for quadrangles containing National Forests. Maps are revised on the basis of jointly developed standards and contain normal features mapped by the USGS, as well as additional features required for efficient management of National Forest System lands. Single-edition maps look slightly different but meet the content, accuracy, and quality criteria of other USGS products.

** Provisional-Edition maps only.

Provisional-edition maps were established to expedite completion of the remaining large-scale topographic quadrangles of the conterminous United States. They contain essentially the same level of information as the standard series maps. This series can be easily recognized by the title "Provisional Edition" in the lower right-hand corner.

*** Topographic Bathymetric maps only.

Topographic Map Information

For more information about topographic maps produced by the USGS, please call 1-888-ASK-USGS or visit us at http://ask.usgs.gov/.

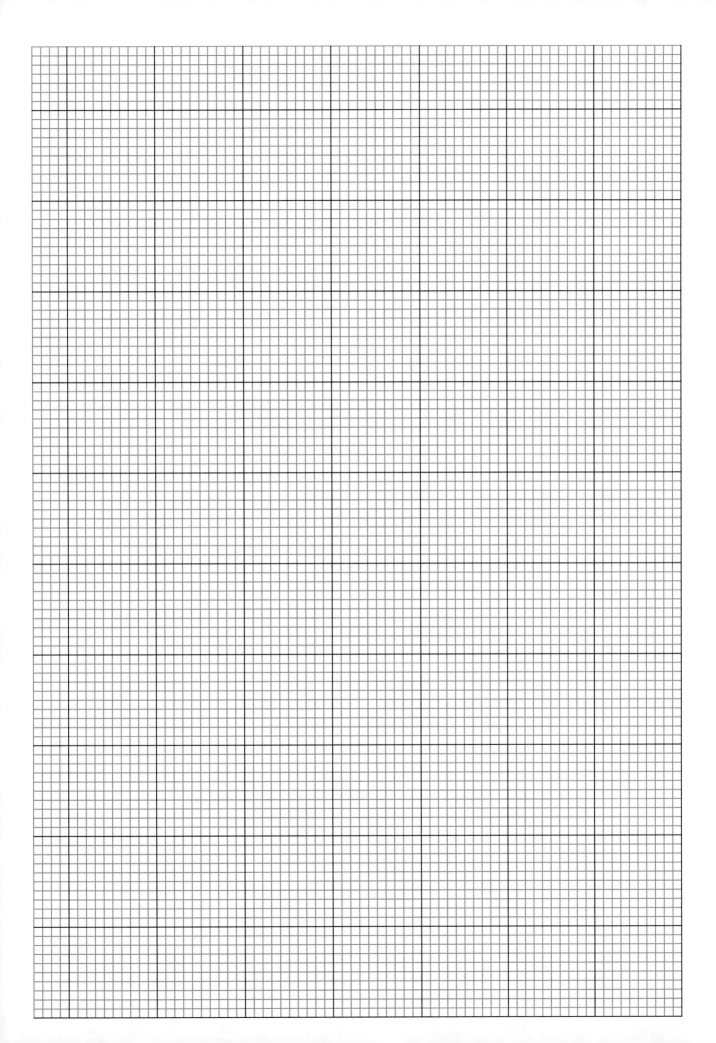

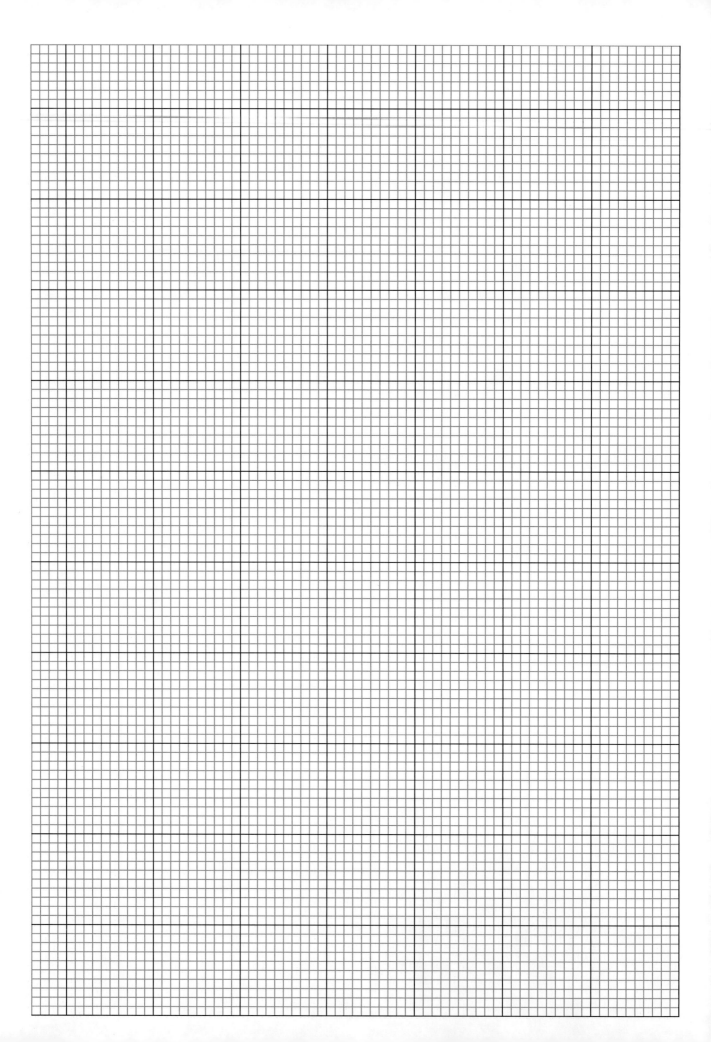

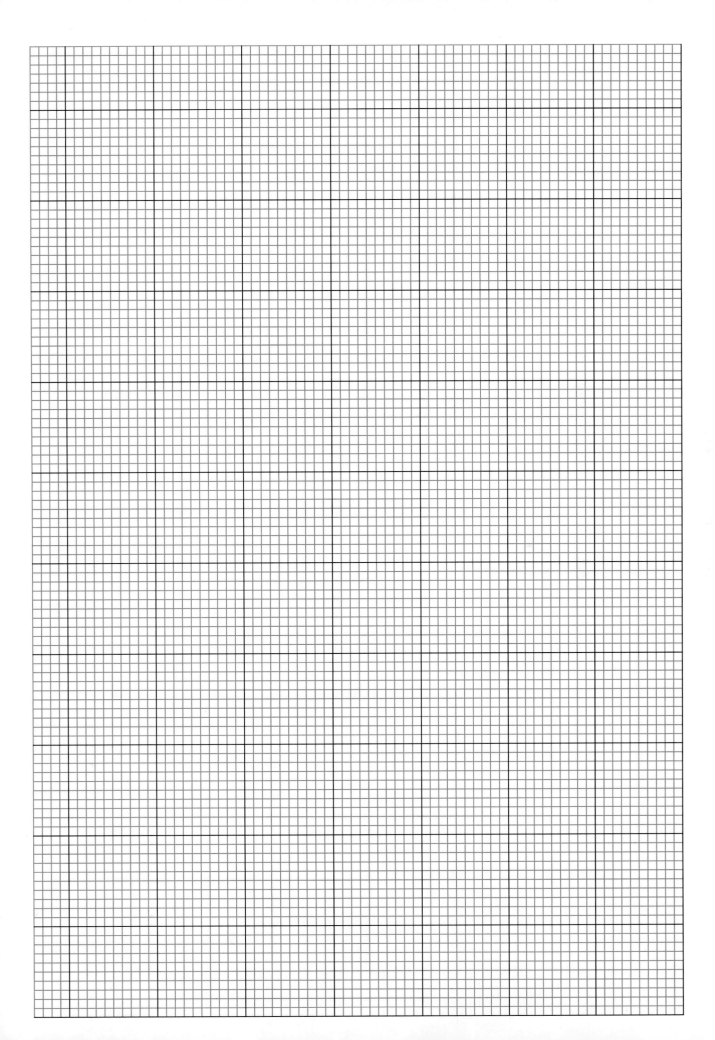

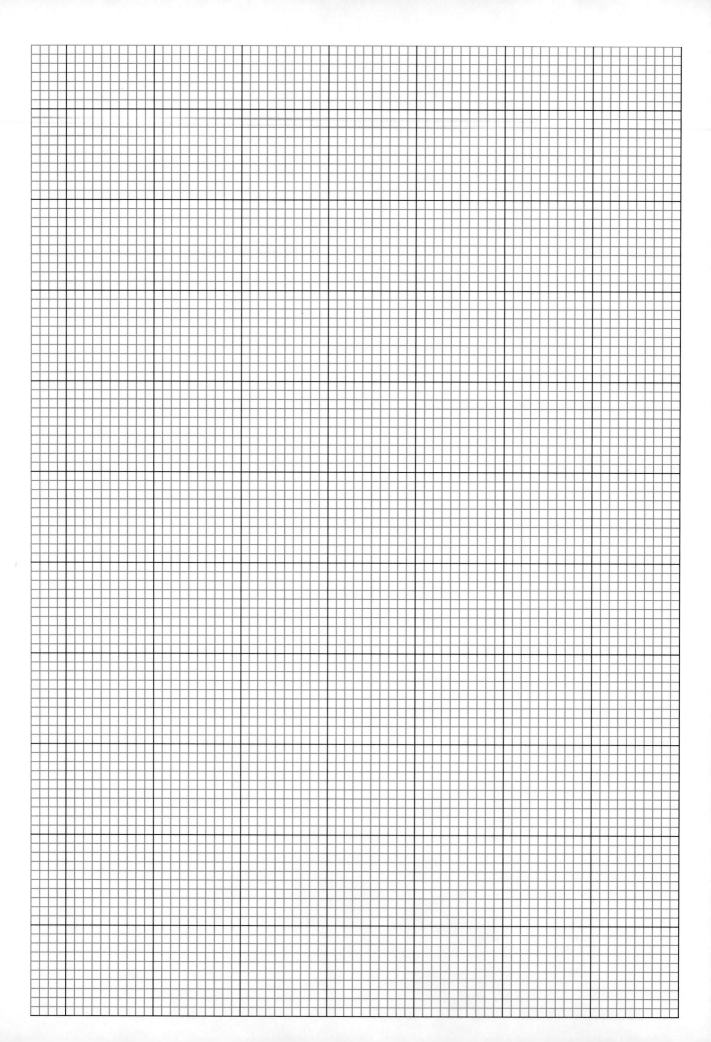

10

Landscapes Formed by Streams

Streams come in different sizes and shapes. They create different landscapes through erosion and by moving and depositing material.

- Learn how streams erode and deposit material
- Become familiar with landforms formed by stream erosion and deposition
- Interpret active and ancient stream processes from landscape features

MATERIALS NEEDED

- Thin string
- Ruler with divisions in tenths of an inch or millimeters
- Graph paper for constructing topographic profiles
- Colored pencils
- Magnifying glass or hand lens to read closely spaced contour lines

10.1 Introduction

Water flowing in a channel is called a **stream** whether it is as large as the Amazon River or as small as the smallest creek, run, rill, or brook. Streams are highly effective agents of erosion and may move more material after one storm in an arid region than wind does all year. This chapter explores why not all streams behave the same way and how streams can produce very different landscapes.

10.2 How Do Streams Work?

All streams operate according to a few simple principles regardless of their size.

- Water in streams flows downhill because of gravity.
- Streams normally flow in a well-defined channel, except during floods when the water overflows the channel and spills out across the surrounding land.
- The motion of water gives a stream kinetic energy, enabling it to do the geologic work of erosion and deposition. The amount of energy depends on the amount (mass) of water and its velocity (remember: kinetic energy = $\frac{1}{2}mv^2$), so big, fast-flowing streams erode more than small, slow-flowing streams.
- Its kinetic energy allows a stream to transport sediment, from the finest mud-sized grain to small boulders. These particles slide or roll on the bed of the stream, bounce along, or are carried in suspension within the water.
- The flow of water erodes unconsolidated sediment from the walls and bed of the channel, and the stream uses that sediment to abrade solid rock.
- Streams deposit sediment when they lose kinetic energy by slowing down (or evaporating). The heaviest particles are deposited first, then the smaller grains, as the energy wanes.

Now for a few geologic terms: A stream **channel** is the area within which the water is actually flowing. A stream **valley** is the region within which the stream has eroded the land. In some cases, the channel completely fills the bottom of the valley; in others, it is much narrower than the broad valley floor. Some valley walls are steep, others gentle.

Exercise 10.1 asks you to describe some differences in streams and Exercise 10.2 begins to look at factors that control stream activity. Streams are complex dynamic systems in which changes in one factor bring about changes in

---

EXERCISE 10.1   Differences between Streams

Name: _____   Section: _____

Course: _____   Date: _____

The basic principles of how streams work, as listed in Section 10.2, are the same for all streams, but the principles can be applied differently, resulting in streams that look very different from one another. In your own words, describe how the streams pictured in **FIGURE 10.1** differ.

_____

_____

_____

FIGURE 10.1 **A tale of two streams.**

(a) Yellowstone River in Wyoming.

(b) River Cuckmere, England.

others, affecting the way the stream looks and behaves. For example, changes in a stream's gradient—the steepness of a stream's channel—can completely change the nature of erosion and deposition, the width of the valley, and the degree to which the channel meanders (its sinuosity).

EXERCISE 10.2 **Getting Familiar with Properties of Streams**

Name: _____   Section: _____

Course: _____   Date: _____

The following questions ask you to compare the streams pictured in Figure 10.1.

(a) Which stream in Figure 10.1 has the wider channel? _____

(b) Which stream has the broader valley? _____

(c) Which stream has the most clearly developed valley walls? _____

(d) Describe the relationship between valley width and channel width for both streams.

_____

_____

_____

(e) Which stream has the straighter channel? _____
Which has a more *sinuous* (meandering) channel? _____

(f) Which stream appears to be flowing faster? _____ What evidence did you use to determine this?

_____

_____

_____

(g) Which stream appears to be flowing more steeply downhill? _____ *Note:* The steepness of a stream channel is called its *gradient* and is a major factor in stream behavior.

FIGURE 10.2 Longitudinal stream profile.

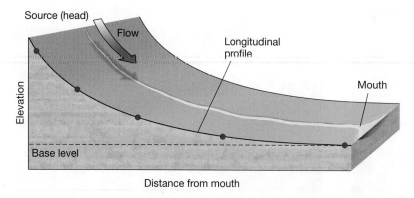

10.2.1 Stream Erosion: Downward or Sideways

A brief lesson in stream anatomy helps to explain stream erosion and deposition. A stream begins at its **headwaters** (or head), and the point at which it ends—by flowing into another stream, the ocean, or a topographic low—is called its **mouth**. The headwaters of the Mississippi River are in Lake Itasca in Minnesota, and its mouth is the Gulf of Mexico in Louisiana. The *longitudinal profile* of a stream from headwaters to mouth is generally a smooth, concave-up curve (**FIG. 10.2**). The gradient (steepness) may vary from a few inches to hundreds of feet of **vertical drop** per mile and is typically steeper at the head than at the mouth.

A stream can erode its channel only as low as the elevation at its mouth, because if it cut deeper it would have to flow uphill to get to the mouth. The elevation at the mouth thus controls erosion along the entire stream and is called the **base level**. Sea level is the ultimate base level for streams that flow into the ocean; base level for a *tributary* that flows into another stream is the elevation where the tributary joins the larger stream.

One difference between the Yellowstone River and the River Cuckmere is the straightness of their channels: the Yellowstone has a relatively straight channel, whereas the Cuckmere channel meanders across a wide valley floor. As noted earlier, the sinuosity of a stream measures how much it meanders, as shown in the following formula. Because sinuosity is a ratio of the two lengths, it has no units. An absolutely straight stream would have a sinuosity of 1.00 (if such a stream existed), whereas streams with many meanders have high values for sinuosity (**FIG. 10.3**).

$$\text{Sinuosity} = \frac{\text{Length of stream channel (meanders and all)}}{\text{Straight-line distance between the same points}}$$

FIGURE 10.3 Stream sinuosity. The straighter a stream, the lower its sinuosity; the more it meanders, the higher its sinuosity.

Low sinuosity (straight)

Low to moderate sinuosity

High sinuosity

EXERCISE 10.3	Why Some Streams Meander but Others Are Straight

Name: _____ Section: _____

Course: _____ Date: _____

In this exercise, you will use the maps in **FIGURES 10.4, 10.5,** and **10.6.** Approximate mile measurements to the nearest 1/10 of a mile.

(a) Compare the course of the Bighorn River between points A and B with that of its tributary between points C and D (Fig. 10.4). Fill in the table below.

	Bighorn River	Unnamed tributary
Channel length (~miles)		
Straight-line length (~miles)		
Sinuosity (no units) Channel length divided by straight-line length		
Highest elevation* (feet)		
Lowest elevation* (feet)		
Vertical drop Highest elevation minus lowest elevation (feet)		
Gradient* (feet per mile) Vertical drop divided by channel length		

*Streams aren't considerate; they don't begin and end on contour lines. Scan the entire stream looking for and estimating the highest and then the lowest point on each stream. Then calculate the vertical drop and gradient.

(b) What is the apparent relationship between a stream's gradient and whether it has a straight or meandering channel?

_____

_____

_____

(c) Test this hypothesis on the Genesee River of New York (Fig. 10.5) and the Casino Lakes area of Idaho (Fig. 10.6). Complete the following table and describe how the Genesee River differs from the Idaho streams.

	Genesee River	Casino Lakes area	
		Stream A–B	Stream C–D
Valley shape (V-shaped or broad with flat bottom)			
Gradient (ft/mile)			
Valley width (miles)		Channel essentially fills valley floor.	
Channel width (feet)			
Valley width/channel width		~1.0	~1.0
Sinuosity Length of stream channel divided by straight-line distance between the same points			

(continued)

Figure 10.4 Part of the Bighorn River in Wyoming (Manderson and Orchard Bench 7.5' quadrangles)

Contour interval = 20 feet

Figure 10.5 The Genesee River south of Rochester, New York.

0   0.5   1 mile
0   0.5   1 km

N

Contour interval = 10 feet

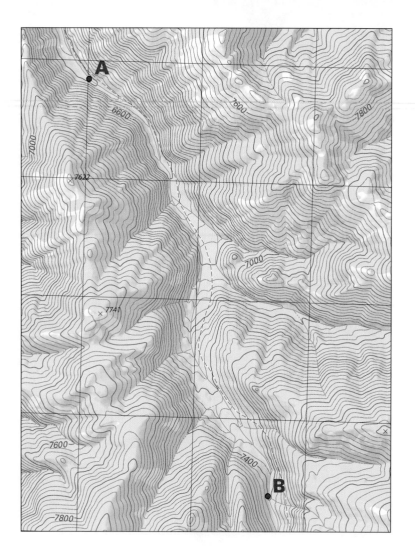

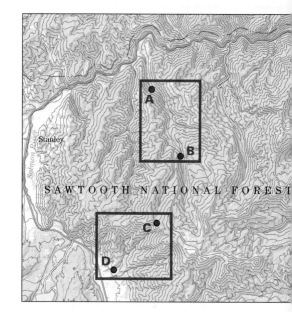

Figure 10.6 Casino Lakes, Idaho (7.5′ quadrangle)

0   0.5   1 mile

0   0.5   1 km

N

Contour interval = 40 feet

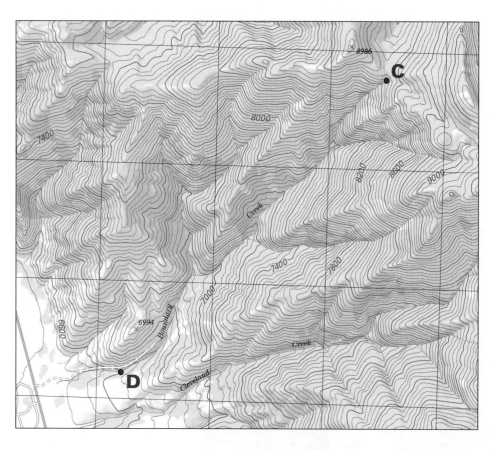

Name: _____ Section: _____

Course: _____ Date: _____

(d) Did these maps support your hypothesis about the relationship between meandering (sinuosity) and gradient? Explain.

_____

_____

_____

(e) What is the apparent relationship between sinuosity and the valley width/channel width ratio?

_____

_____

_____

(f) What is the apparent relationship between stream gradient and the shape of a stream valley?

_____

_____

_____

Now apply what you've learned to the photographs of the streams in Figures 10.1 and 10.3.

(g) Which probably has the steeper gradient–the River Cuckmere or the Yellowstone River? Explain your reasoning.

_____

_____

_____

(h) Which of the streams in Figure 10.3 probably has the steepest gradient? The gentlest gradient? Explain your reasoning.

_____

_____

_____

10.3 Stream Valley Types and Features

The Yellowstone River and River Cuckmere in Figure 10.1 illustrate the two most common types of stream valleys: steep-walled, V-shaped valleys whose bottoms are occupied fully by the channel, and broad, flat-bottomed valleys much wider than the channel and within which the stream meanders widely between the valley walls. **FIGURE 10.7** shows how these valleys form.

FIGURE 10.7 Evolution of stream valleys.

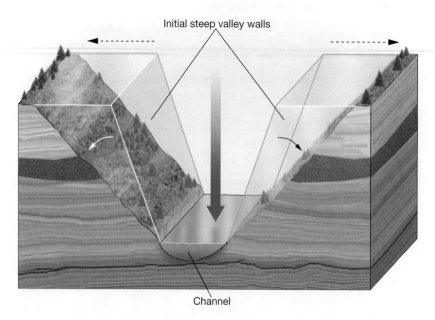

Initial steep valley walls

Channel

(a) Steep, V-shaped valley: Vertical erosion carves the channel and valley downward vertically (large blue arrow), producing steep valley walls. Mass wasting (slump, creep, landslides, and rock falls) reduces slope steepness to the angle of repose (curved arrows) and widens the top of the valley (dashed arrows).

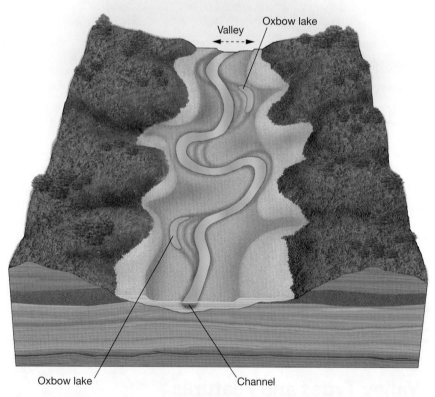

Valley **Oxbow lake**

Oxbow lake **Channel**

(b) Broad, flat-bottomed valley: As the stream meanders, it widens the valley (arrows). Mass wasting gentles the slope of the valley walls as in part (a). Oxbows mark the position of former meanders.

FIGURE 10.8 Floodplain and associated features.

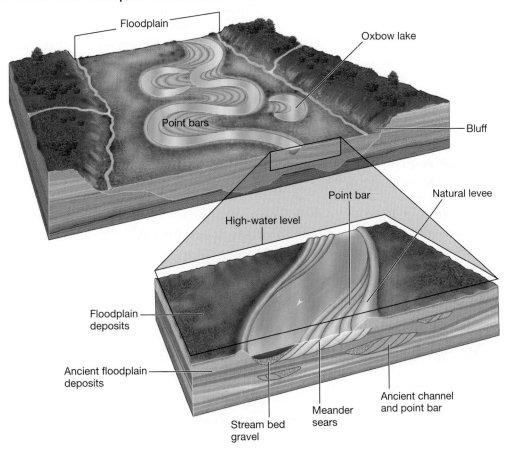

When water is added to a stream in a V-shaped valley, the channel expands and fills more of the valley. When more water enters a stream with a broad, flat valley and a relatively small channel, it spills out of the channel onto the broad valley floor in a **flood**. Sediment carried by the floodwater is deposited on the **floodplain**, and other depositional and erosional features can be recognized easily on topographic maps or photographs (**FIG. 10.8**).

10.3.1 Features of Floodplains

Natural levees are ridges of sediment that outline the channel, and they form when a stream overflows its banks and deposits its coarsest sediment next to the channel. Several generations of natural levees are visible in Figure 10.8, showing how the meanders changed position with time. **Point bars** form when water on the inside of a meander loop slows down, causing sediment to be deposited. At the same time, erosion occurs on the outside of the meander loop because water there moves faster. The result is that meanders migrate with time, moving outward (toward their convex side) and downstream. Sometimes a stream cuts off a meander and straightens itself. The levees that formerly flanked the meanders help outline the former position of the river, leaving **meander scars**. Meander scars that have filled with water are called **oxbow lakes**.

10.4 Changes in Streams over Time

Streams erode *vertically* by leveling the longitudinal profile to the elevation of the mouth. Streams also erode *laterally*, broadening their valleys by meandering. Most streams erode both laterally and vertically at the same time, but the balance between vertical and lateral erosion commonly changes as the stream evolves.

Headwaters of a high-gradient stream are much higher than its mouth, and stream energy is used largely in vertical erosion, lowering channel elevation all along its profile. Over time, the gradient lessens as erosion lowers the headwater area. The stream still has enough energy to alter the landscape and uses some of that energy to erode laterally. The valley then widens by a combination of mass wasting and meandering. Even when headwaters are lowered to nearly the same elevation as the mouth, a stream still has energy for geologic work; but because it cannot cut vertically below its base level, most energy at this stage must be used for lateral erosion.

Initially, a stream meanders within narrow valley walls; but with time it erodes those walls farther and farther, eventually carving a very wide valley. As its gradient decreases, a stream redistributes sediment that it had deposited, moving it back and forth across the floodplain.

Although most streams follow this sequence of progressively lowering their gradients and therefore their erosional behavior, not all do so. Some streams have very gentle gradients and meander widely from the moment they begin. Streams on the Atlantic and Gulf coastal plains are good examples of this kind of behavior.

| EXERCISE 10.4 | Interpreting Stream Behavior |

Name: _____   Section: _____
Course: _____   Date: _____

FIGURES 10.9a–c on the following pages show three meandering streams, each of which balances energy use differently between vertical and lateral erosion.

(a) From their valley width/channel width ratios alone, which stream would you expect to have the steepest gradient? The gentlest gradient? Explain your reasoning. (You don't have to calculate these; but if it helps, feel free to do so.)

_____

_____

_____

(b) Which stream do you think is doing the most vertical erosion? The least? Explain.

_____

_____

_____

(c) For each of the three maps, describe features that indicate former positions of the river channels. Do all three maps show this information? If not, explain why.

_____

_____

_____

(d) Label one example of the following stream erosional and depositional features on Figure 10.9a: valley; channel; meander; point bar; oxbow lake; meander scar.

(e) Indicate on Figure 10.9c where the velocity of the Arkansas River is the greatest and where it is the least.

FIGURE 10.9 (a) The St. Francis River in Arkansas and Mississippi.

Contour interval = 50 feet

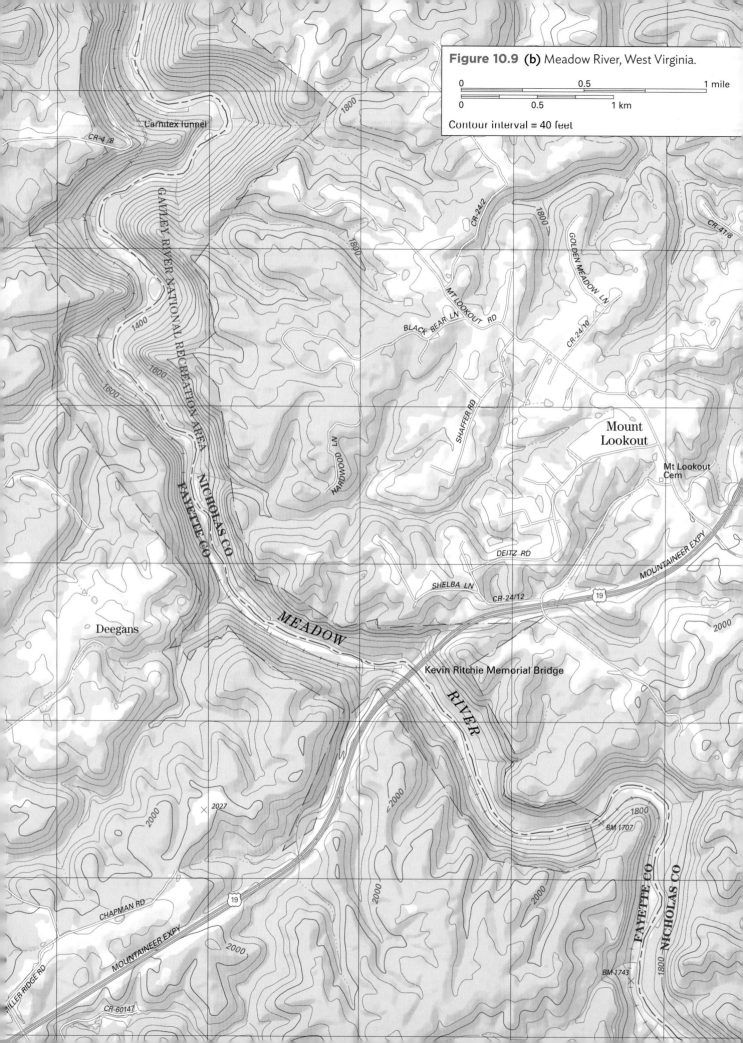

Figure 10.9 (b) Meadow River, West Virginia.

Contour interval = 40 feet

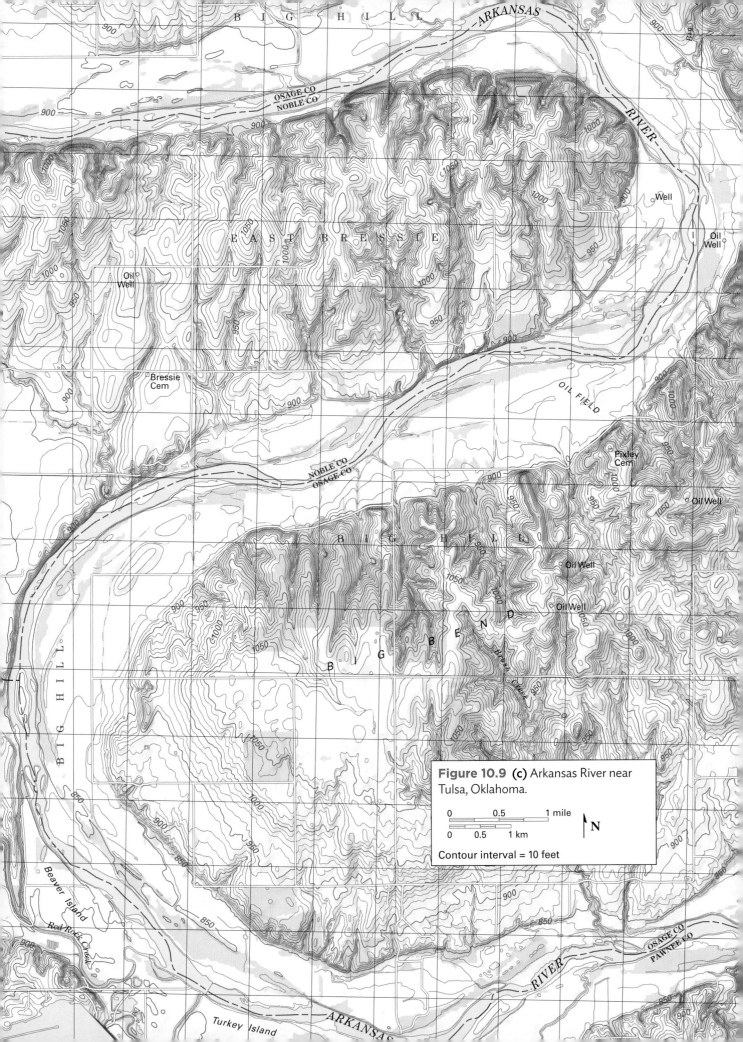

Figure 10.9 **(c)** Arkansas River near Tulsa, Oklahoma.

0   0.5   1 mile

0   0.5   1 km

N

Contour interval = 10 feet

10.5 Stream Networks

Streams are particularly effective agents of erosion because they form networks that cover much of Earth's surface. Rain falling on an area runs off into tiny channels that carry water into bigger streams and eventually into large rivers. Each stream—from tiniest to largest—expands headward over time as water washes into its channel, increasing the amount of land affected by stream erosion. Understanding the geometric patterns of stream networks and the way they affect the areas they drain is the key to understanding how to prevent or remedy stream pollution, soil erosion, and flood damage.

10.5.1 Drainage Basins

The area drained by a stream is its **drainage basin**, which is separated from adjacent drainage basins by highlands called **drainage divides**. The drainage basin of a small tributary may cover a few square miles, but that of the master stream may be hundreds of thousands of square miles. **FIGURE 10.10** shows the six major drainage basins of North America. Five deliver water to an ocean (Pacific, Arctic, Hudson Bay/Atlantic, Gulf of Mexico/Atlantic), but the Great Basin is bounded by mountains and there is no exit for the water.

FIGURE 10.10 Major drainage basins of North America.

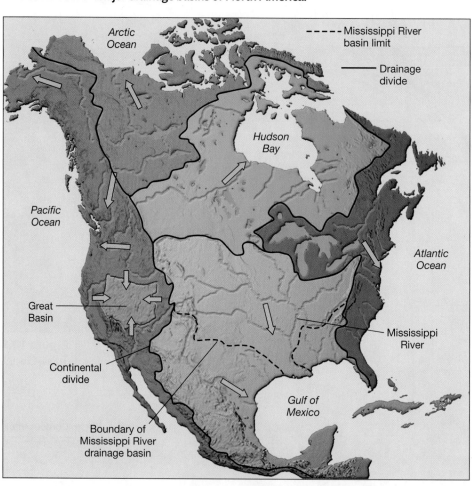

The yellow arrows show the dominant flow direction—note that some of these drain *northward.* The Mississippi drainage basin is the largest and drains much of the interior of the United States. The *continental divide* separates streams that flow into the Atlantic Ocean from those that flow to the Pacific. The Appalachian Mountains are the divide separating the Gulf of Mexico and direct Atlantic Ocean drainage; the Rocky Mountains separate Gulf of Mexico and Pacific drainage. A favorite tourist stop in Alberta, Canada, is a *triple* divide that separates waters flowing north to the Arctic Ocean, west to the Pacific, and south to the Gulf of Mexico.

10.5.2 Drainage Patterns

Master and tributary streams in a network typically form recognizable patterns (**FIG. 10.11**). **Dendritic** patterns (from the Greek *dendros,* for veins in a leaf), develop where surface materials are equally resistant to erosion. This may mean horizontal sedimentary or volcanic rocks; loose, unconsolidated sediment; or igneous and metamorphic areas where most rocks erode at the same rate. **Trellis** patterns form where ridges of resistant rock alternate with valleys underlain by weaker material. **Rectangular** patterns indicate zones of weakness (faults, fractures) perpendicular to one another. In **radial** patterns, streams flow either outward (**centrifugal**) from a high point (e.g., a volcano) or inward (**centripetal**) toward the center of a large basin. **Annular** drainage patterns occur where there are concentric rings of alternating resistant and weak rocks—typically found in structures called domes and basins. Parallel drainage patterns occur on a uniform slope when several streams with parallel courses develop simultaneously.

FIGURE 10.11 Common drainage patterns.

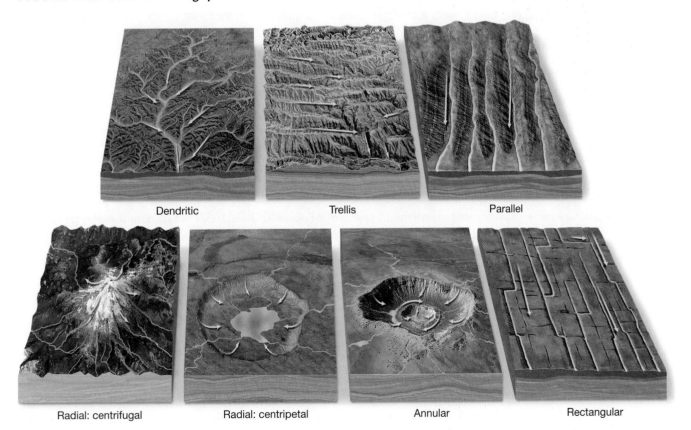

Dendritic

Trellis

Parallel

Radial: centrifugal

Radial: centripetal

Annular

Rectangular

Drainage Basins and Stream Divides

Name: _____ Section: _____
Course: _____ Date: _____

FIGURE 10.12 is a map showing several tributaries on the north and south sides of the Missouri River near Jefferson City, Missouri. One large tributary, the Osage River, joins the Missouri from the south, near the eastern margin of the map, but most of the tributaries on the north are much smaller.

(a) With a colored pencil, trace one of the tributary creeks feeding *directly* into the Missouri from the north. With the same pencil, trace tributaries that flow directly into this creek, and then the tributaries of these smaller streams. Remember the "rule of V" as you trace the smaller streams to their headwaters.

(b) With a different color, trace an adjacent tributary of the Missouri and its tributaries. Repeat for more streams and their tributaries on the north side of the Missouri River, using a different color for each stream.

(c) You have just outlined most of the drainage on the north side of the Missouri. Now, with again a different colored pencil, trace the divides that separate the individual drainage basins for each master stream. This should be easy because you've already identified streams in each drainage basin with a different color.

Note that some divides are defined sharply by narrow ridges, but others are more difficult to locate within broad upland areas where most of the headwaters are located.

(d) Local residents are worried that a recent toxic spill at an electrical substation (asterisk in Fig. 10.12) will work its way into the drainage system. Based on your drainage basin analysis, shade in areas that might be affected. Be conservative: if there is any doubt, err on the side of including areas rather than excluding them.

Recognizing Drainage Patterns

Name: _____ Section: _____
Course: _____ Date: _____

(a) What drainage pattern is associated with the Mississippi River drainage basin in Figure 10.10? What does that tell you about the materials that underlie the central part of the United States?

_____

_____

_____

(b) What drainage patterns are associated with the areas shown in Figures 10.12 and 10.17? What do these patterns indicate about the rocks underlying those areas?

_____

_____

_____

_____

_____

_____

_____

_____

_____

_____

Figure 10.12 Drainage divides of the Missouri River near Jefferson City, Missouri.

Contour interval = 65 feet

10.6 Changes in Stream-Carved Landscapes with Time

Just as a single stream or entire drainage network changes over time, so too do **fluvial** (stream-created) *landscapes*. Consider a large block of land uplifted to form a plateau. Several things will change with time: the highest elevation, the number of streams, the stream gradients, and the amount of flat land relative to the amount of land that is part of valley walls. These changes reflect the different ways in which stream energy is used as the landscape is eroded closer to base level. **FIGURE 10.13** summarizes these idealized changes. In the real world, things rarely remain constant long enough for this cycle to reach its end: sea level may rise or fall due to glaciation or tectonic activity, the land may be uplifted tectonically, and so on. Nevertheless, the stages in Figure 10.13 are typical of landscapes produced by stream erosion and can be recognized on maps and other images.

FIGURE 10.13 Idealized stages in the evolution of a stream-carved landscape.

Stage 1: A fluvial landscape is uplifted, raising stream channels above base level.
- Few streams, but with relatively steep gradients.
- Broad, generally flat divides between streams.
- Channels are relatively straight.
- Relief is low to modest (not much elevation difference between divides and channels).

Stage 2: Main streams cut channels downward, and tributaries form drainage networks.
- Numerous tributaries develop with moderate gradients, dissecting most of the area.
- Stream divides are narrow and sharp; most of the area is in slope, with little flat ground.
- Main streams and some tributaries meander moderately.
- Relief is high, and little if any land is at the original uplifted elevation.

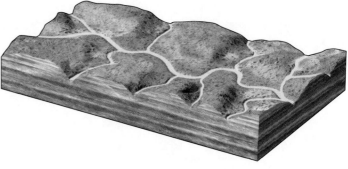

Stage 3: Stream erosion has lowered land surface close to base level.
- Few streams, as in Stage 1, but with gentle gradients.
- Stream divides are broad and flat; relief is low.
- Streams meander broadly.

EXERCISE 10.7 Recognizing Stages of Landscape Erosion

Name: _____ Section: _____

Course: _____ Date: _____

(a) Examine the topographic maps of **FIGURES 10.14**, **10.15**, and **10.16**. Fill in the following table comparing aspects of the three areas.

	Southeast Texas (Fig. 10.14)	Colorado Plateau (Fig. 10.15)	Appalachian Plateau (Fig. 10.16)
Approximate relief: elevation difference between highest and lowest points			
Number of streams (few, intermediate, most)			
Estimated amount of land area that is valley slopes as a percentage			
Estimated stream gradients			
Stream divides: broadly rounded, angular, can't see divides			
Stage of stream dissection (Stage 1, 2, or 3)			

(b) With this practice, look at the fluvial landscapes in Figures 10.1a, 10.6, and 10.9a and suggest which stages of erosion each map represents.

_____

_____

_____

_____

_____

_____

_____

_____

_____

_____

_____

Figure 10.14 Stream dissection in southeast Texas (Fred, Spurger, Magnolia Springs, and Potato Patch Lake quadrangles).

Contour interval = 10 feet

N

0   1   2 miles

0   1   2 km

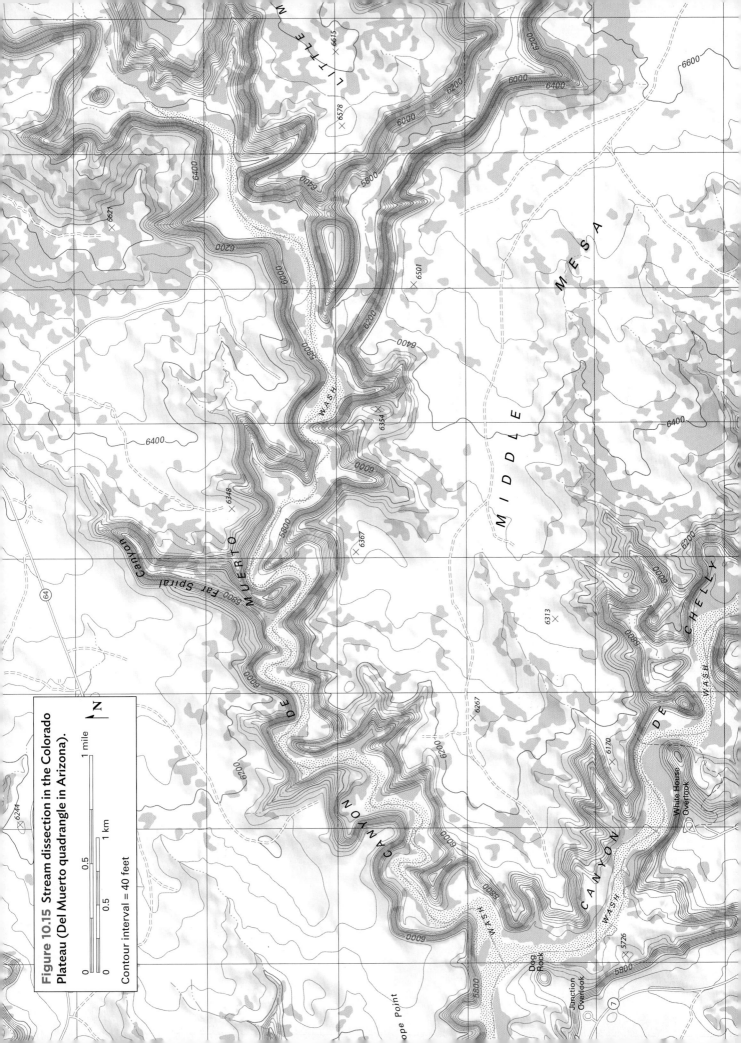

Figure 10.15 Stream dissection in the Colorado Plateau (Del Muerto quadrangle in Arizona).

Contour interval = 40 feet

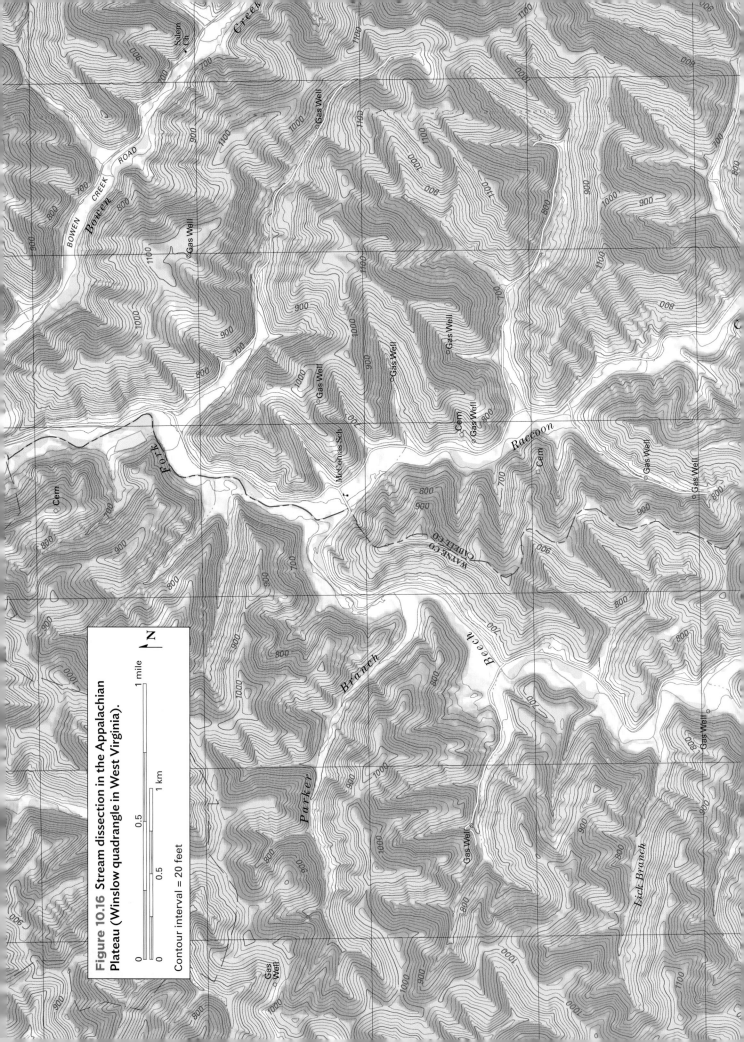

Figure 10.16 Stream dissection in the Appalachian Plateau (Winslow quadrangle in West Virginia).

Contour interval = 20 feet

10.7 When Streams Don't Seem to Follow the Rules

Most stream erosion and deposition follow the principles you just deduced, but there are some notable exceptions. Actually, these streams aren't violating any rules; they are following them to the letter, but their situations are more complex than the basic ones we have examined.

Consider, for example, the Susquehanna River as it flows across the Pennsylvania landscape shown in **FIGURE 10.17**. This area is part of the Valley and Ridge Province of the Appalachian Mountains and is characterized by elongate ridges and valleys made of resistant and nonresistant rocks, respectively. Streams are the dominant agent of erosion in the area. Also look at the Green River, illustrated in **FIGURE 10.18**, where meanders are incised in bedrock. In the following two exercises, you will deduce and suggest what the origins of these two rivers may have been.

EXERCISE 10.8   **Deducing the History of the Susquehanna River**

Name: _____   Section: _____
Course: _____   Date: _____

(a) Figure 10.17 shows a part of the Valley and Ridge Province in Pennsylvania. What is unusual about the relationship between the Susquehanna and Juniata rivers and the valley-and-ridge topography?

_____

_____

_____

Most of the small streams flow in the elongate valleys; but the Susquehanna and its tributary, the Juniata River, cut across the ridges at nearly right angles. It is tempting to think that the big streams had enough energy to cut through the ridges while the small ones couldn't, but this is not the case. The answer lies in a multistage history of which only the last phase is visible today.

(b) Why do most of the smaller streams flow in the elongate valleys?

_____

_____

_____

(c) Suggest as many hypotheses as you can to explain why the two larger rivers cut across the valley-and-ridge topography. *Hint*: How might the landscape have been different at an earlier time?

_____

_____

_____

_____

_____

_____

Rivers with enough energy to cut through the ridges should certainly have been able simply to meander around them, but the Susquehanna and Juniata rivers didn't take the easy way out. It's almost as if they didn't even know the ridges and valleys were there.

(continued)

Name: _____   Section: _____

Course: _____   Date: _____

(d) With that clue, suggest a series of events that explains the behavior of the Susquehanna and Juniata rivers. *Hint*: This type of stream is called a *superposed* stream.

_____

_____

_____

_____

FIGURE 10.17 The Susquehanna River cutting across the Appalachian Valley and Ridge Province in Pennsylvania.

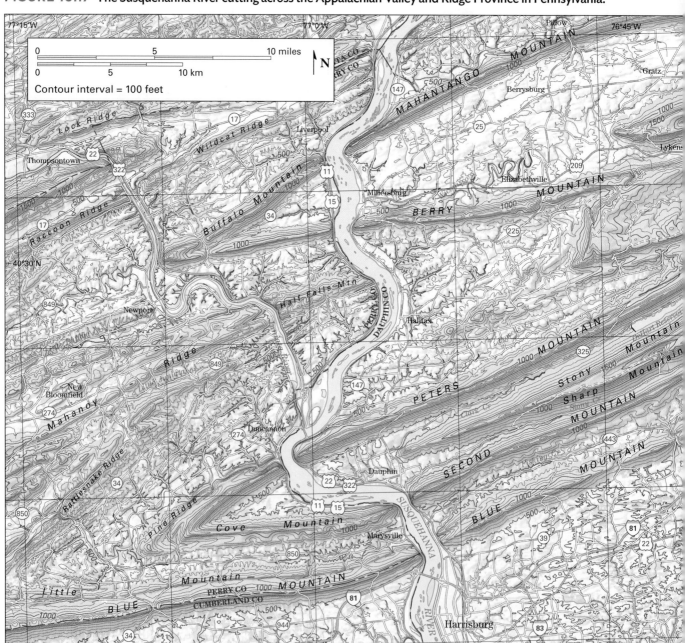

Name: _____   Section: _____

Course: _____   Date: _____

The Green River is in the Colorado Plateau, an area where meanders of many rivers cut deeply (are *incised*) into the bedrock (**FIG. 10.18**). The most famous is the Colorado River itself, particularly where it flows through the Grand Canyon. This behavior is totally unlike that of the meandering streams encountered earlier in this chapter. The Green River seems to violate rules of stream behavior; but, as with the Susquehanna, it is following them perfectly. Some geologic detective work will let you figure out the difference.

(a) Describe the path of the Green River as shown in Figure 10.18.

_____

_____

_____

(b) What is the sinuosity of the Green River? _____

(c) When a river meanders with such sinuosity, how is it using most of its energy—in lateral or vertical erosion?

_____

_____

(d) What evidence is there that the Green River is eroding laterally?

_____

_____

_____

(e) What evidence is there that the Green River is eroding vertically?

_____

_____

_____

(f) What is the probability that the Green River will straighten its path by cutting through the walls of the Bowknot Bend? Explain.

_____

_____

_____

(g) The Green River flows into the Colorado, which flows into the Gulf of Southern California. How far above base level is the river in this area? Is this what you expect for a meandering stream? Explain.

_____

_____

_____

(h) Suggest an origin for the incised meanders. (Don't forget possible effects of tectonic activity.)

_____

_____

_____

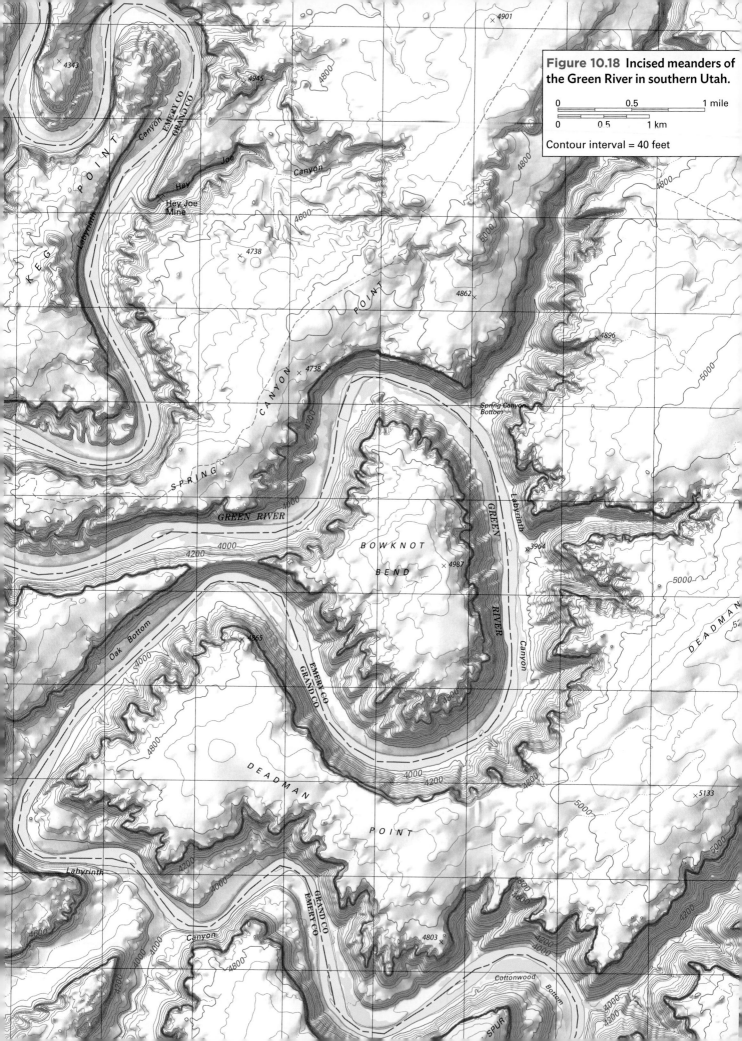

Figure 10.18 Incised meanders of the Green River in southern Utah.

Contour interval = 40 feet

10.8 When There's Too Much Water: Floods

A flood occurs when more water enters a stream than its channel can hold. Many floods are seasonal, caused by heavy spring rains or melting of thick winter snow. Others, called **flash floods**, follow storms that can deliver a foot or more of rain in a few hours. **FIGURE 10.19** is a map compiled by the Federal Emergency Management Agency (FEMA) showing the estimated flood potential in the United States, based on the number of square miles that would be inundated. It might appear that states lightly shaded in Figure 10.19 would have little flood damage, but that would be an incorrect reading of the map because two of the worst river floods in U.S. history occurred in Rapid City, South Dakota, and Johnstown, Pennsylvania. Indeed, these two cities have been flooded many times. The *area* of potential flooding may not be as large as in some other states, but the *conditions* for flooding may occur frequently. South Dakota's Rapid Creek has flooded more than 30 times since the late 1800s, including a disastrous flash flood on June 9, 1972, triggered by 15 inches of rain in 6 hours. The creek overflowed or destroyed several dams, ruined more than 1,300 homes and 5,000 cars, and killed more than 200 people in Rapid City (**FIG. 10.20**).

FIGURE 10.19 Flood risk in the United States.

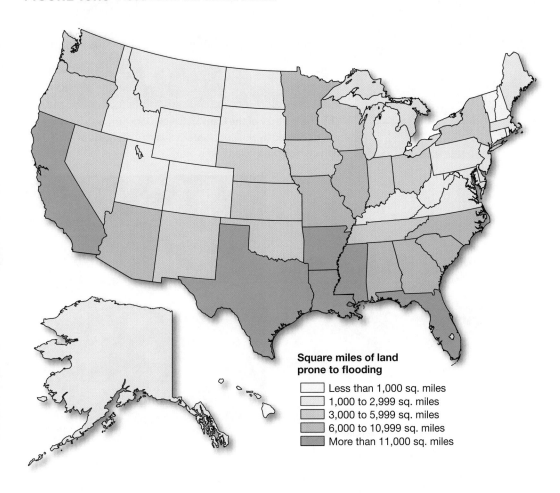

Square miles of land prone to flooding

- Less than 1,000 sq. miles
- 1,000 to 2,999 sq. miles
- 3,000 to 5,999 sq. miles
- 6,000 to 10,999 sq. miles
- More than 11,000 sq. miles

FIGURE 10.20 Effects of the June 9, 1972, flood in Rapid City, South Dakota.

(a) Rapid Creek has a narrow floodplain where it follows through Dark Canyon, 1 mile upstream of Rapid City. Floodwaters filled the entire floodplain. Concrete slabs (arrows) are all that is left of the homes built in the floodplain.

(b) The spillway of the Canyon Creek Dam (arrow) was clogged by debris carried by the floodwater, causing water to flow over the dam, which then failed completely.

(c) Not the usual lineup of cars for gasoline.

(d) This railroad trestle was washed away, along with highway bridges, slowing relief efforts.

Name: _____   Section: _____

Course: _____   Date: _____

Streams shown on maps earlier in the chapter are all subject to flooding, but the potential problems for cities along their banks are not the same. In the following questions, you will use the maps from this chapter to estimate problems caused by potential floods.

(a) Refer to Figures 10.9a, b, and c. For which of these would a flood 20-feet higher than the current stream channels cause damage to human structures and roads? In each case, describe what would happen.

_____

_____

_____

_____

_____

(b) Describe the effect of a flash flood that raised the level of the Bighorn River (Fig. 10.4) 20 feet above its banks. Estimate and describe the affected areas.

_____

_____

_____

_____

_____

(c) Would the Geneseo airport on the east bank of the Genesee River (Fig. 10.5) have to be evacuated if water rose 20 feet? Explain.

_____

_____

_____

_____

(d) Compare the map of the Missouri River with a recent satellite image of the same area in **FIGURE 10.21**, below. What kind of information does the map add? What information does the satellite image add?

_____

_____

_____

_____

(e) What things have been built in the Missouri River floodplain that could be destroyed or made unusable in a flood where the river rose 20 feet? How would the loss of these affect relief efforts?

_____

_____

_____

_____

FIGURE 10.21 Aerial image of the floodplain of the Missouri River near Jefferson City, Missouri.

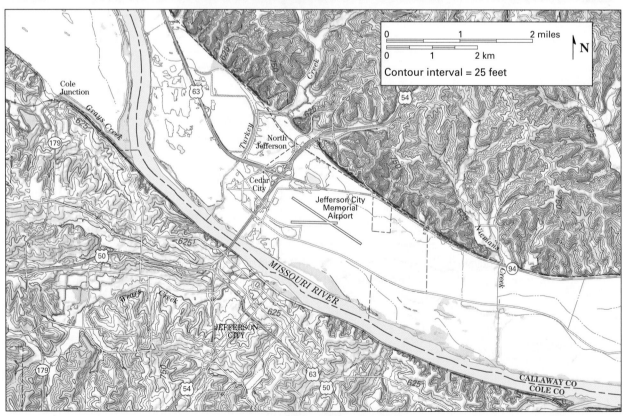

Name: _____   **Section:** _____

Course: _____   **Date:** _____

It may be surprising to learn that normal stream behavior can have "interesting" effects on international relationships. For example, consider the boundary between the United States and Mexico in the vicinity of Brownsville, TX, and Matamoros, Coahuila State, Mexico. The boundary is the center of the Rio Grande River, clearly a meandering stream with significant sinuosity.

You have seen earlier in this chapter that the channels of meandering streams change over time.

(a) How do you think the Rio Grande River might change its course over the next 50-100 years?

_____

_____

(b) Indicate on the map places where these changes might cause parts of the U.S. to become part of Mexico, and parts of Mexico to become part of the U.S.

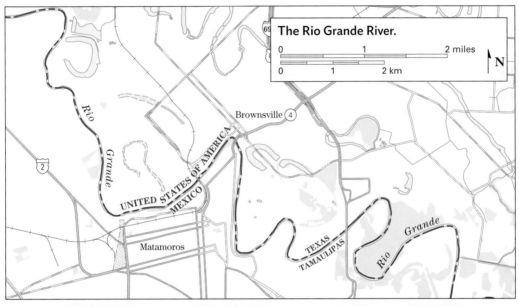

The Rio Grande River.

(continued)

Name: _____   Section: _____

Course: _____   Date: _____

? What Do You Think You have been hired by the U.S.-Mexico
International Boundary and Water Commission (IBWC) as a consultant
for issues related to the Rio Grande River. Based on your answers to (a) and (b)
above, what do you recommend the IBWC do to anticipate changes in the river
channel? Should the river remain the international border? How can property
rights be preserved? How will business and agriculture be affected? Use a
separate piece of paper for your report.

11

Glacial Landscapes

These jagged mountains are typical of features carved by glaciers, which are powerful and aggressive agents of erosion.

11.1 Introduction

Glaciers are broad sheets or narrow "rivers" of ice that last year round and flow slowly across the land. They hold more than 21,000 times the amount of water in all streams and account for about 75% of Earth's freshwater. Today, huge continental glaciers cover most of one continent (Antarctica) and nearly all of Greenland. Smaller glaciers carve valleys on the slopes of high mountains, causing a characteristically sharp, jagged topography (**FIG. 11.1**). Distinctive landforms show that continental glaciers were even more widespread during the Pleistocene Epoch—the so-called ice ages—covering much of northern Europe and North America.

Today, glaciers are receding at rates unprecedented in human history. As they shrink, the balance of the water cycle changes among glaciers, oceans, streams, and groundwater, altering our water supply and modifying ecologic systems worldwide. Glaciated landscapes contain clues to past climate changes and help us understand what is happening now and plan for the future. We know, for example, that continental glaciers advanced and retreated several times during the Pleistocene, and we know how fast those changes were. This enables us to measure human impact on the rates of these processes.

FIGURE 11.1 The Grand Teton Mountains in Wyoming display jagged topography typical of areas affected by alpine glaciers.

11.1.1 Types of Glaciers

Some small glaciers form on mountains and flow downhill, carrying out their erosion and depositional work in the valleys they carve. These are called, appropriately, **mountain**, **valley**, or **alpine glaciers** (**FIG. 11.2a**). Although they flow in valleys like streams, mountain glaciers work differently and form distinctly different landscapes. The Antarctic and Greenland **continental ice sheets** described earlier are not confined to valleys. They flow across the countryside, as masses of ice thousands of feet thick, dwarfing hills and burying all but the tallest peaks (**FIG. 11.2b**). The North American and European ice sheets did the same during the Pleistocene. Exercise 11.1 compares the ways in which glaciers and water do their geologic work.

FIGURE 11.2 Mountain and continental glaciers.

(a) The Crowfoot Glacier in Banff National Park, Canada.

(b) Continental ice sheet, Antarctica.

EXERCISE 11.1	Comparison of Glaciers and Streams

Name: _____ Section: _____

Course: _____ Date: _____

Glaciers behave differently from stream water—even glaciers that occupy valleys. Based on what you know about streams and have learned about glaciers in lecture and from your geology text, complete the following table.

		Streams	Glaciers
	State of matter	Liquid	
	Composition	H_2O	
	Areal distribution	In channels with tributaries flowing into larger streams	
	Rate of flow	Fast or slow, depending on gradient	
	Erodes by . . .	1. Abrasion using sediment load 2. Dissolving soluble minerals, rocks	
	Maximum depth to which erosion can occur	Base level: sea level for streams that flow into the ocean; master stream for tributaries	
	Deposits when . . .	Stream loses kinetic energy either by slowing down or evaporating	
Type(s) of sediment	Maximum clast size range	Small boulders to mud	
	Sorting	Generally well sorted	
	Clast shape	Moderately to well rounded, depending on distance transported	

11.1.2 How Glaciers Create Landforms

Glaciers erode destructional landforms and deposit constructional landforms. A glacier can act as a bulldozer, scraping the regolith from an area and plowing it ahead. When it encounters bedrock, a glacier uses its sediment like grit in sandpaper to abrade the rock and carve unique landforms. Glacial erosion occurs wherever ice is flowing, whether the glacier front is advancing or retreating.

FIGURE 11.3 Till and outwash.

(a) Till with characteristic poor sorting (small boulders through sand, silt, and mud) and angular rock fragments.

(b) Layers of outwash. Each layer is very well sorted; most contain well-rounded, sand-sized grains and some show cross bedding. Note the coarser, pebbly layer (arrow) resulting from an increase in meltwater discharge.

Depositional landforms are equally distinctive—some are ridges tens or hundreds of miles long, and others are broad blankets of sediment. All form when the ice in which the sediment is carried melts, generally when the glacier front is retreating. Some sediment, called **till**, is deposited directly from melting ice; but some is carried away by meltwater and is called **outwash** (or *glaciofluviatile* [glacial + stream] *sediment*). Till and outwash particles differ in size, sorting, and shape (**FIG. 11.3**) because of the different properties of the water and ice that deposited them.

11.2 Landscapes Produced by Continental Glaciation

Landscapes formed by continental glaciation differ from those formed by streams in every imaginable way: the shapes of hills and valleys are different, and both continental and mountain glaciers produce unique landforms not found in fluvial landscapes. **FIGURE 11.4** shows an area affected by continental glaciation for comparison with fluvial landscapes, and Exercise 11.2 explores the differences more fully.

Continental glaciers scour the ground unevenly as they flow, producing an irregular topography filled with isolated basins that often become lakes when the ice melts. Many of these post-Pleistocene lakes are still not incorporated into normal networks of streams and tributaries, even though the ice melted thousands of years ago. Glaciers also deposit material unevenly, clogging old streams and choking new ones with more sediment than they can carry. *Poorly integrated drainage, with many swampy areas, is thus characteristic of recently glaciated areas.* In addition, continental glaciers carve bedrock into characteristic streamlined shapes that indicate the direction the ice flowed.

FIGURE 11.4 Comparison of valleys carved by glaciers and streams.

(a) Glacial landscape: an unnamed valley in Glacier National Park, Montana.

(b) Fluvial landscape: Yellowstone River valley in Wyoming.

EXERCISE 11.2	Comparison of Landscapes Formed by Glaciers and Streams

Name: _____ Section: _____
Course: _____ Date: _____

(a) What difference do you notice in the shape of the valleys pictured in Figure 11.4a and b? In particular, compare the overall shape of the valleys pictured and the characteristics of each valley's base.

_____

_____

(continued)

Name: _____   Section: _____
Course: _____   Date: _____

(b) Suggest a reason for the difference in valley shape. *Hint:* Use a colored pencil to fill in the portion of both valleys in which active erosion occurred. (We will return to this issue below.)

_____

_____

(c) Now compare the extensively glaciated area (below) with the photographs, maps, and digital elevation models of fluvial landscapes (in Chapter 10). Describe the differences without worrying about technical terms for the glacial features. *Consider:* Is there a well-developed stream network?

_____

_____

_____

Aerial view of part of the Canadian Shield after extensive continental glaciation.

11.2.1 Erosional Landscapes

When a continental glacier passes through an area, it carves the bedrock into characteristic shapes called *sculptured bedrock*. The DEM in Exercise 11.3 shows the topography of an area of sculptured bedrock eroded by continental glaciers in southern New York State. The hills are underlain by gneisses that were resistant to erosion and the valleys by marbles and schists that were more susceptible to glacial erosion.

11.2.2 Depositional Landscapes

Landscapes formed by continental glaciers contain several depositional landforms. Those composed of till deposited directly from the melting ice are called **moraines** or **drumlins**; those made of outwash carried by meltwater streams include **outwash plains** and **eskers**.

Moraines: Moraines form when melting ice drops the boulders, cobbles, sand, silt, and mud that it has been carrying. Some moraines are irregular ridges; others are broad carpets pockmarked by numerous pits separated by isolated hills. The type of moraine—ridge or carpet—depends on whether the front (**terminus**) of the glacier is *retreating* (i.e., the ice is melting faster than it is being replaced) or *stagnating* (i.e., the terminus is in a state of dynamic equilibrium in which new ice is replenishing what is melting so that the position of the terminus doesn't change).

Till deposited from a retreating glacier forms an irregular carpet of debris called a **ground moraine** that may be hundreds of feet thick. Large blocks of ice are isolated as a glacier retreats, and some may be buried by outwash or till. When a block melts, the sediment subsides and forms a depression in the ground moraine called a **kettle hole**. For this reason, the irregular surface of a ground moraine is often referred to as "knob and kettle" topography.

The terminus of a stagnating glacier may remain in the same place for hundreds or thousands of years, and the till piles up in a ridge that outlines the terminus. This ridge is called a **terminal moraine** and shows the maximum extent of the glacier.

FIGURE 11.5 shows depositional features created at the termini of continental glaciers. Use these models to identify landforms in the following DEMs and maps. A glossary of glacial depositional and erosional features can be found in **TABLE 11.1** (on page 293) at the end of this chapter.

Drumlins: Drumlins are streamlined, asymmetric hills made of till that occur in four large swarms in Nova Scotia, southern Ontario/New York State, southeastern New England (most famously Bunker Hill), and the Midwestern states of Wisconsin, Iowa, and Minnesota. Their origin is debated, but their shapes clearly reflect the dynamics of ice flowing through till, and their shape and orientation are good indicators of glacial direction.

Outwash Plains: A melting continental glacier generates an enormous amount of water, which flows away from the terminus carrying all the clasts the water is powerful enough to carry, usually sand, gravel, and small boulders. Meltwater streams round and sort this sediment and eventually deposit it beyond the terminal moraine as a generally flat outwash plain (Fig. 11.5).

Eskers: Tunnels may form at the base of a continental glacier and act as pipelines to carry meltwater even as the glacier ice surrounding the tunnels continues to flow. Well-sorted sediment carried by these subglacial streams may build up to form ridges 50 feet high or more. These ridges, called *eskers*, trace the courses of meandering subglacial streams and stand above the unsorted till of the adjacent ground moraine (Fig. 11.5).

FIGURE 11.5 Formation of depositional landforms after continental glaciation.

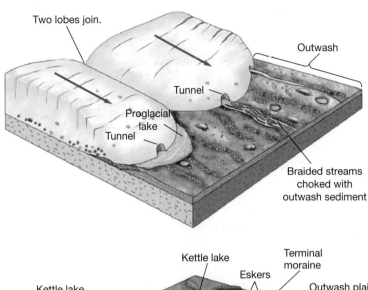

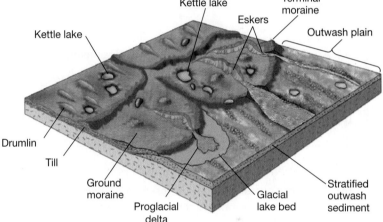

Name: _____   **Section:** _____

Course: _____   **Date:** _____

(a) Describe the topography in the DEM below in your own words, paying particular attention to the shapes of the bedrock hills. When features in a glacial landscape display a strong alignment as in **FIGURE 11.6**, geologists say that the landscape has a "topographic grain" related to the direction in which ice moved. Describe the topographic grain in the map and the direction the ice probably moved.

_____

_____

_____

Look again at the topographic map (Fig. 11.6) for more detail.

(b) Construct a topographic profile, using the graph paper provided at the end of this chapter, from St. John's Church to Lake Rippowam (Line A–B) on the topographic map in Figure 11.6, using 10 × vertical exaggeration. Are the hills symmetrical or asymmetrical?

_____

(c) Are the steep and gentle slopes distributed randomly or systematically? Describe the distribution.

_____

_____

(d) What is controlling the steep slopes of these asymmetric hills?

_____

_____

(e) Based on your description and observations, draw arrows on the DEM below and on Figure 11.6 showing the direction in which the glacier moved across the map area. Explain your reasoning.

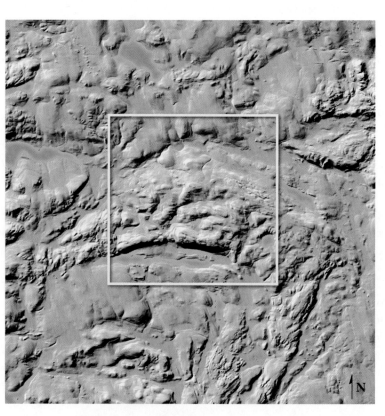

_____

_____

_____

_____

_____

_____

_____

(f) Fill in the blanks to complete the following sentence that describes how sculptured rock records glacial movement: The direction of ice flow indicated by sculptured rock is *from* the _____ side of the hill *toward* the_____ side.

DEM of the Peach Lake quadrangle in New York (area in rectangle shown in Fig. 11.6).

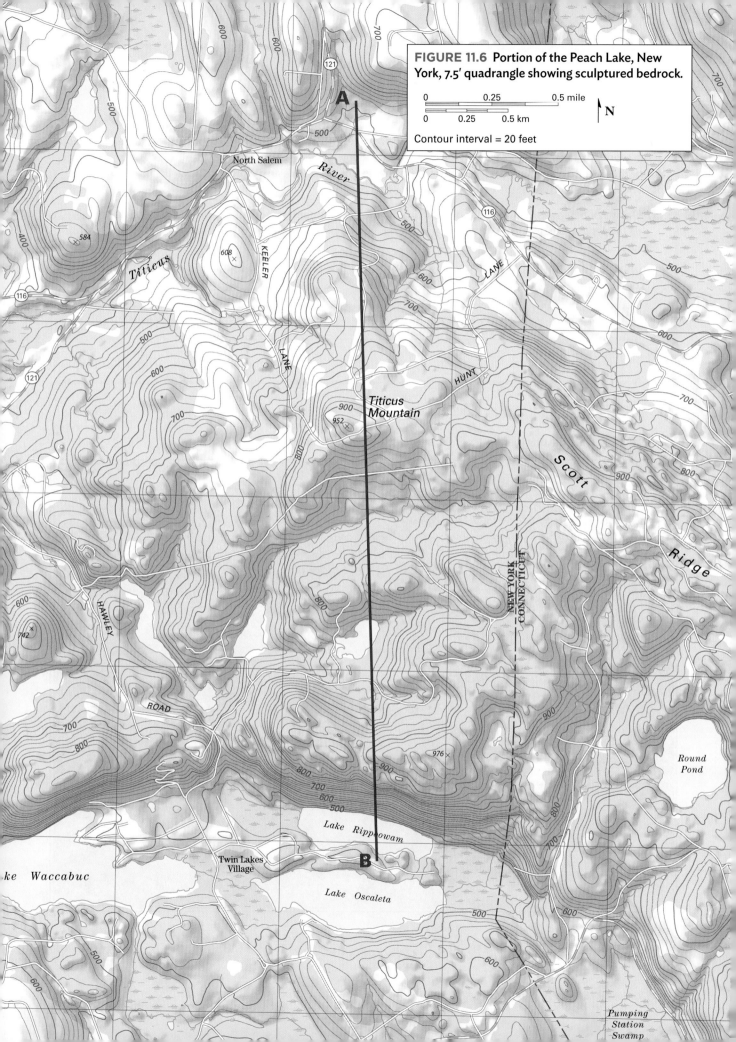

FIGURE 11.6 Portion of the Peach Lake, New York, 7.5' quadrangle showing sculptured bedrock.

Contour interval = 20 feet

N

Name: _____   Section: _____

Course: _____   Date: _____

FIGURE 11.7 is a DEM of southeastern New York and southwestern Connecticut showing the irregular glacial topography at the southernmost extent of glaciers in the United States. Long Island is composed largely of till and outwash deposited during the advance and retreat of *two* continental ice sheets. This makes the area more complex than that in Figure 11.5, but landforms that formed at the two glacial termini are recognizable.

(a) Outline or otherwise identify the two terminal moraines, the southernmost outwash plain, and the outwash and glacial deltas deposited between the terminal moraines during a time of glacial retreat. Use different colors to separate features deposited during the two different glacial advances.

(b) Which of the two moraines is younger? Explain your reasoning.

_____

_____

(continued)

FIGURE 11.7 DEM of Long Island, New York, showing two terminal moraines.

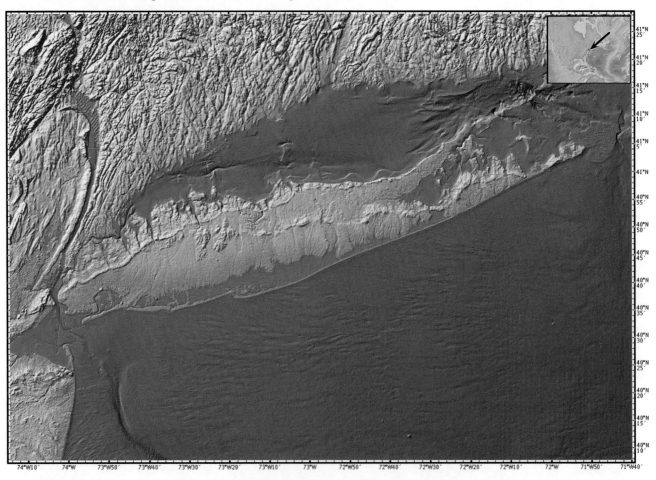

Name: _____   Section: _____
Course: _____   Date: _____

FIGURE 11.8 is a map of the Arnott moraine in Wisconsin, formed approximately 14,000 years ago. It shows the classic irregular "knob and kettle" topography of terminal moraines. Compare this morainal area with stream-produced landscapes as seen in the maps from Chapter 10.

(a) Are the stream networks and divides as well developed and clearly defined in this area?

_____

(b) Suggest an explanation for the absence of a well-integrated drainage system.

_____

_____

_____

(c) Suggest an explanation for the numerous swampy areas.

_____

_____

_____

(d) Compare the sizes and shapes of the hills (knobs) in this area with those in the fluvial landscapes.

_____

_____

(e) Suggest an origin for the several small lakes in the moraine. What will be their eventual fate?

_____

_____

_____

(f) There are several gravel pits in the area. Why might gravel be more easily and profitably mined from these specific areas than from others?

_____

_____

(g) What difficulties do you think might be involved in farming in ground moraine?

_____

_____

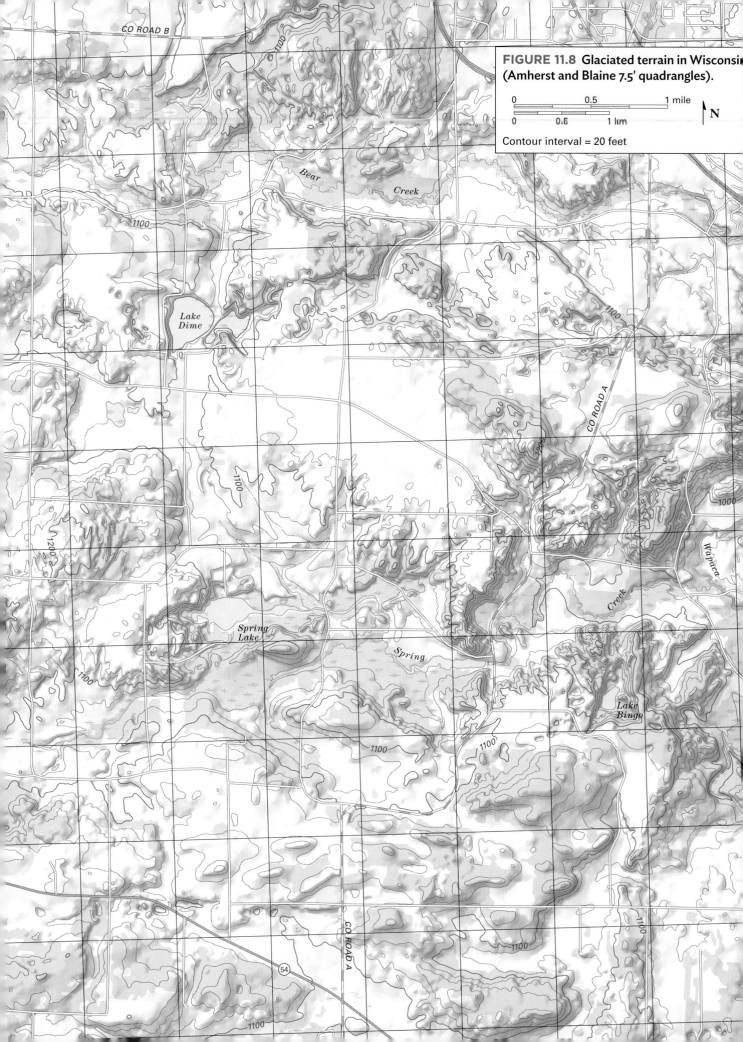

FIGURE 11.8 Glaciated terrain in Wisconsin (Amherst and Blaine 7.5′ quadrangles).

Contour interval = 20 feet

Name: _____

Course: _____

Section: _____

Date: _____

Not all ridges in glaciated areas are moraines (some may be eskers) and not all hills are knobs in a ground moraine (some may be drumlins). Eskers are quite long. **FIGURE 11.9** shows part of an esker that extends for more than 40 miles. You can't tell from a map or DEM that eskers are made of outwash rather than till, but they are much narrower than terminal moraines and lack the kettle holes associated with melting ice. The asymmetric shape of drumlins makes them easy to recognize.

(continued)

FIGURE 11.9 Part of the Passadumkeag, Maine, 7.5′ quadrangle showing the Enfield Horseback esker (the DEM inset shows the esker in its regional setting).

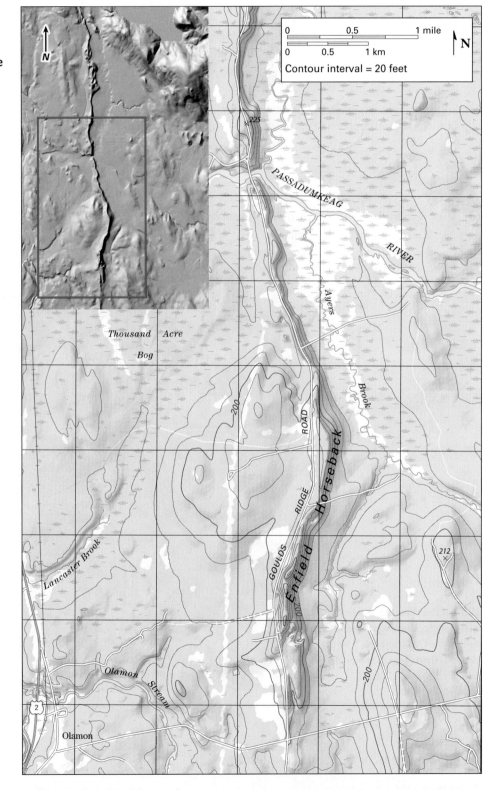

Name: _____   Section: _____

Course. _____   Date: _____

Examine Figures 11.7 and 11.8, and suggest other ways to differentiate eskers from moraines.

(a) Now examine the DEM (below) and the map of the Weedsport, New York area (**FIG. 11.10**) and describe the topography.

_____

_____

_____

_____

(b) Sketch the long axis of one of the hills on the DEM. What is its orientation? _____ Do the same for the others. What was the direction of glacial movement in this area? _____ Show this with a large arrow on the map.

(c) Sketch or construct profiles of the five hills numbered on Figure 11.10. Are the hills symmetrical or asymmetrical? Describe the distribution of their steep and gentle slopes.

_____

_____

_____

(d) The direction of ice flow indicated by a drumlin is *from* the _____ side of the drumlin *toward* the _____ side.

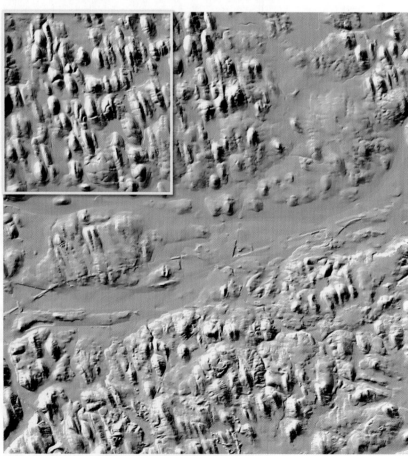

DEM of the Weedsport, New York 7.5′ quadrangle, showing the southern end of the Ontario/New York State drumlin field (area in rectangle shown in Fig. 11.10).

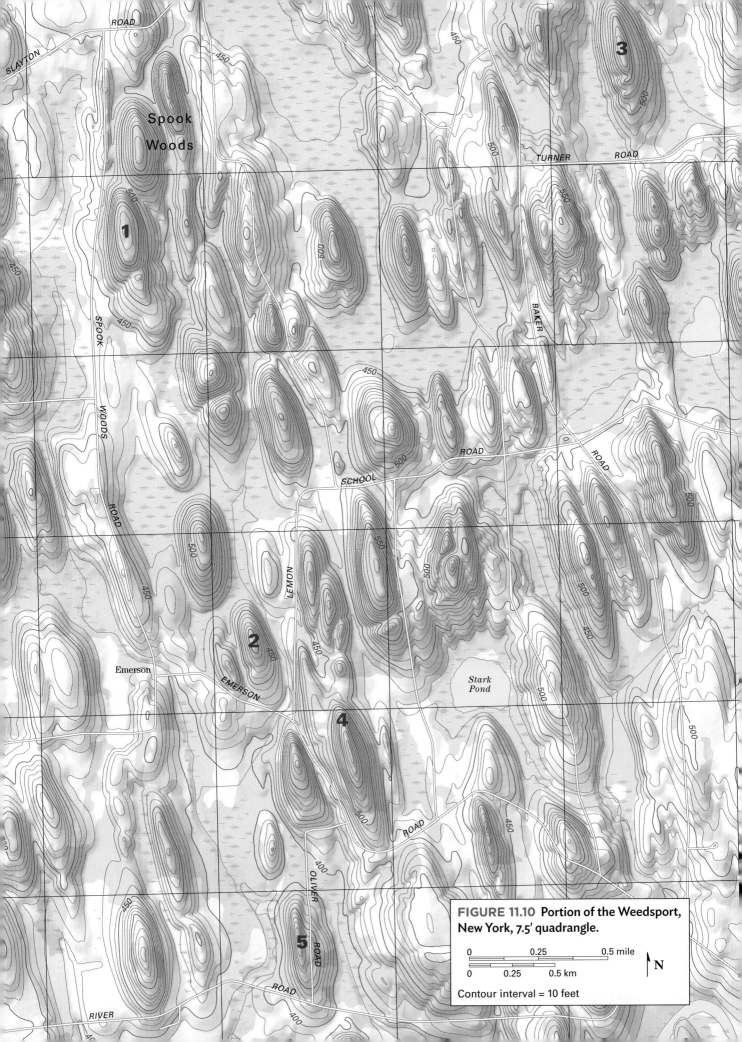

FIGURE 11.10 Portion of the Weedsport, New York, 7.5′ quadrangle.

Contour interval = 10 feet

11.3 Landscapes Produced by Mountain Glaciation

Alpine, Himalayan, Rocky Mountain, or Sierra Nevadan vistas with their jagged peaks, steep-walled valleys separated by knife-sharp divides, and spectacular waterfalls are the result of valley glacier erosion. Small glaciers form in depressions high up in the mountains and flow downhill, generally following existing stream valleys. The depressions are deepened and expanded headward by *plucking*, which eats into the mountainside, and the ice flows downhill, filling the former stream valleys and modifying them by abrasion (**FIG. 11.11**). When the ice melts, the expanded depressions become large, bowl-shaped amphitheaters (**cirques**); formerly rounded stream divides are replaced by knife-sharp ridges (**arêtes**); and the headward convergence of several cirques leaves a sharp, pyramidal peak (**horn peak**, or **horn**, named after the Matterhorn in the Alps).

Valley glaciers transform stream valleys in unmistakable ways, including their cross-sectional and longitudinal profiles, and the way that streams flow into one another. Figure 11.11c shows the characteristic U-shaped cross-sectional profile. Why are stream valleys V-shaped whereas glaciated valleys are U-shaped? Let's see if your ideas in Exercise 11.2 were right.

The answer is in how the two agents of erosion operate (**FIG. 11.12**). A stream actively erodes only a small part of its valley at any given time, cutting its channel downward or laterally (on the left in Fig. 11.12). Mass wasting gentles the valley walls, creating the V shape. In contrast, a valley glacier fills the entire valley, abrading the walls not only at the bottom, but everywhere it is in contact with the bedrock (center). When the ice melts, it leaves a U-shaped valley (right).

FIGURE 11.11 Erosional features formed by valley glaciers.

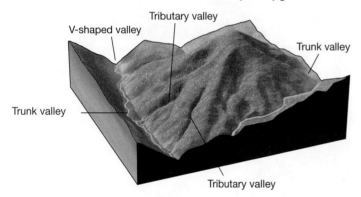

(a) Before valley glaciation, the area has a normal fluvial topography.

(b) The extent of the glaciers during valley glaciation.

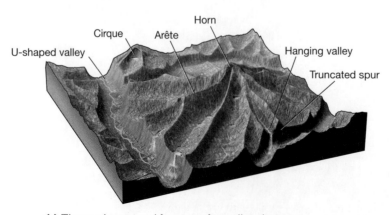

(c) The area's erosional features after valley glaciation.

FIGURE 11.12 Glacial modification of a fluvial valley.

Ice fills the entire valley, eroding everywhere, not just at the bottom like a stream. As it erodes, the valley is deepened and widened everywhere, resulting in the U shape on the right.

Most streams have smoothly concave **longitudinal profiles**. Valley glaciers, being solid, can overcome gravity locally and flow uphill for short distances. Their longitudinal profile is more irregular than a stream's, commonly containing a series of shallow scooped-out basins. Water can collect in some of these basins, producing a chain of lakes called **paternoster** (or **cat-step**) **lakes** (**FIG. 11.13**).

(a) Smooth, concave stream profile.

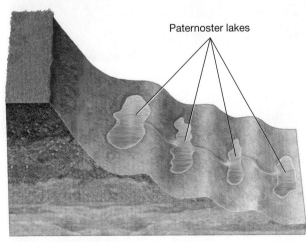

Paternoster lakes

(b) Irregular, scalloped glacial valley profile with paternoster lakes filling the basins.

FIGURE 11.13 Longitudinal profiles of stream and glacial valleys.

The base level for a tributary stream is the elevation of the stream, lake, or ocean into which it flows, and most tributary streams flow smoothly into the water body at their base level. In contrast, glaciers have no base level, so large glaciers can carve downward more rapidly than smaller ones. Previously existing main stream valleys, with large glaciers, can therefore be carved more deeply than their tributary valleys. When the ice melts, the result is a **hanging valley**, where the tributary stream hangs over the deeper main valley (**FIG. 11.14**). Today, Bridal Veil Creek falls almost 200 m to the floor of Yosemite Valley. The mouth of the creek and the floor of the valley were once at the same elevation, but the small valley glacier that filled Bridal Veil Creek could not keep pace with the main glacier that eroded downward more rapidly.

FIGURE 11.14 The hanging valley at Bridal Veil Falls, Yosemite National Park, California.

EXERCISE 11.6   **Recognizing Features of Valley Glaciation**

Name: _____   Section: _____
Course: _____   Date: _____

(a) The following DEM shows part of Glacier National Park and several valley glacier erosional features. Label examples of horn peaks, arêtes, cirques (with a C), and U-shaped valleys (with a U), and look also for flat places (uniform color) that could be filled with lakes (label with an L).

(continued)

DEM (false color) of a portion of Glacier National Park showing classic features of valley glaciation.

Lowest elevation   Highest elevation

(b) Look also at the chapter-opening photograph (on page 269) and identify as many features of valley glaciation as you can.

EXERCISE 11.7   Case Histories in Geologic Reasoning: A Glacial Dilemma

Name: _____   Section: _____
Course: _____   Date: _____

Mt. Katahdin is the highest peak in Maine, reaching almost a mile above sea level. At first glance, the glacial history of the area seems to be straightforward. A closer look suggests it is a bit more complicated. Examine the shaded relief map of the Mt. Katahdin area (**FIG. 11.15**).

(a) What glacial landform is represented by North Basin, South Basin, and Little North Basin? _____

(b) What glacial feature is represented by the Knife Edge, Hamlin Ridge, and Keep Ridge? _____

(c) What kind of glacial feature is Mt. Katahdin? _____

(d) What type of glacier produced these features? _____

The regional setting of Mt. Katahdin is shown in **FIGURE 11.16**. The area of northern New England, Quebec, and New Brunswick shown in this map was covered by continental ice sheets that pushed as far south as Long Island. The Passadumkeag esker in Figure 11.9 is located just a bit south of Mt. Katahdin, as shown by the green dot in Figure 11.16.

Suggest an explanation for the origin of the alpine glacial features on Mt. Katahdin and surrounding peaks in the midst of an area affected by continental glaciation. There is more than one possible answer.

_____

_____

_____

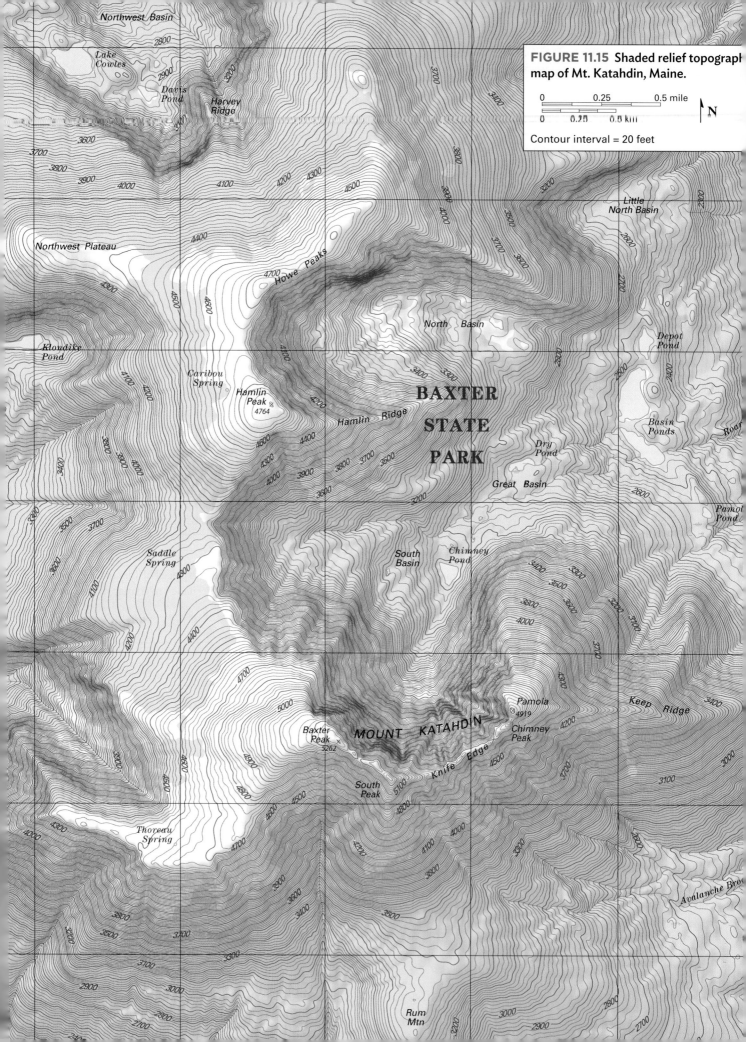

FIGURE 11.15 Shaded relief topograph[ic] map of Mt. Katahdin, Maine.

Contour interval = 20 feet

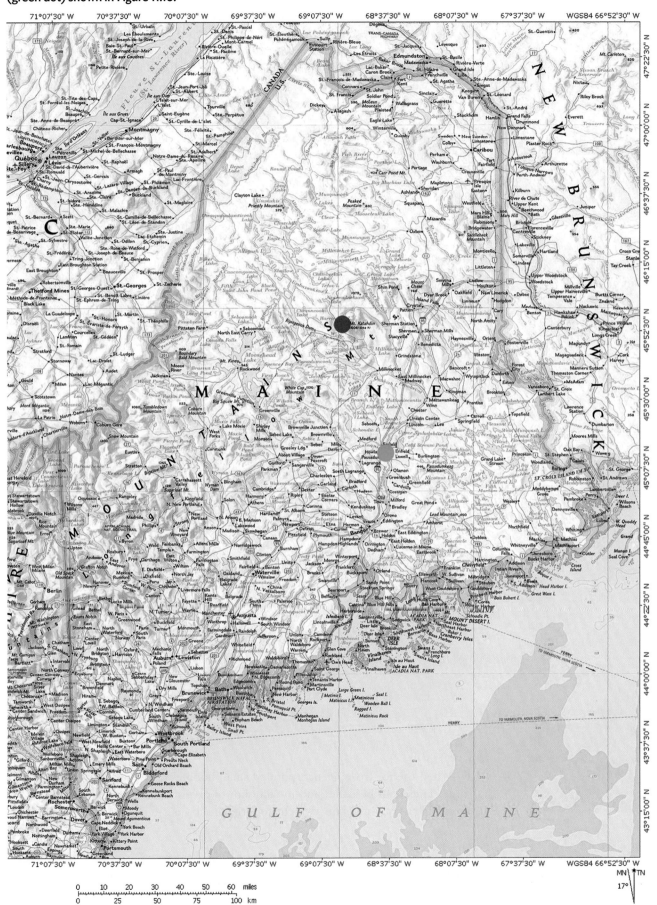

11.4 Glaciers and Climate Change

Intense weather events such as tornadoes, hurricanes, and typhoons, the rise of sea level, and persistent drought have brought the issue of global climate change from the field and laboratories of geologists and meteorologists to media headlines and the world political arena. Glacial landforms and sophisticated studies of polar ice provide the best record of atmospheric changes over the past several hundred thousand years, enabling us to base hypotheses about future changes on solid fact rather than guesswork.

Terminal moraines and outwash plains provide evidence that what movies call "the ice age" and geologists call the Pleistocene Epoch involved four intervals (glacials), in which continental glaciers advanced across the northern hemisphere, each followed by an "interglacial," a period in which the ice retreated or melted completely. Carbon dating of wood from glacial and interglacial sediments gives us a time frame for these events: the first glacial began around 700,000 years ago, and the last continental ice sheets in Europe and North America disappeared around 12,000 years ago.

Although the Pleistocene glaciers have disappeared from North America and Europe, ice sheets remain in Antarctica and Greenland, and they record temperature and greenhouse gas levels (CO_2, methane) over the hundreds of thousands of years that they have existed. Geologists have drilled into these ice sheets and collected ice cores (**FIG. 11.17**) for study in refrigerated laboratories.

FIGURE 11.17 An ice core showing seasonal layering and a dark layer of volcanic ash.

When the ice froze, it incorporated small bubbles of the air that existed at the time. These can be dated and the amount of greenhouse gases can be measured, providing a timeline against which modern conditions can be compared (**FIG. 11.18**).

FIGURE 11.18 Atmospheric temperature and CO_2 concentration from Vostok ice cores, Antarctica.

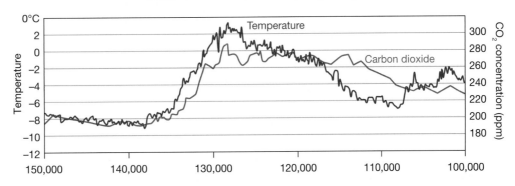

Vostok ice cores 150,000 to 100,000 years ago

EXERCISE 11.8   Retreat of the Athabasca Glacier, Alberta, Canada

Name: _____   Section: _____

Course: _____   Date: _____

We don't need sophisticated instruments to understand how glaciers respond to climate change. All we have to do is look. The Athabasca Glacier in Jasper National Park, Alberta, Canada, is one of the most popular tourist destinations in the province and *the* most visited glacier in North America. The glacier has been retreating since 1843, and the Alberta tourist bureau is concerned that if the retreat continues, it won't be long before there won't be a glacier to visit. The photographs below dramatically document the glacier's retreat — the black and white photograph from 1906 and the color photograph from 2006 are of approximately the same location. Notice how the glacier is now only in the far left-hand portion of the image.

The map given locates the changes in the end of the glacier — what is known as the glacier's terminus — from 1843 to 1999. You have been asked to advise the tourist bureau and National Parks Canada on how much longer the glacier will be a tourist attraction.

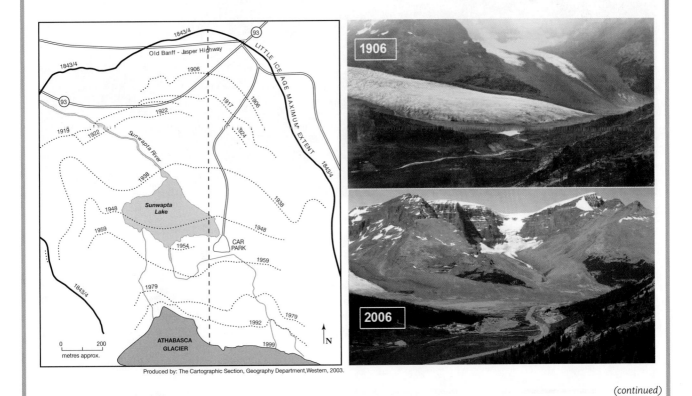

Produced by: The Cartographic Section, Geography Department, Western, 2003.

(continued)

Name: _____   Section: _____
Course: _____   Date: _____

(a) You will note on the map that the glacier's terminus does not follow a straight line but flows around the contours of the landscape. But, to obtain an estimate, you decide to base your estimate along a straight-line (the red, dashed line that intersects the glacial termini for different years).

Measure the amount of retreat from 1843 to 1999 along the dashed line. It is _____ m.
Determine the average rate of the retreat during that period (_____ m/156 years = _____ m/yr).

You note that the rate has not always been consistent. In some time periods it appears to retreat more quickly than others. You decide to compute the rates during different shorter time periods.

- Determine the rate of retreat along the dashed line from 1906 to 1999. _____ m/yr
- Determine the rate of retreat along the dashed line from 1843 to 1906. _____ m/yr

(b) You have computed three rates of retreat over three different time periods. The following picture shows a recent view of this glacier. It is now approximately 6.2 km from its terminus to its top, or head.

First, determine estimates for the potential life of this glacier using the three rates you determined in part (a). Divide 6.2 km by each rate (remember to be careful to *use the same units*) to give you three different time periods. These are

1 _____ years   2 _____ years   3 _____ years

Head of
Athabasca Glacier   2010 terminus   1938 terminus

? **What Do You Think** Using the information that you have, what are your recommendations to the Alberta tourist bureau? Specifically, how long can they expect the glacier to last and why? What might happen that could cause the rate of retreat either to accelerate or to slow? To the best of your knowledge, propose reasons for both (on a separate sheet of paper).

TABLE 11.1 Glossary of glacial landforms.

Erosional Landforms		
Landform	**Type of glacier**	**Description**
Arête	Valley	Sharp ridge separating cirques or U-shaped valleys
Cirque	Valley	Bowl-shaped depression on mountainside; site of snowfield that was source of valley glacier
Hanging valley	Valley	Valley with mouth high above main stream valley; formed by tributary glacier that could not erode down as deeply as larger main glacier
Horn peak	Valley	Steep, pyramid-shaped peak; erosional remnant formed by several cirques
Sculptured bedrock	Continental	Asymmetric bedrock with *steep* side facing the direction toward which the ice flowed; also called roche moutonnée
U-shaped valley	Valley	Steep-sided valley eroded by a glacier that once filled the valley
Depositional Landforms		
Landform	**Type of glacier**	**Description**
Drumlin	Continental	Streamlined (elliptical), asymmetric hill composed of till; the long axis parallels glacial direction, the gentle side faces the direction toward which the ice flowed
Esker	Continental	Narrow, sinuous ridge made of outwash deposited by a subglacial stream in a tunnel at the base of a glacier
Ground moraine	Valley and continental	Sheet of till deposited by a retreating glacier
Kettle hole	Valley and continental	Depression left when a block of ice that had been isolated from a melting glacier is covered with sediment and melts
Lateral moraine	Valley	Till deposited along the walls of a valley when a valley glacier retreats
Outwash plain	Valley and continental	Outwash deposited by meltwater beyond the terminus of a glacier
Proglacial delta	Continental	Outwash deposited into a proglacial lake
Terminal moraine	Valley and continental	Wall of till deposited when the terminus of a glacier remains in one place for a long time
Lakes Associated with Glacial Landscapes		
Landform	**Type of glacier**	**Description**
Ice-marginal lake	Valley and continental	Lake formed by disruption of local drainage by a glacier
Kettle lake	Valley and continental	Lake filling a kettle hole
Paternoster lake	Valley	Series of lakes filling basins in a glaciated valley
Tarn	Valley	Lake occupying part of a cirque

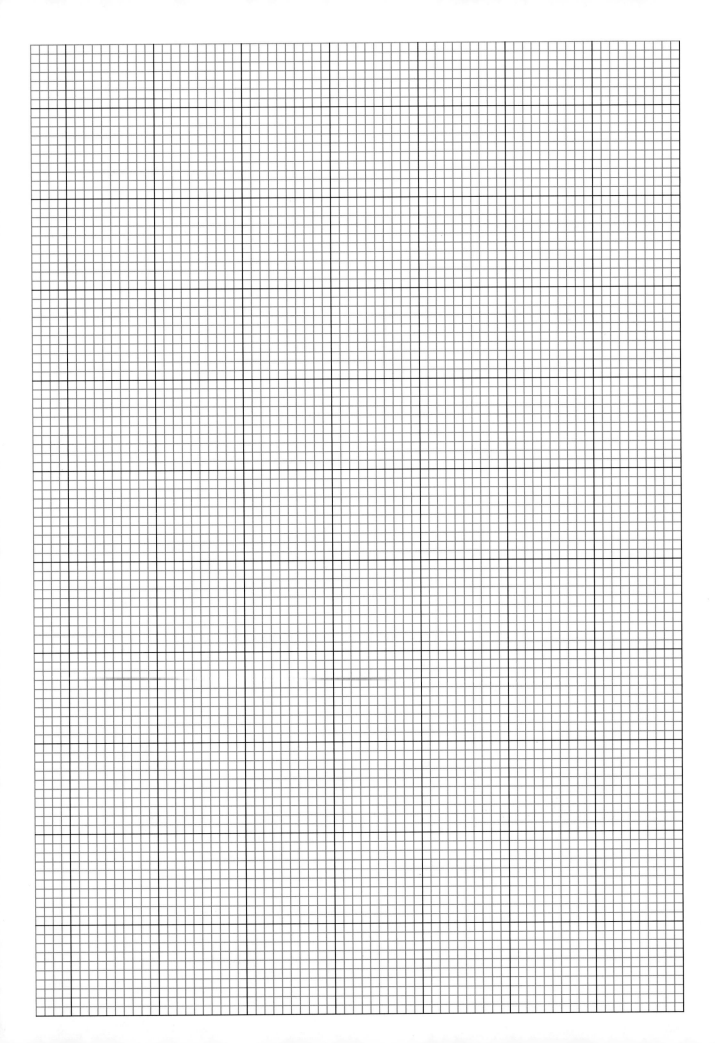

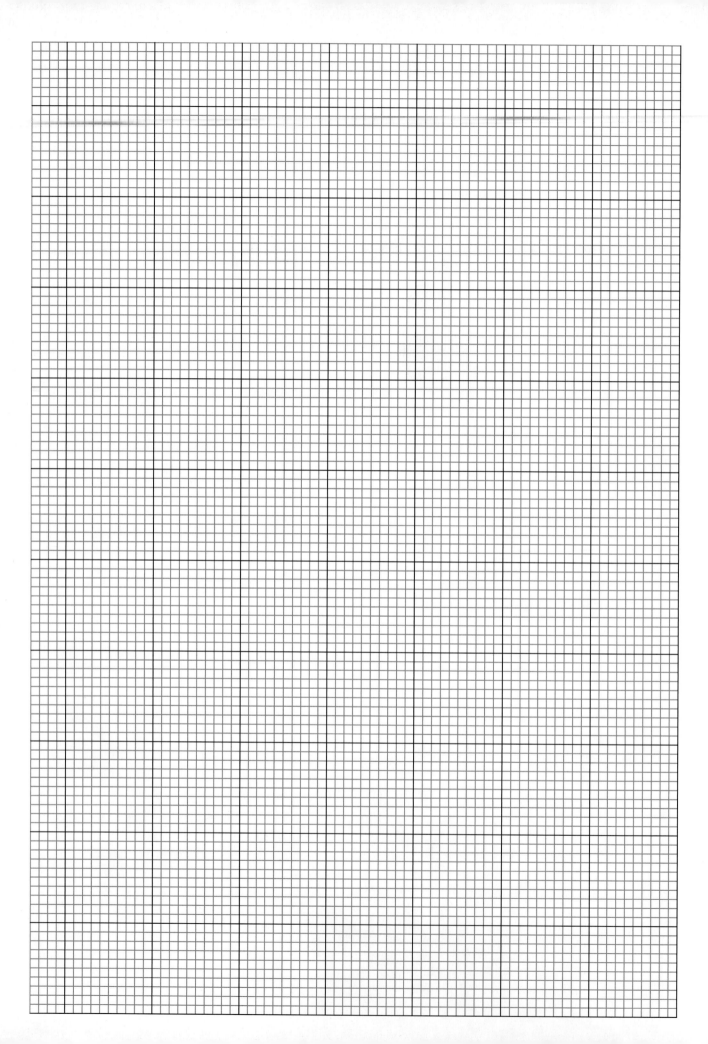

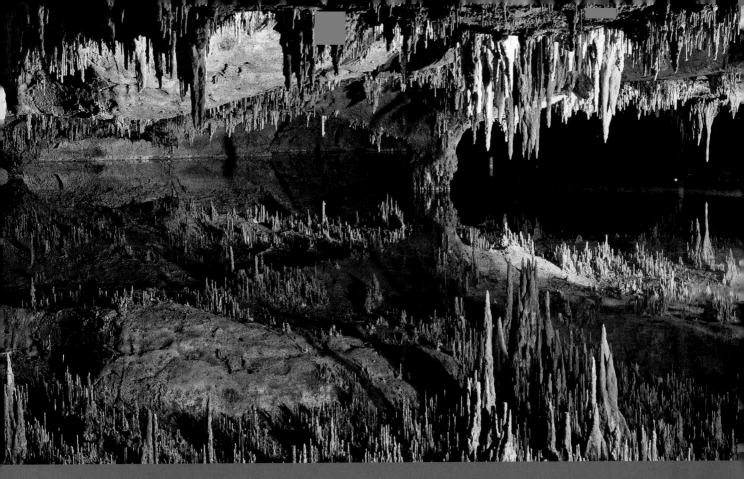

12

Groundwater as a Landscape Former and Resource

- Understand how
  groundwater infiltrates and
  flows through Earth materials
- Explain why groundwater
  erodes and deposits
  differently from streams and
  glaciers
- Learn to recognize
  landscapes formed by
  groundwater and to interpret
  groundwater flow direction
  from the topographic features
- Understand how geologists
  carry out groundwater
  resource and pollution
  studies

MATERIALS
NEEDED

- Specimens of four different
  materials for the infiltration
  exercise
- Tracing paper and colored
  pencils

12.1 Introduction

Some rain and snow that falls on the land runs off into streams, some evaporates into the air, and some is absorbed by plants. The remainder sinks into the ground and is called **groundwater**. Water moves much more slowly underground than in streams because it must drip from one pore space to another rather than flow in a channel. Groundwater therefore has much less kinetic energy than stream water and cannot carry the particles with which streams scrape and wear away at bedrock. Instead, groundwater erodes chemically, by dissolving soluble rocks such as limestone, dolostone, and marble, which are largely made of carbonate minerals like calcite and dolomite.

If groundwater erodes underground, how can it form landscapes at the surface? As groundwater erodes rocks from below, it undermines their support of the surface above, and the land may collapse to produce very distinctive landscapes, often pockmarked with cavities called sinkholes (**FIG. 12.1a**). Landscapes in areas of extreme groundwater erosion are among the most striking in the world, with narrow, steep-sided towers unlike anything produced by streams or glaciers (**FIG. 12.1b**). Groundwater-eroded landscape is called **karst topography,** after the area in Slovenia and northeastern Italy where geologists first studied it in detail, but spectacular examples are also found in Indiana, Kentucky, Florida, southern China, Puerto Rico, and Jamaica.

Groundwater is also a vital resource, used throughout the world for drinking and washing and for irrigating crops. Strict rules govern its use in many places because it flows so slowly that renewal of the groundwater supply takes a long time. In addition, its slow flow and underground location make it difficult to purify once it has been polluted. We examine both roles of groundwater in this chapter: as landscape former and as resource. Exercise 12.1 starts by introducing some factors that affect groundwater flow.

FIGURE 12.1 Landscapes carved by groundwater.

(a) Karst topography characterized by sinkholes and underground caves in southern Indiana.

(b) Karst towers in the Guilin area of southern China.

12.2 Aquifers and Aquitards

Materials that transmit water readily are called **aquifers** (from the Latin, meaning "to carry water"), and those that prevent water from infiltrating are called **aquitards** or **aquicludes** (they re*tard* or ex*clude* water). Groundwater flows through pores and fractures in bedrock as well as through unconsolidated sediment. Some rocks, like a poorly cemented sandstone, contain pores when the rock forms (**primary porosity**), but others that have no pores at all when they form (e.g., granite, obsidian) become porous and permeable when broken by faulting or fracturing (**secondary porosity**).

EXERCISE 12.1 Factors Affecting Infiltration and Groundwater Flow

Name: ______________________ **Section:** __________
Course: ______________________ **Date:** __________

Let's look first at factors that control the flow of water underground. Your instructor will provide four containers screened at the bottom and filled with materials that have different grain shapes and sizes like in the figure below.

(a) If equal amounts of water were poured into each of the four containers, which material do you think would permit the greatest amount of water to pass through to the graduated cylinder? Rank your predictions here:

Greatest amount *Least amount*

____

____

In a few sentences, explain your reasoning for these choices.

____

____

____

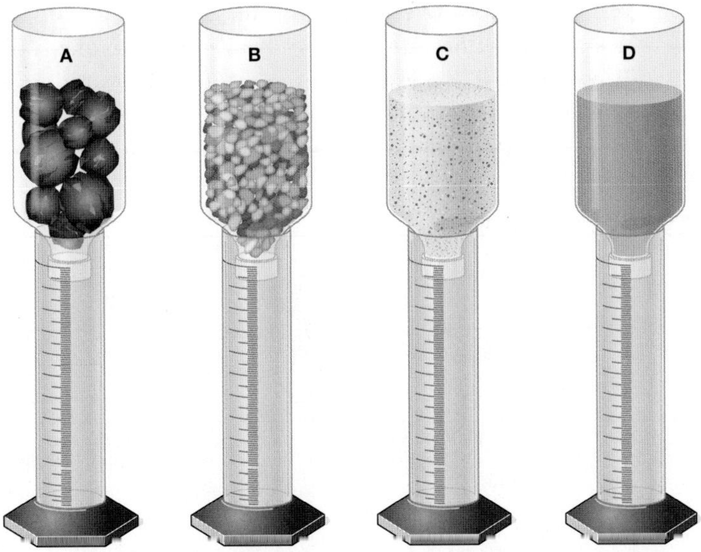

(b) Which material will transmit water fastest? Slowest? Explain.

Fastest water transmission *Slowest water transmission*

____

____

In a few sentences, explain your reasoning for these choices.

____

____

____

(continued)

Name: _____   Section: _____
Course: _____   Date: _____

In the top section of the table, describe the indicated properties of the contents in each container.

Now let's see how good your predictions are. Pour equal amounts of water into the four bottles and measure the amount of water that passes through and its rate of flow. Record your observations in the lower parts of table below.

	Observation	Container A	Container B	Container C	Container D
Properties of material that might affect water flow	Grain size				
	Sorting				
	Grain shape				
	Porosity				
	Permeability				
Volume of water transmitted (ml)	30 seconds				
	60 seconds				
	90 seconds				
	120 seconds				
	150 seconds				
	180 seconds				
	210 seconds				
	240 seconds				
Other observations	Water color				
	Amount of water retained				

(continued)

EXERCISE 12.1 Factors Affecting Infiltration and Groundwater Flow (continued)

Name: ______________________ **Section:** ____________
Course: ______________________ **Date:** ____________

(c) Which materials transmitted the most water? Which retained the most water?

____

____

(d) What properties of the materials are correlated with good water transmission? Which are correlated with water retardation and retention?

____

____

____

Being porous is not necessarily enough for a material to be a good aquifer. Pumice and scoria are very porous, but their pores are not connected. Pore spaces must be connected for water to move from one to another—a property called **permeability**. The materials in Exercise 12.1 that transmitted water easily had to be both porous (there was room for the water between grains) *and* permeable (the water could move through the material). *An aquifer must be both porous and permeable.* Exercise 12.2 explores these concepts.

EXERCISE 12.2 The Difference between Porosity and Permeability

Name: ______________________ **Section:** ____________
Course: ______________________ **Date:** ____________

Are all porous rocks aquifers? Hold pieces of highly porous pumice and scoria above two beakers or rest them on the rims as shown in the following figure. Using a water dropper, slowly add water to the top of the rock and carefully observe how much water passes into the beaker.

Porosity and permeability in pumice and scoria.

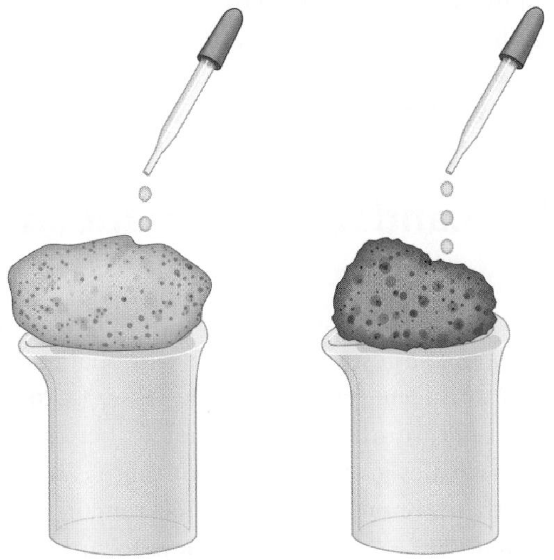

(continued)

Name: _____   Section: _____

Course: _____   Date: _____

(a) Are pumice and scoria porous? Permeable? Explain.

_____

_____

_____

(b) Is a material that is similar to scoria more likely to function as an aquifer or an aquitard? Explain your answer.

_____

_____

_____

(c) Is a material that is similar to pumice more likely to function as an aquifer or an aquitard? Explain your answer.

_____

_____

_____

_____

Groundwater's slow pore-to-pore movement helps chemical erosion because the longer water sits in contact with minerals, the more it can dissolve them. Similarly, even a few drops of water placed on a sugar cube will eventually dissolve their way through the cube. Unfortunately, chemical erosion can cause dangerous health problems if humans or animals drink water in which poisonous material is dissolved. For example, when groundwater wells were drilled in Bangladesh to provide safer drinking water than that from streams contaminated with bacteria, the groundwater turned out to contain high concentrations of arsenic. An international effort is under way to identify the source of the arsenic and devise methods for removing it from the groundwater.

12.3 Landscapes Produced by Groundwater

Groundwater is the only agent of erosion that creates landforms from below the surface of the Earth, and all karst topographic features are destructional. Caves and caverns, the largest groundwater erosional features, are underground, unseen, and often unsuspected at the surface (**FIG. 12.2a**). In some areas, extensive cave networks result from widespread chemical erosion. **Sinkholes** form when these underground cavities grow so large that there is not enough rock to support the ground above them. **Karst towers** are what remain when the rock around them has been dissolved (**FIG. 12.2b**). **Karst valleys** form when several sinkholes develop along an elongate fracture (**FIG. 12.2c**).

Unlike streams and glaciers, deposition by groundwater does not produce landforms. Groundwater deposition occurs in caverns when drops of water evaporate, leaving behind a tiny residue of calcite—sometimes on the roof of the cavern, sometimes on the floor. These build up slowly over time to produce stalactites and stalagmites, respectively (**FIG. 12.3**). In some instances, a stalactite and stalagmite may grow together to form a column. In Exercise 12.3, you will explore karst topography.

FIGURE 12.2 Groundwater landscape erosion from beneath the Earth's surface.

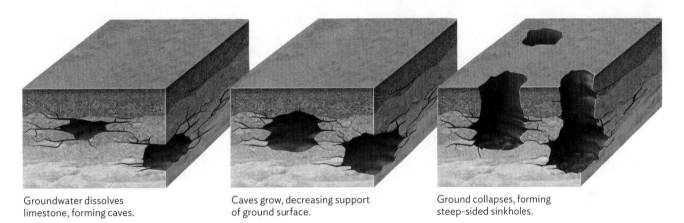

Groundwater dissolves limestone, forming caves.

Caves grow, decreasing support of ground surface.

Ground collapses, forming steep-sided sinkholes.

(a) Origin of sinkholes by solution and collapse.

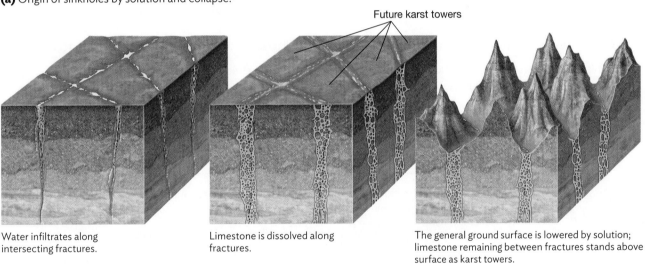

Future karst towers

Water infiltrates along intersecting fractures.

Limestone is dissolved along fractures.

The general ground surface is lowered by solution; limestone remaining between fractures stands above surface as karst towers.

(b) Origin of karst towers as erosional remnants.

(c) Origin of karst valleys along fractures.

FIGURE 12.3 Groundwater deposition in caverns.

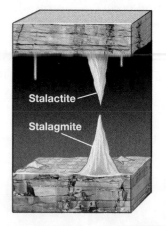

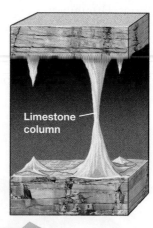

Time

Stalactites grow *downward* from the roof; stalagmites grow *upward* from the floor. Columns form when stalactites and stalagmites merge.

EXERCISE 12.3 **Karst Topography**

Name: _____   Section: _____
Course: _____   Date: _____

This chapter examines two classic karst areas in the United States to help you become familiar with groundwater erosional landscapes—one in Kentucky (**FIG. 12.4**) and one in Florida (**FIG. 12.7** on page 307). In this exercise, we will focus on the location in Kentucky.

Compare the map of the Mammoth Cave area (Fig. 12.4) with the diagrams showing how karst features develop (Fig. 12.2).

(a) This area receives a moderate amount of rainfall, yet there are very few streams. Suggest an explanation for this phenomenon. (This is a characteristic of nearly all karst regions.)

_____

_____

_____

(b) Look for changes in topographic grain (texture) across the map area. Draw lines separating places with different topographic textures.

(c) Describe the topography in the southernmost quarter of the area. What karst features are present?

_____

_____

_____

_____

(d) Some of the circular depressions in this part of the map are partially filled with water. Suggest an origin for these lakes (we will return to this point later).

_____

_____

_____

(continued)

FIGURE 12.4 Karst topography near Mammoth Cave in Kentucky.

Contour interval = 10 feet

Name: _____   Section: _____

Course: _____   Date: _____

(e) Describe the topography in the northern three-quarters of the map. What karst feature is represented by The Knobs? What feature is represented by the Woolsey Valley?

_____

_____

_____

_____

_____

_____

(f) Karst features are found throughout the entire map but have a different appearance in the southern quarter compared to the north. Describe the differences and suggest a possible explanation.

_____

_____

_____

_____

_____

_____

(g) Some of the sinkholes contain lakes, but others are dry. Suggest an explanation for the difference. *Hint:* Determine the elevation of lakes in the sinkholes and write it in each lake. Compare those to the elevations of the bottoms of the sinkholes that do not have lakes.

_____

_____

_____

_____

_____

12.4 The Water Table

Look at Figure 12.7, a map showing a karst area in northern Florida in which sinkholes are filled with water. The level of water in the lakes in Figure 12.7 is controlled by a feature called the **water table**. Gravity pulls groundwater downward until it reaches an aquitard that it cannot penetrate, and the water begins to fill pores in the aquifer just above the aquitard. More water percolating downward finds pores at the bottom of the aquifer already filled, and so it saturates those even higher. The water table is the boundary between the **zone of saturation** (where all pore spaces are filled with water) and the **zone of aeration,** where some pores are partly filled with air (**FIG. 12.5**). You can model the water table by adding water to a beaker filled with sand and watching the level of saturation change.

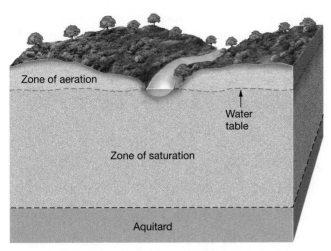

A stream or lake forms where the water table intersects the land surface.

12.5 Groundwater Resources and Environmental Problems

Groundwater is vital for human existence, both for drinking and to irrigate crops. We drill wells to depths below the water table and use pumps to extract the water from an aquifer. But groundwater flows so slowly that it can rarely replenish itself as rapidly as our needs dictate. Indeed, some water that we drink today entered its aquifer system hundreds of years ago. If we take too much groundwater, our wells run dry—a serious problem for many farms between the Rocky Mountains and the Mississippi River that depend on the same aquifer (the Dakota sandstone aquifer) for their water.

When we pump water from an aquifer, it comes from the saturated pore spaces below the water table, and water must flow downward to replace it from other parts of the aquifer (**FIG. 12.6**). This lowers the water table, especially in the area around the well. Each well creates a **cone of depression** in the water table, as shown in Figure 12.6. You model this phenomenon every time you order a thick milk shake from your local fast-food restaurant. Next time, take the top off the shake and insert the straw just below the surface of the shake. Sip slowly and look at what happens to the surface. Because the shake flows sluggishly (not unlike groundwater), the area

FIGURE 12.6 Formation of a cone of depression by removal of water from an aquifer.

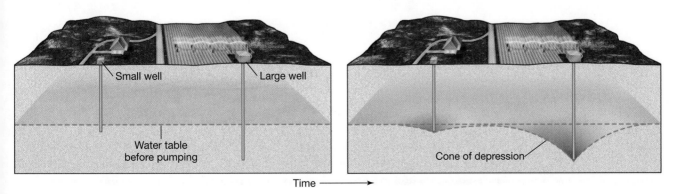

below the straw is drawn down more than the area near the edges of the cup. You've just made a small cone of depression.

Two types of problems are associated with groundwater resources: issues of *quantity* (whether there is enough groundwater to meet our needs) and *quality* (whether the groundwater is pure enough for us to drink or whether it has become contaminated through natural or human causes). We will explore these issues with two case histories, both based on real-world situations, in Exercises 12.6 and 12.7. But first, let's learn more about how to find the water table in Exercise 12.4 and how to solve a geologic puzzle in Exercise 12.5.

EXERCISE 12.4   Locating the Water Table

Name: _____   Section: _____

Course: _____   Date: _____

Look at the map of the Interlachen area in Florida (Fig. 12.7), a karst landscape underlain by highly soluble limestone. In addition to showing another karst region, this map illustrates other important concepts about groundwater—its value as a resource and environmental and other problems associated with living in karst topography.

The water table generally mimics the surface topography in a slightly subdued shape (Fig. 12.5). Groundwater flows downhill from higher parts of the water table to lower parts.

(a) Construct a profile of the Interlachen area from Point A to Point B in Figure 12.7. From your data of lake level elevations, sketch the water table on this profile.

(continued)

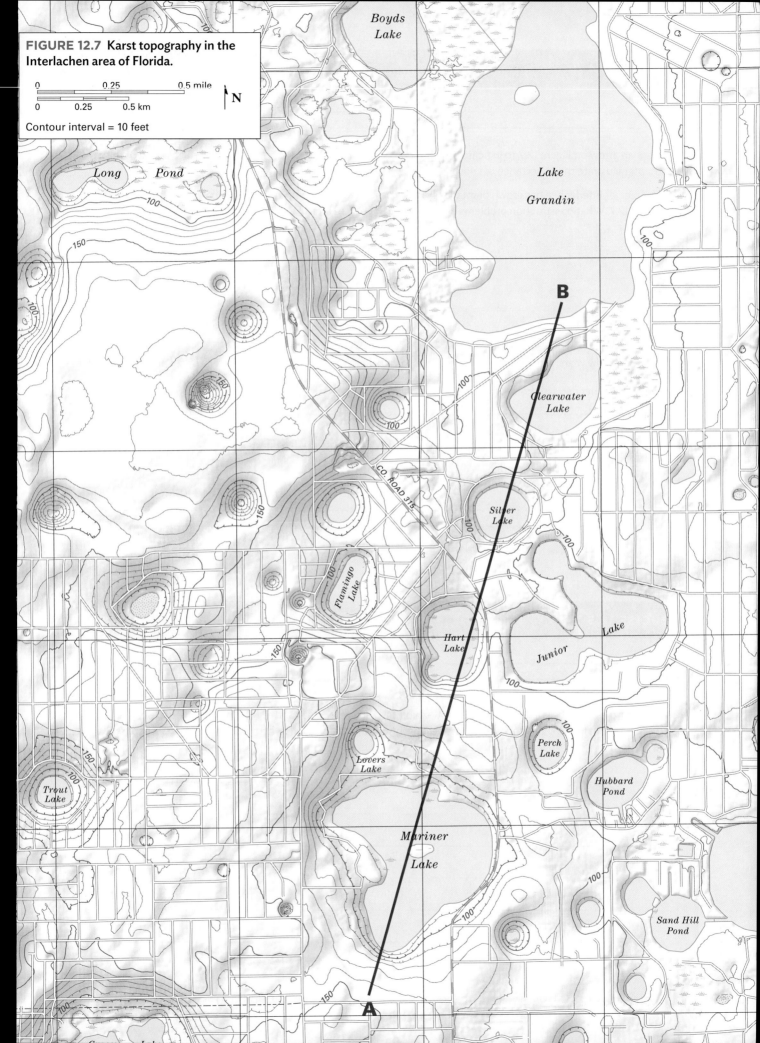

FIGURE 12.7 Karst topography in the Interlachen area of Florida.

0   0.25   0.5 mile
0   0.25   0.5 km

N

Contour interval = 10 feet

Boyds Lake

Lake Grandin

Long Pond

B

Clearwater Lake

Silver Lake

CO. ROAD 315

Flamingo Lake

Hart Lake

Junior Lake

Perch Lake

Lovers Lake

Hubbard Pond

Trout Lake

Mariner Lake

Sand Hill Pond

A

Name: _____   Section: _____
Course: _____   Date: _____

(b) Draw an arrow on Figure 12.7 to indicate the direction of groundwater flow. The case histories at the end of this chapter illustrate the importance of knowing which way groundwater is flowing.

(c) The area of Interlachen has grown dramatically in the past 20 years, and part of its suburban street grid can be seen in Figure 12.7. What construction problems do developers face in an area like this?

_____

_____

_____

_____

(d) In the past few years, some lakes in this area have shrunk dramatically or dried up entirely. Suggest possible reasons for this change.

_____

_____

_____

_____

Name: _____   Section: _____
Course: _____   Date: _____

Carlsbad Caverns National Park in New Mexico (**FIG. 12.8**) contains spectacular caverns with stalactites, stalagmites, and columns, yet the area where it is located does not exhibit the karst features found at Mammoth Cave and southern Florida. Suggest an explanation. What conditions might be causing caverns to form below ground without a karst topography above ground?

_____

_____

_____

_____

_____

_____

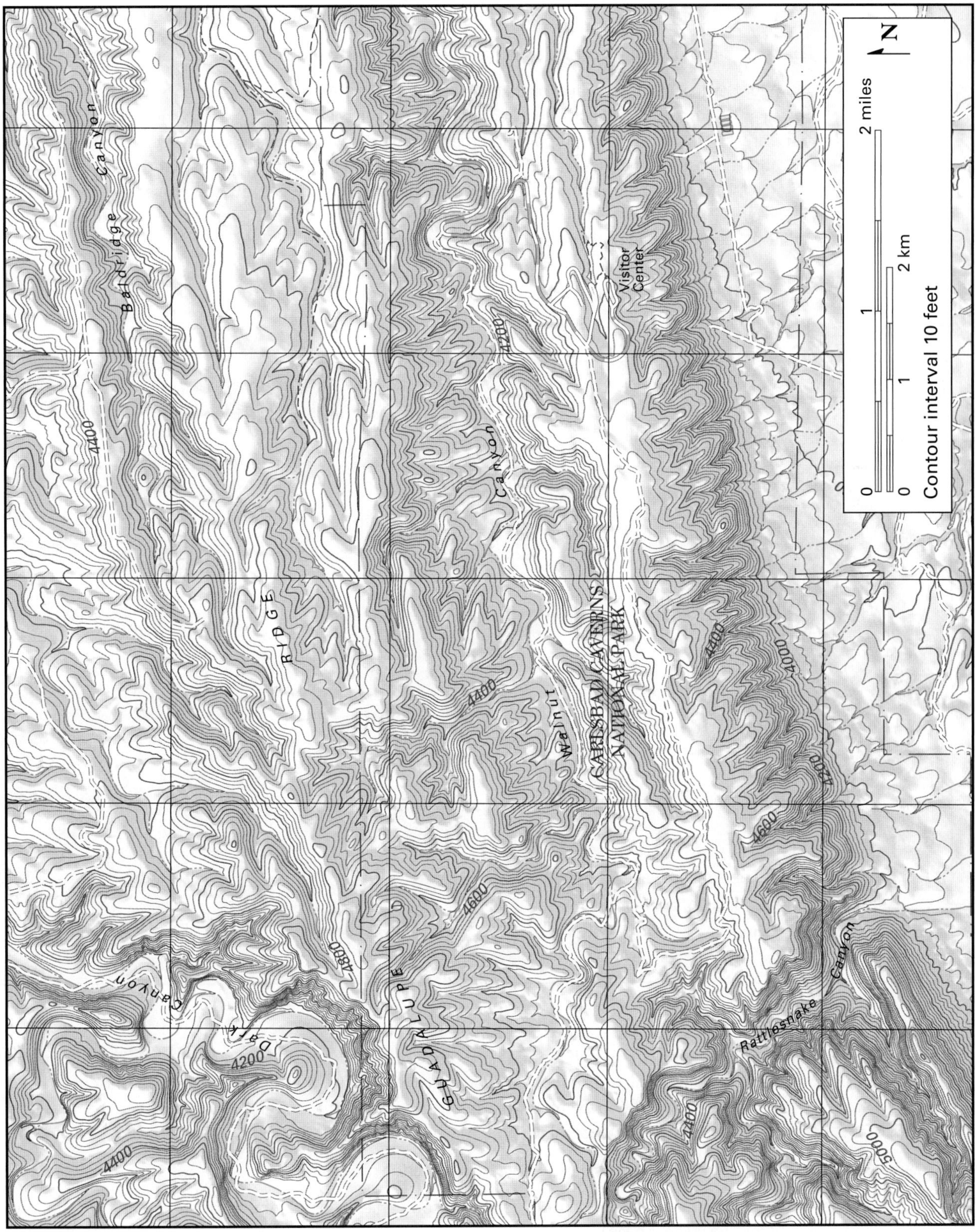

FIGURE 12.8 Topography of the Carlsbad Caverns area of New Mexico.

Name: _____ Section: _____

Course: _____ Date: _____

This exercise presents a real-life problem, although details have been changed to protect the privacy of the parties involved.

The figure below is a map showing an area in which farmers have relied on groundwater for generations for drinking and to irrigate their crops. In an attempt to diversify the local economy, the town selectmen offered tax incentives to bring two new industries into the area: Grandma's Soup Company and the Queens Sand and Gravel Quarry. Trouble arose when Farmer Jones (see location on map) found that his well had run dry even though it was the rainy season. Having taken Geology 101 at the State University several years ago, he suspected that pumping by the new companies had lowered the water table. He hired a lawyer and sued Grandma's Soup Company to prevent it from pumping "his" water. Grandma's lawyers responded that the company had nothing to do with his problem.

Depth to water table *before* commercial pumping.

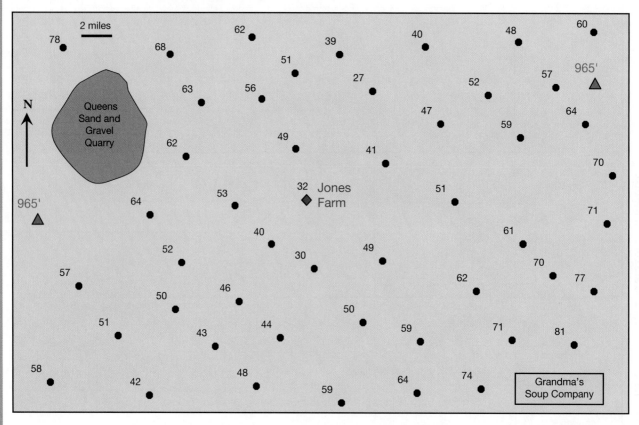

The judge, who has never taken a geology class, has appointed *you* to resolve the problem. Your tasks are (1) find out why Farmer Jones has lost his water, and (2) figure out how to keep everyone happy. After all, Jones and his friends have been benefiting from the fact that Grandma's Soup Company has been paying an enormous amount of taxes. Your assistants have already gathered a lot of useful data.

- A survey shows that the land is as flat as a pancake, with an elevation of 965 feet.

- Records show the depth to the water table in wells *before* the companies opened (above) and *after* the two companies had been pumping for a year (next page).

(continued)

Name: _____   Section: _____
Course: _____   Date: _____

Depth to water table *after* **commercial pumping.**

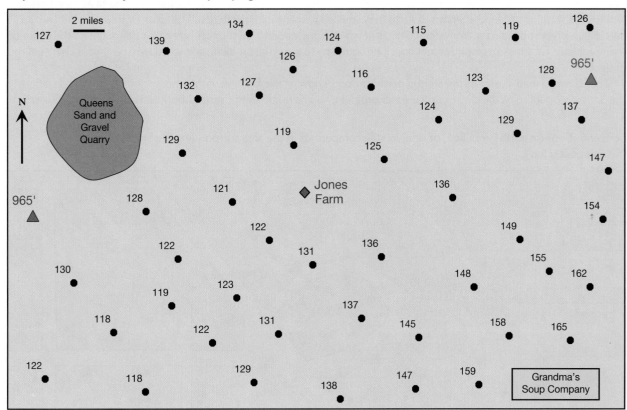

**What Do You Think** Based on the data in the two maps, answer the following questions on a separate sheet of paper:

- Why has Farmer Jones's well run dry?
- What steps do you recommend could be taken to keep the Jones Farm and the new industries in business?

Hint 1: Use your contouring skills to make a topographic map of the water table before and after the quarry and soup company began pumping.

Hint 2: Consider what the quarry and soup company do with the water they pump.

Name: _____   Section: _____
Course: _____   Date: _____

Groundwater *quality* is becoming a critical issue as growing populations place stress on the environment. Commercial pollution or leakage from residential septic tanks may make water unsuitable for drinking. The following case history happens all too commonly.

(continued)

Name: _____   Section: _____
Course: _____   Date: _____

The Smiths bought a home in a suburban development where the water supply was a shallow groundwater aquifer. A few months ago, they noticed a strange smell when they took showers, and the drinking water has begun to taste foul. Last week, while washing dishes, the Smiths thought they smelled gasoline in their kitchen. The local TV news channel ran an investigative report on leakage from storage tanks at gas stations, and the Smiths suspected that this might be the cause of their problem. They hired a groundwater consulting company to solve the problem, and you have been put in charge of the project.

The following map illustrates the Smiths' neighborhood, showing the location of their home and two local gas stations the Smiths suspect to be the cause of their water problem. You have surveyed the neighborhood using state-of-the-art

Concentrations (in parts per billion) of semivolatile compounds in the area surrounding the Smith home (nd = not detected).

🔲 Goodmile Gas Co.

⛽ GoFast Gas Co.

|← 1 mile →|

▲ Smith home

(continued)

Name: _____   Section: _____
Course: _____   Date: _____

detection instruments to measure the amount of volatile compounds in the soil around the Smiths' home. These chemicals are released into the ground when corroded gas tanks leak and are indicators of soil and groundwater contamination. Concentrations greater than 50 parts per billion (ppb) are considered dangerous. The data are shown in red numbers on the map.

Hint: Make a contour map of the soil contamination level using a 10-ppm contour interval. This is not a map showing elevation, but the rules are very much the same. Color-code the map to show levels of potential danger: red for values above the danger threshold (50 ppb); a second color for probable future threat (between 30 and 50 ppb); and a third color for possible future threat (between 10 and 30 ppb).

(a) Does either gas station have a leakage problem? Explain your conclusion.

_____

_____

(b) In what direction does the local groundwater flow? How do you know?

_____

_____

(c) Which homes will be the next to feel the effects of the gasoline leakage?

_____

_____

(d) What additional problems has your research discovered?

_____

_____

_____

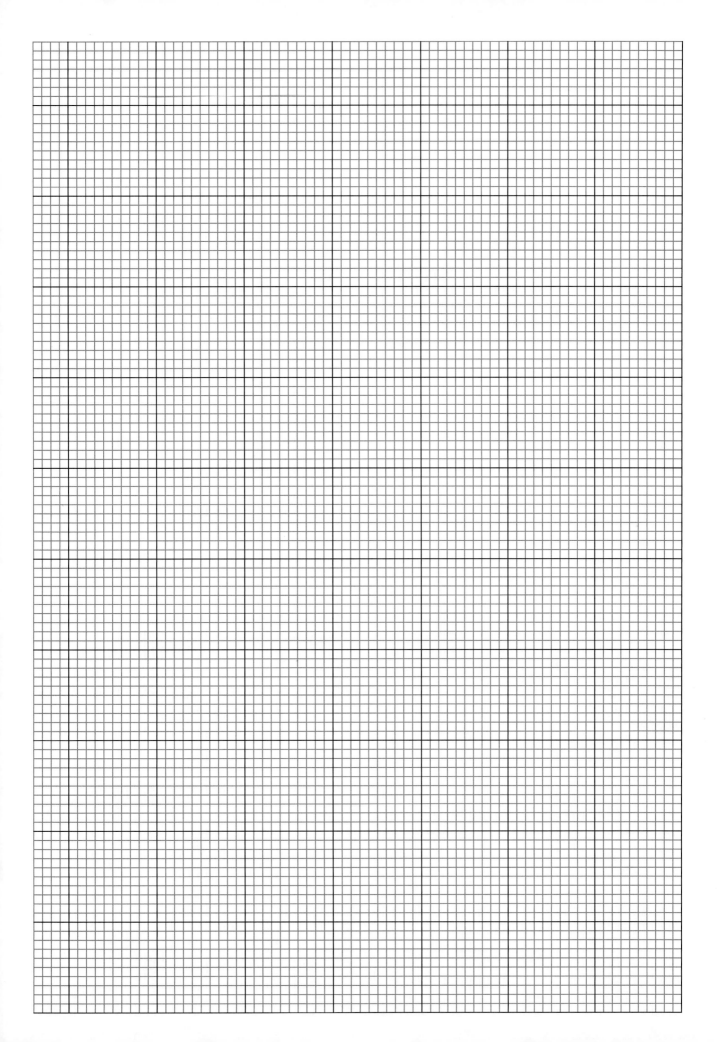

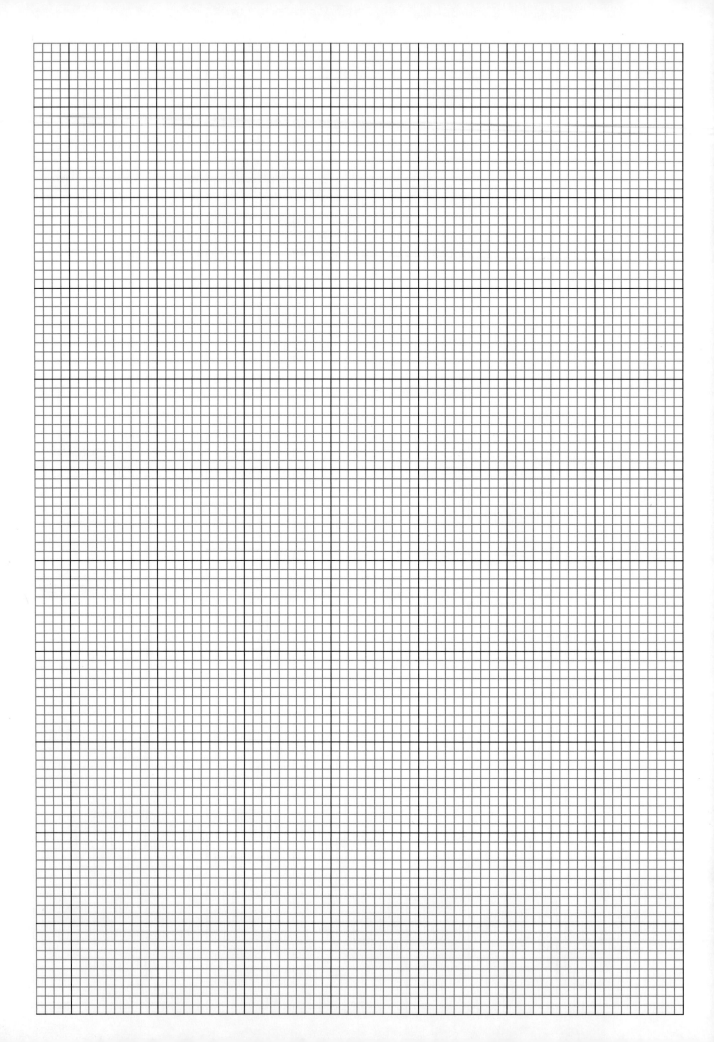

13

Processes and Landforms in Arid Environments

Buttes are common landforms in arid environments. The buttes pictured in this photo are from Monument Valley along the Arizona-Utah state line.

13.1 Introduction

Arid regions are defined as those that receive less than 25 cm (10 in.) of rain per year, which provides very little water for stream and groundwater erosion or for chemical and physical weathering. Weathering, mass wasting, and wind and stream erosion are still the major factors in arid landscape formation, but because they are applied in very different proportions than in humid regions, the results are strikingly different from those in temperate and tropical landscapes (**FIG. 13.1**).

Movies and television portray arid regions with camels resting at oases amid mountainous sand dunes. Some arid regions, like parts of the Sahara and Mojave deserts, do look like that (minus camels in the Mojave, of course). But some arid regions actually have more rock than sand, while others lie next to shorelines where there is no shortage of water—and even the North and South Poles rank among Earth's most arid regions (**FIG. 13.2**).

With these factors in mind, complete Exercise 13.1, comparing features of arid and humid landscapes.

FIGURE 13.1 Arid landforms from the southwestern United States.

(a) Erosional remnants in Monument Valley, Arizona.

(b) Hoodoo panorams in Bryce Canyon, Utah.

(c) Delicate Arch in Arches National Park, Utah.

FIGURE 13.2 Types of arid regions.

(a) Landscape in Death Valley, in the Mojave Desert, California.

(b) Polar desert, Norway.

(c) Rocky desert, Utah.

EXERCISE 13.1	Comparing Arid and Humid Landscapes

Name: _____   Section: _____

Course: _____   Date: _____

Compare arid landscapes in a part of Canyonlands National Park and the Grand Canyon, located in Utah and Arizona, respectively, with humid landscapes of the Delaware Water Gap at the New Jersey/Pennsylvania border and an area in Utah (see photos on the next page).

(continued)

Name: _____   Section: _____
Course: _____   Date: _____

(a) In which areas are the landforms more angular? In which are they more rounded? Suggest an explanation for the difference.

_____

_____

_____

(b) In which areas do you think landscape changes are more rapid? Are slower? Explain your answer.

_____

_____

_____

(c) What other differences do you see between the arid and humid regions?

_____

_____

_____

(d) What type of erosion appears to be dominant in the arid regions? In the humid regions? How is this possible?

_____

_____

Comparison of arid and humid landscapes.

(a) Canyonlands National Park, Utah.

(b) Grand Canyon, Arizona.

(c) Delaware Water Gap, New Jersey/Pennsylvania.

(d) Humid region in Utah.

13.2 Processes in Arid Regions

Now let's examine the underlying causes for the different evolution of landscapes in arid and humid areas. Features tend to survive longer in arid areas than in humid ones because weathering, erosion, and deposition are slowed by the scarcity of water needed to abrade, dissolve, and carry debris away. In the absence of water, soluble rocks like limestone, dolostone, and marble—easily weathered and eroded in humid regions to form valleys—become ridge formers.

Both physical and chemical weathering occur in arid regions. The dominant processes of physical weathering are usually different from those in humid areas; for example, where do you think root wedging is more common? In areas where rain is rare, the major source of water for chemical weathering may be the thin film of dew formed on rocks each morning. The scarcity of water limits chemical weathering severely, and soil formation is slow. Exercise 13.2 explores some of these differences.

EXERCISE 13.2	Comparing Processes in Arid and Humid Regions

Name: _____ Section: _____
Course: _____ Date: _____

In a few short sentences, describe how the following processes are different in arid regions and humid regions? Consider the nature and intensity of the processes and the role that water plays in each. What are the dominant processes in each category in the two different areas?

	Arid regions	Humid regions
Physical weathering		
Chemical weathering		
Mass wasting		
Soil formation		
Stream erosion		
Wind erosion		
Groundwater activity		

13.3 Progressive Evolution of Arid Landscapes

The evolution of arid landscapes differs from that in humid regions. **FIGURE 13.3** shows the stages in the development of an arid landscape, starting with a block made of resistant and nonresistant rocks that has been uplifted by faulting (Fig. 13.3a). Physical weathering takes advantage of fractures, causing cliffs to retreat and forming a rubble-strewn plain (Fig. 13.3b).

Streams—perhaps flowing after only a few storms during the rainy season but still very effective—erode materials from the highlands and redeposit them as **alluvial fans** that begin to fill in the lowlands (Fig. 13.3c). With further deposition, alluvial fans coalesce to form a broad sand and gravel deposit (a **bajada**) flanking the uplifted block (Fig. 13.3c).

Where several fault blocks and valleys are present, bajada sediment from neighboring blocks eventually fills in the intervening valley, burying all but a few eroded remnants of the original blocks (Fig. 13.3d). These remnants stand out as isolated peaks, called **inselbergs**, above the sediment fill. Streams flowing into the basin from surrounding blocks have no way to leave and so form **playa lakes**, which may evaporate to form **salt flats**. Monument Valley (Fig. 13.3e) is a classic example of remnants of resistant rock isolated from their source layers by arid erosion.

Arid landscapes are typically more angular than humid landscapes. Compare, for example, photos (b) and (c) in Exercise 13.1. In the Grand Canyon, there is little soil. Rockfall is the dominant process of mass wasting, and sharp, angular slope breaks indicate contacts between resistant and nonresistant rocks. In Exercise 13.3, you will practice interpreting these features.

Nearly vertical slopes characterize the resistant sandstones, and more gently sloping surfaces characterize the less resistant shales. In contrast, a well-developed soil profile has formed above the bedrock in the Delaware Water Gap, partially masking erosional differences in the underlying rock. Soil creep is the dominant mass wasting process, and this smooths the topography.

EXERCISE 13.3	**Interpreting Arid Landscapes**

Name: _____   Section: _____

Course: _____   Date: _____

In the following series of exercises you will practice interpreting aspects of different arid landscapes.

Grand Canyon, Arizona

(a) Why hasn't soil creep smoothed the slopes of the Grand Canyon as pictured in this photo?

_____

_____

_____

Death Valley, California/Nevada

Examine the topographic overview (**FIG. 13.4**) of the Death Valley area. This is one of the driest places in North America and, with an elevation of more than 250 feet *below* sea level, one of the lowest spots on the continent. It also contains classic examples of arid landscapes.

(continued)

FIGURE 13.3 Stages in the evolution of an arid landscape.

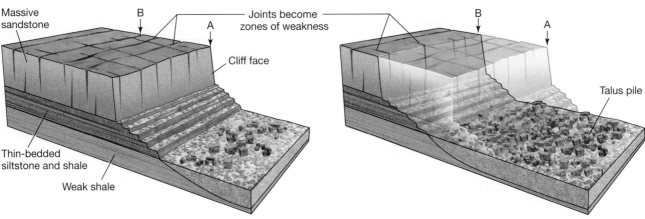

(a) An uplifted block of resistant sandstone and more easily eroded shales.

(b) Idealized cliff retreats from A to B as underlying rocks weather physically. Rockfall produces talus piles at the base of the cliff.

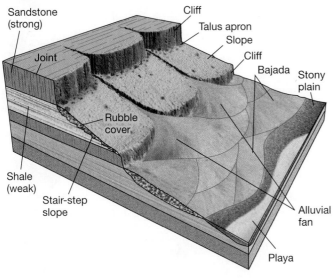

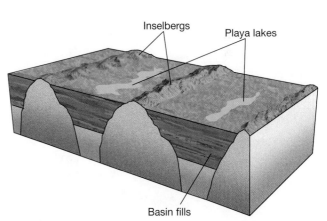

(c) A debris apron covers the lowlands and protects the bedrock slope from further erosion.

(d) A late stage of landscape development involving several uplifted blocks.

(e) Buttes in Monument Valley, Arizona. Buttes are small- to medium-sized erosional remnants with flat tops. Similar but larger features are called *mesas*, from the Spanish word for table.

FIGURE 13.4 Topographic overview and profile of part of the Death Valley area (Death Valley National Monument in dark pink at center of map).

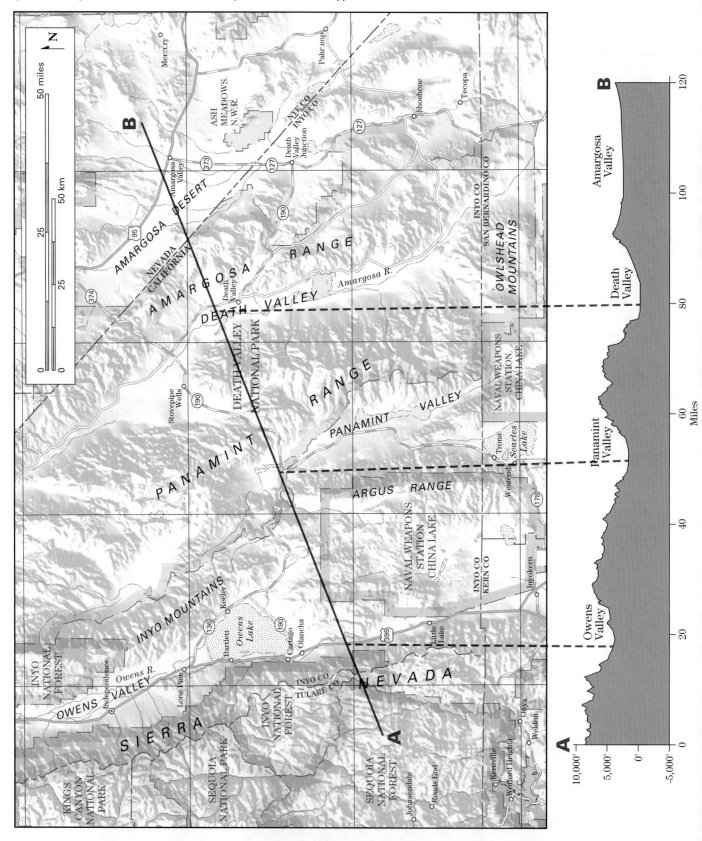

Name: _____   Section: _____

Course: _____   Date: _____

(b) Compare this region in Figure 13.4 with Figure 13.3. Is this a single fault block like in Figure 13.3a or several blocks like in Figure 13.3d? Explain your reasoning.

_____

_____

_____

(c) Sketch in the border fault(s) with a colored pencil on both the map view and the profile in Figure 13.4.

(d) Why is Death Valley lower in elevation than the other major valleys shown in the topographic profile in Figure 13.4? Consider factors that played roles in forming the valley: stream erosion, tectonic activity.

_____

_____

_____

Now examine Death Valley more closely, starting with the Stovepipe Wells area (**FIG. 13.5**).

(e) What are the broad, rounded landforms that project into Death Valley from the mountains adjacent on the west and southeast? How did they form?

_____

_____

_____

(f) Several streams flow into Death Valley from these mountains. Where do they go? What is their base level?

_____

_____

_____

(g) A stippled pattern identifies areas of rapidly moving sand dunes. Where did this sand come from?

_____

_____

_____

(h) A more detailed view of the sand areas would show individual dunes, but these would not be accurate representations of the current land surface. Why not?

_____

_____

_____

(continued)

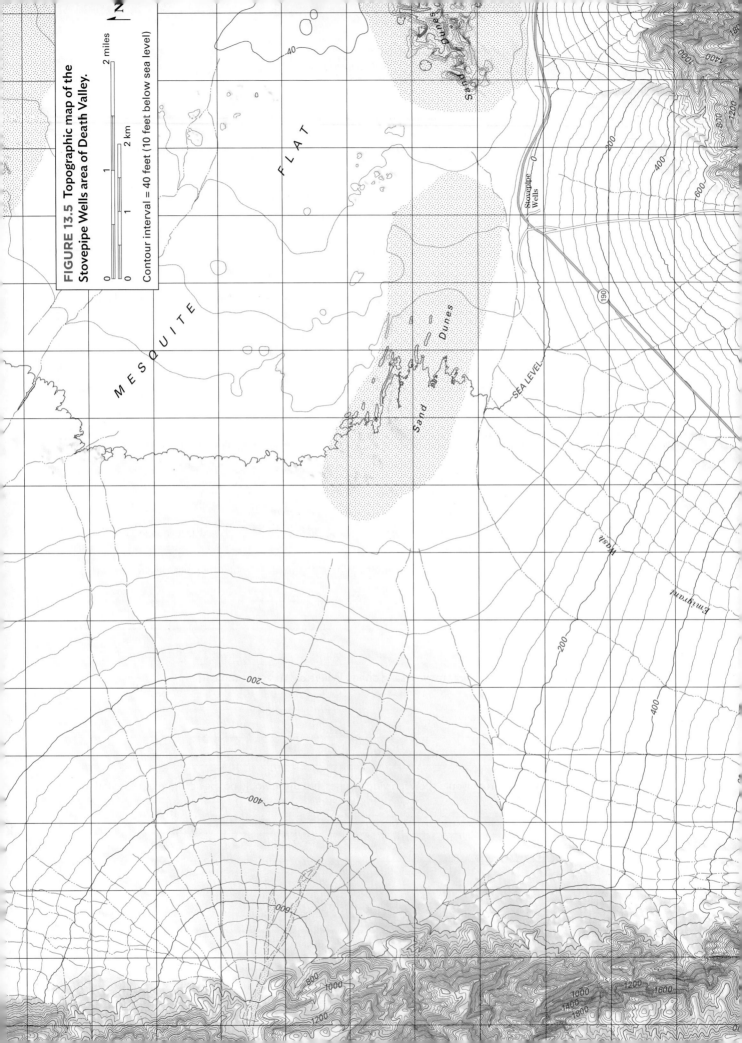

FIGURE 13.5 Topographic map of the Stovepipe Wells area of Death Valley.

Contour interval = 40 feet (10 feet below sea level)

Name: _____   Section: _____
Course: _____   Date: _____

Examine the map of the area surrounding the town of Death Valley, including the Death Valley Airport and Furnace Creek Inn (**FIG. 13.6**).

(i) What arid environment depositional landforms are present? Label these on the map.

(j) A large lake that is not present throughout the year is shown with a blue-stipple pattern in Cotton Ball Basin. What is this type of lake called and how does it form?

_____

_____

_____

(k) What evidence on the map suggests that the water in the streams and lakes is unsafe to drink?

_____

_____

_____

(l) Is all of the water in these lakes brought in by the streams shown on the map? Explain your reasoning.

_____

_____

_____

(m) Compare Figures 13.4, 13.5, and 13.6 with Figure 13.3. How far advanced in the arid erosion cycle is the Death Valley area? What changes can we expect to take place in the future?

_____

_____

_____

_____

Buckeye Hills, Arizona

The Buckeye Hills southwest of Phoenix are in an arid area similar to Death Valley, but this area differs in some ways. A topographic overview of the Buckeye Hills is shown in **FIGURE 13.7**.

(n) Describe the landscape shown in this overview.

_____

_____

_____

_____

(o) How are the Buckeye Hills similar to the part of Death Valley shown in Figure 13.4?

_____

_____

_____

_____

(continued)

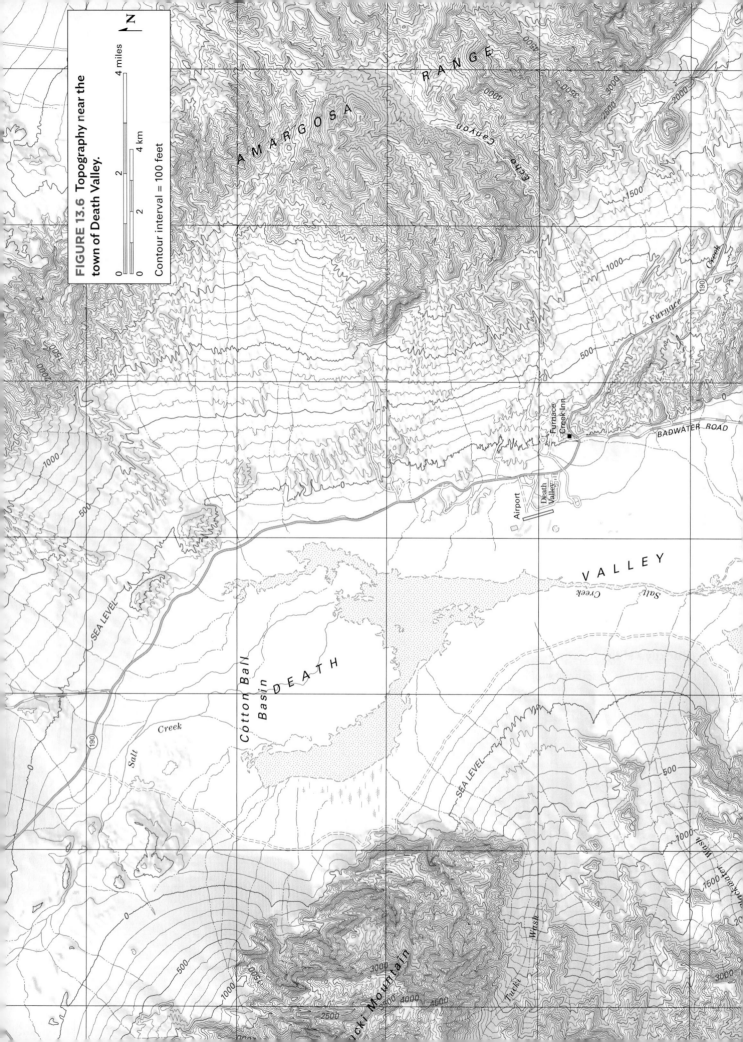

FIGURE 13.6 Topography near the town of Death Valley.

Contour interval = 100 feet

4 miles
4 km

N

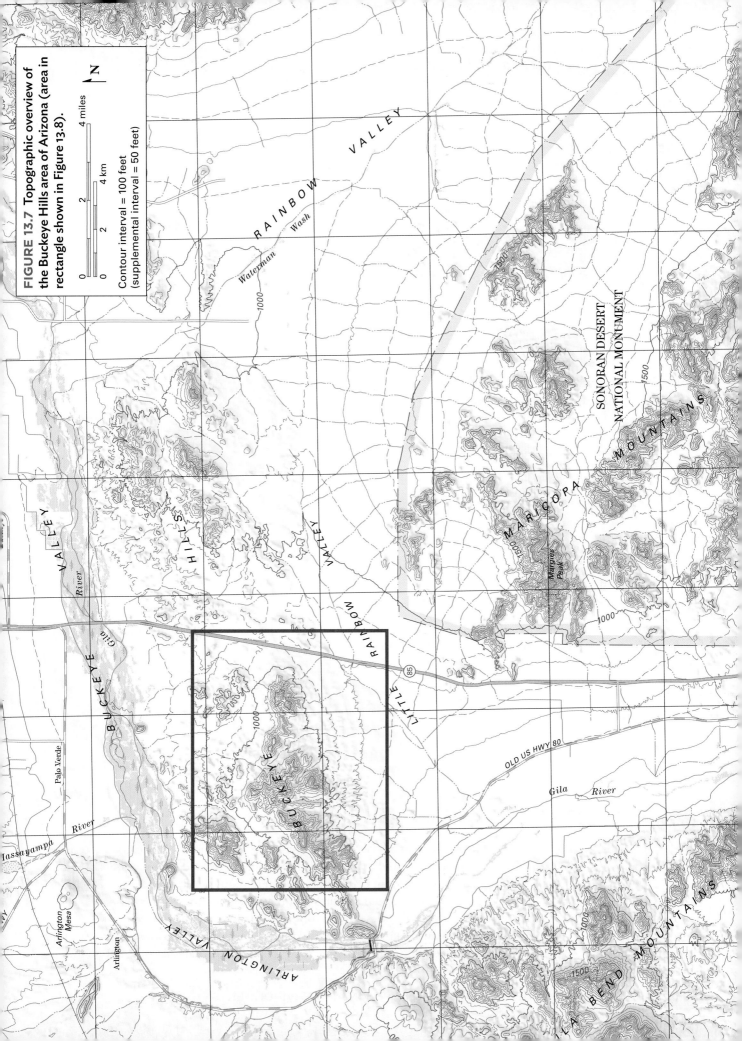

FIGURE 13.7 Topographic overview of the Buckeye Hills area of Arizona (area in rectangle shown in Figure 13.8).

N

4 miles

4 km

Contour interval = 100 feet
(supplemental interval = 50 feet)

RAINBOW VALLEY

Waterman Wash

1000

SONORAN DESERT
NATIONAL MONUMENT

1500

1500

MARICOPA MOUNTAINS

Margies Peak

1500

1000

VALLEY

River

Gila

HILLS

BUCKEYE

Palo Verde

Hassayampa River

Arlington Mesa

Arlington

BUCKEYE VALLEY

ARLINGTON VALLEY

RAINBOW VALLEY

LITTLE

85

1000

OLD US HWY 80

Gila River

1000

GILA BEND MOUNTAINS

1500

Name: _____   Section: _____
Course: _____   Date: _____

(p) How are they different?

_____

_____

_____

_____

Now look at the detailed map of the central part of the Buckeye Hills (**FIG. 13.8**).

(q) Identify three erosional and depositional features of arid landscapes. List them here and label them on the map.

_____

_____

_____

(r) What is the probable origin of the small hills just east of Highway 85 that are pictured on this map?

_____

_____

_____

(s) What evidence is there of current stream activity in the Buckeye Hills?

_____

_____

_____

_____

(t) Refer to the diagram showing the evolution of arid landscapes (Fig. 13.3). Which area, Death Valley or the Buckeye Hills, is in a more advanced state of development? Explain your reasoning.

_____

_____

_____

_____

13.4 Wind and Sand Dunes

Wind has a greater impact in arid areas than in humid areas because there is less cohesion between sand grains and little vegetation to hold the sand in place. This is because a small amount of moisture between grains at the surface holds the grains together weakly in humid climates, but that phenomenon is absent in arid regions. There are also fewer trees or other obstacles to block the full force of the wind. Wind erodes the way streams do—using its kinetic energy to pick up loose grains and then using this sediment load to abrade solid rock—literally sandblasting cliffs away.

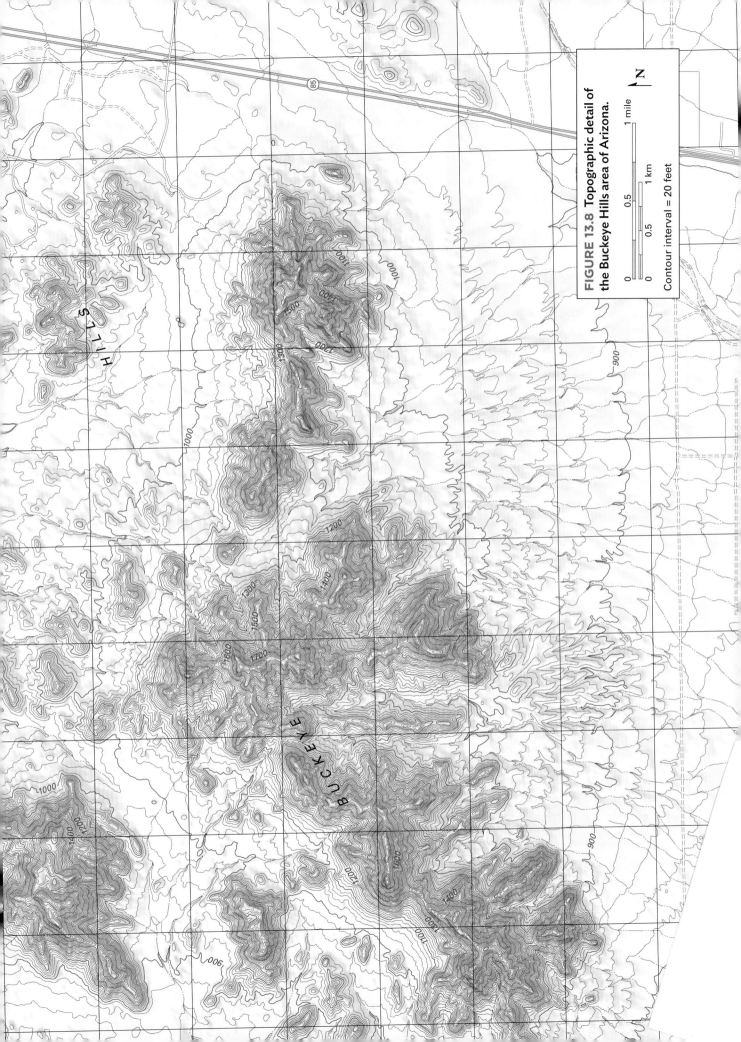

FIGURE 13.8 Topographic detail of the Buckeye Hills area of Arizona.

Contour interval = 20 feet

Dunes are the major depositional landform associated with wind activity in deserts, but they also form along shorelines where there is an abundant source of sand. There are several kinds of dunes, depending on the amount of sand available, the strength of the wind, and how constant the wind direction is (**FIG. 13.9**). **FIGURE 13.10** shows the ripple-marked surface of a large transverse dune from the stippled area on the Stovepipe Wells map in Death Valley (Fig. 13.5). Exercise 13.4 explores the history of sand dunes.

| EXERCISE 13.4 | Interpreting the History of Sand Dunes |

Name: _____   Section: _____
Course: _____   Date: _____

The Nebraska Sand Hills

The Nebraska Sand Hills are the remnant of a vast mid-continent sea of sand that existed beyond the Pleistocene ice front approximately 30,000 years ago. More humid conditions today permit grasses, trees, and shrubs to stabilize the dunes so that they no longer move across the countryside like those in Death Valley or the Sahara, but the hills preserve evidence of the more arid conditions under which they formed. Examine the overview of the Nebraska Sand Hills (**FIG. 13.11**).

(a) What evidence is there that this is still a relatively arid region?

_____

_____

_____

_____

(b) Describe the general topographic grain of the area, including the size and shape of the ridges and valleys present.

_____

_____

_____

_____

(c) These ridges are the remnants of the Pleistocene sand sea. What features do they represent?

_____

_____

_____

Examine the map detail (**FIG. 13.12**) of the area outlined in red in Figure 13.11.

(d) Describe the shape of the ridges. Draw a topographic profile, using the graph paper provided at the end of this chapter, along the line indicated to help answer the next question.

_____

_____

(e) What does the shape of the dunes tell you about the wind during the Pleistocene?

_____

_____

(continued)

FIGURE 13.9 Types of sand dunes (red arrows indicate wind direction).

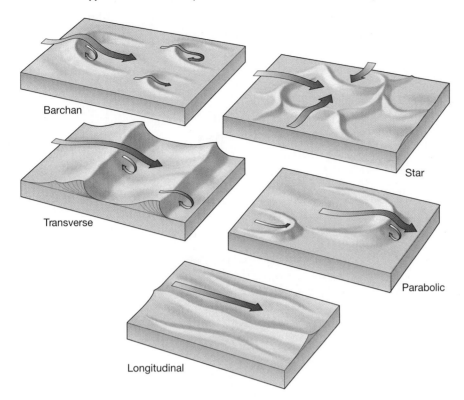

Barchan

Star

Transverse

Parabolic

Longitudinal

FIGURE 13.10 Origin and movement of transverse dunes.

Wind

Windward side

Slip face

Dune

Surface ripples

Slip face

Cross-bed surface

Main bedding surface

(a) Transverse dunes at Stovepipe Wells in Death Valley. The shape of a sand dune depends on wind direction. Small ripples may form on the dune surface.

(b) Movement of sand grains in a transverse dune. Sand is moved from the upwind dune face to the slip face, causing the dune to migrate downwind.

FIGURE 13.11 Topographic overview of part of the Nebraska Sand Hills (area in rectangle shown in Figure 13.12).

Contour interval = 100 feet

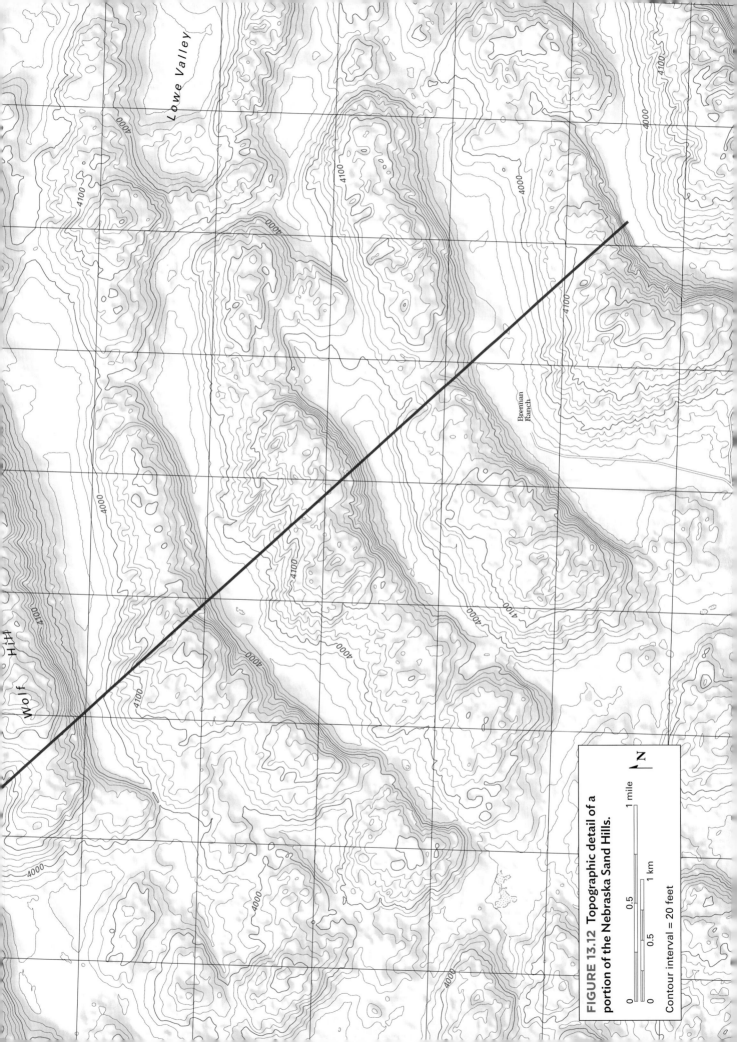

FIGURE 13.12 Topographic detail of a portion of the Nebraska Sand Hills.

Contour interval = 20 feet

Lowe Valley

Wolf Hill

Brennan Ranch

4000

4100

N

0   0.5   1 km

0   0.5   1 mile

Name: _____   Section: _____

Course: _____   Date: _____

? What Do You Think   Nebraska ranks #4 in the United States in generating electricity with wind–the same wind that forms sand dunes. The wind turbines shown in the first photograph below are only some of the over 500 turbines that generate electricity for 250,000 homes across the state.

In this exercise, suppose that a local initiative is exploring the building of a series of 500 windmills across the Sand Hills area (as seen in Figure 13.12). The windmills would supply low cost electricity for the area, and also allow for future development in the region. However, in this project, the windmills and their arrangement on the landscape will be more similar to those seen in the second photograph below, showing an area outside of Palm Springs, California.

As an informed citizen, you want to start a blog covering this initiative, so that your neighbors understand its pros and cons. To start, on a separate sheet of paper, create two columns, one labeled "Pro" and the other "Con." Then, in each column, briefly describe at least three positive and negative aspects of this project. Consider what you have learned about arid regions, how the turbines might impact other uses of the land, and the possible environmental risks. One thing that many people don't consider is how will the electricity get from the turbines to cities and towns? Then describe your current position either for, against, or possibly somewhere in between, based on your current information.

Wind turbines in Nebraska.

Wind turbines in Palm Springs, CA.

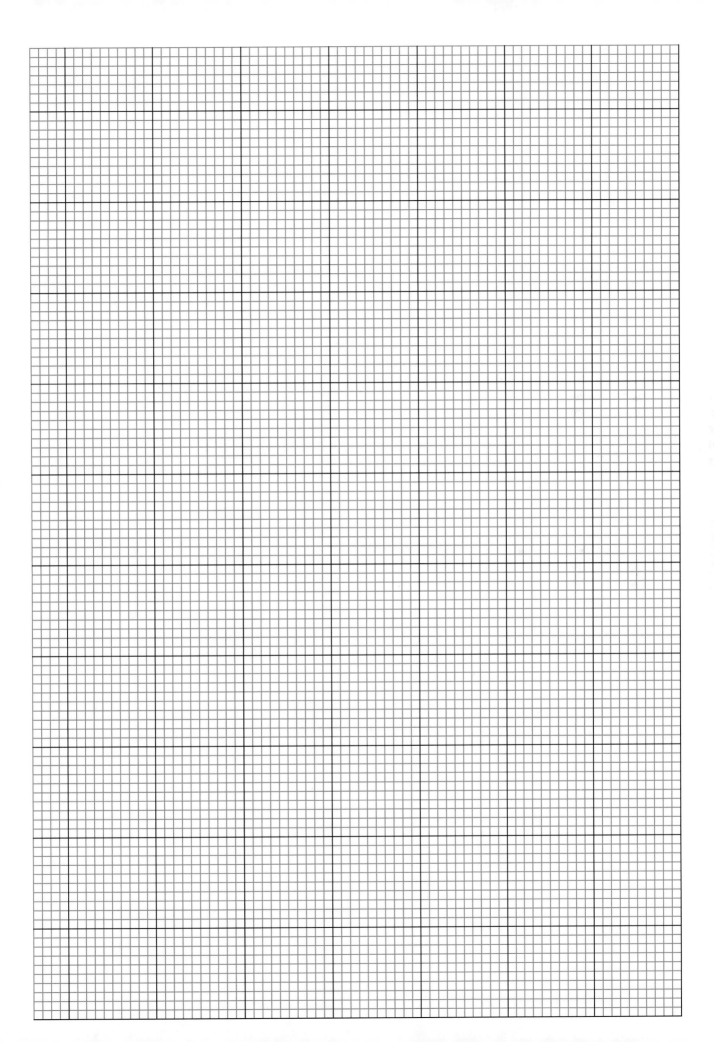

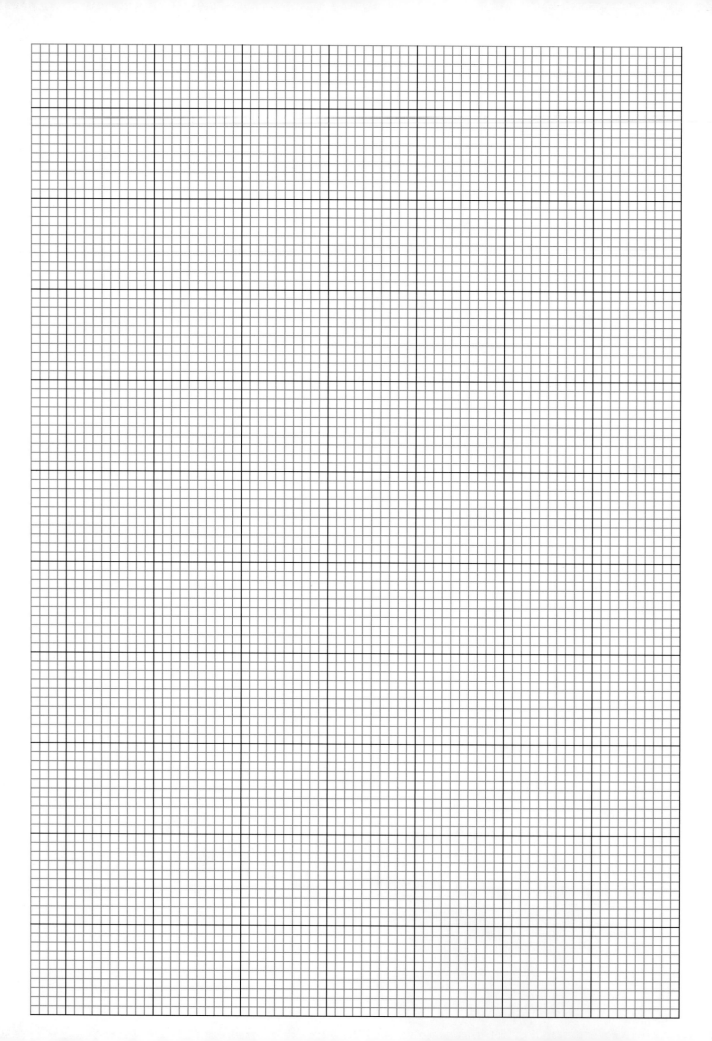

14

Shoreline Landscapes

Coastlines come in a wide variety of types, including rocky, sandy, and mangrove swamps.

LEARNING OBJECTIVES

- Recognize the different kinds of shorelines and how they evolve

- Understand that processes of shoreline erosion and deposition produce unique landforms

- Understand the impact that human activity has on shorelines

- Recognize potential shoreline hazards and ways to lessen their impact

MATERIALS NEEDED

- Colored pencils

- Magnifying glass or hand lens to help read close-spaced contour lines

14.1 Introduction

Hundreds of millions of people live along North America's Atlantic, Gulf of Mexico, Pacific, and Great Lakes shorelines, and millions more vacation there every year. Recent hurricanes like Katrina and Sandy remind us that shorelines are the most dynamic places on Earth, able to change dramatically in a few hours or days and endanger lives and property. Shoreline landscapes are unique places where the relatively stable geosphere interacts with the much more rapidly changing hydrosphere and atmosphere: wind generates waves that crash onshore, eroding and transporting materials.

Humans will always live along shorelines, but we must do so intelligently to lessen the impact of the rapid changes and must plan wisely as global climate change increases the risk of future catastrophic events. This chapter examines the nature of shorelines, the processes that create and destroy their distinctive landforms, the effects of human activity on those processes, and the disastrous events to which shorelines are uniquely subjected.

14.2 Factors Controlling Shoreline Formation and Evolution

There are many types of shorelines, with examples ranging from rocky cliffs to gentle sandy beaches, untouched coral reefs, marshes, and mangrove groves to heavily urbanized coastlines (**FIG. 14.1**). But the nature and rapidity of shoreline change are controlled by four important factors, regardless of the type of shoreline:

- Material of which the shoreline is made
- Weather, wind, climate, and climate change
- Tidal range
- Tectonic activity

14.2.1 Shoreline Materials

The type of material present is one of the most important factors in the effectiveness of erosional and depositional processes and therefore shoreline stability. Not all beaches are made of tan sand—consider Hawaii's green, black, and red sand beaches and Bermuda's pink sands—and some shores have little or no sand at all and are made of materials like corals, rocks or boulders, marshes or mangroves, or man-made piers or jetties. And, of course, some shorelines like the southern end of Manhattan Island in New York City, parts of San Francisco, and other harbor cities are now completely man-made. In Exercise 14.1, you will examine the images of shorelines pictured in Figure 14.1.

14.2.2 Weather, Wind, Climate, and Climate Change

Waves are the dominant force in shoreline erosion, and the height and velocity of waves in a region are controlled by weather factors such as the types of weather systems, the strength and direction of winds, and the frequency of storms. Wind also controls how loose materials will be moved along the shoreline in a process called longshore drift (discussed in section 14.3.3) and can have devastating effects on man-made parts of the shoreline.

Climate change has a profound effect on shorelines. For example, during an ice age, glaciers advance across the continents locking enormous amounts of water in ice that would otherwise have been in the oceans. Today, global warming is having the opposite effect, melting the Greenland and Antarctic ice caps as well as most mountain glaciers. It isn't just adding meltwater: as the oceans get warmer, the water in the upper 700 m expands. Between 1961 and 2003, thermal expansion accounted for about 25% of sea-level rise. Between 1993 and 2003, that proportion had risen to nearly 50%.

FIGURE 14.1 Types of shorelines.

(a) Rocky shoreline at West Quoddy Head in Maine.

(b) Sandy and rocky shoreline, Hawaii.

(c) Coral reef along the Hawaiian shoreline.

(d) Salt marsh along the coast of Cape Cod in Massachusetts.

(e) Mangrove swamp in northeastern Brazil.

(f) A seawall built at the southern tip of Manhattan (New York City).

EXERCISE 14.1 Erodability and Stability of Shoreline Materials

Name: ____________________ Section: ____________

Course: ____________________ Date: ____________

Review the types of shoreline shown in Figure 14.1 and answer the following questions.

(a) Which of these shorelines do you think would be the most difficult to erode? Which would be the easiest? Explain your choices in a few sentences.

(b) Compare the bedrock cliffs in Figure 14.1 a and b. What factors will determine how resistant to erosion these cliffs are? How can waves erode the bedrock cliffs?

(c) Consider man-made shorelines like the one in Figure 14.1f as compared to the natural shorelines in the other photographs of the figure. Identify one shoreline that may be stronger than the man-made shoreline in (f) and one that may be weaker. Explain your reasoning for each choice.

(d) Indicate with arrows the wind directions that generate the waves shown in Figure 14.1b. Explain your reasoning.

(e) Which area do you think would be most damaged by high winds and high waves? Explain why you made your choice.

FIGURE 14.2 Tidal range at Mont-Saint-Michel, France.

(a) At low tide.

(b) At high tide.

14.2.3 Tidal Range

The gravitational attraction of the Sun and Moon cause the tides—water rising and falling along the shore twice a day in most coastal areas. Tidal range is typically a few feet, but in some places coastal geometry results in a far higher range. The Bay of Fundy between the Canadian provinces Nova Scotia and New Brunswick has the highest range, as much as 53 feet between high and low tides. Mont-Saint-Michel in France ranks "only" fourth at 46 feet, but as **FIGURE 14.2** shows, the result is dramatic.

The tides move enormous amounts of sediment all along the shoreline, and fluvial processes extend across exposed tidal flats at low tide. As we shall see later, tidal stage is a major factor in determining how much damage coastal storms will cause when they come onshore.

14.2.4 Tectonic Activity

Tectonic activity creates ocean basins, enlarges or shrinks them, and may uplift or lower the land along the coast. In addition, when submarine volcanoes build their cones above sea level, they create new lands (**FIG. 14.3**), including heavily populated island nations such as Japan, Indonesia, the Philippines, and the state of Hawaii. These islands are in a constant state of change, as waves erode the land that the new lavas build. When an ocean widens or begins to close, the water must occupy the greater or smaller volume of the ocean basin, lowering or raising sea level along the ocean shoreline.

FIGURE 14.3 Growth of the island of Hawaii by addition of lava from Kilauea Volcano to the shoreline.

(a) Red-hot lava entering the ocean.

(b) New land created by lava flows in the 1940s.

14.2.5 Emergent and Submergent Shorelines

Short-term fluctuations in sea level occur sporadically during major storms and twice daily in the tidal cycle, but long-term sea-level changes are caused by tectonic activity and climate change. Geologists commonly group shorelines into two categories based on how they respond to these long-term changes. If the land sinks or sea level rises, the shoreline appears to be drowned, with irregular coastline, prominent bays, and abundant islands, marshes, and lagoons. These are called **submergent shorelines**. Land along **emergent shorelines** appears to have risen from the sea by tectonic uplift or a drop in sea level. Emergent shorelines are typically straight and bounded by steep cliffs; where tectonic uplift has occurred, remnants of former shoreline features may be found well above sea level.

The following exercises will give you experience recognizing different types of shorelines, the processes that create them, and how they have changed over time.

EXERCISE 14.2	Effects of Climate Change on Shorelines

Name: _____   Section: _____
Course: _____   Date: _____

(a) What would be the effect on the oceans if continental glaciers worldwide expanded by 10%? Explain.

_____

_____

_____

(b) What would be the effect on the world's shorelines if continental glaciers worldwide expanded by 10%?

_____

_____

_____

(c) Conversely, what would be the effect on the world's oceans and shorelines if continental glaciers shrunk by 10%?

_____

_____

_____

(d) Starting about 2 million years ago, much of northern North America, Europe, and Asia were covered with continental glaciers. In what way was the location of the world's shorelines at that time different from that of today's shorelines? Explain your reasoning.

_____

_____

_____

_____

_____

Effects of Plate Tectonic Processes on Shorelines

Name: _____   Section: _____
Course: _____   Date: _____

(a) What effect will continued sea-floor spreading in the Atlantic Ocean have on east and gulf coast sea level? Explain your reasoning. *Hint*: Look at Exercise 2.7.

_____

_____

_____

_____

(b) What effect would partial closing of the Atlantic Ocean have on sea level? Explain.

_____

_____

_____

_____

EXERCISE 14.4   **Recognizing Emergent and Submergent Shorelines**

Name: _____   Section: _____
Course: _____   Date: _____

Emergent shorelines look very different from submergent shorelines. Examine the following maps: part of the Atlantic Coast in Maine (**FIG 14.4**), and part of the Pacific Coast in California (**FIG 14.5**). One of these is a typical submergent shoreline, the other a classic emergent shoreline. Apply your geologic reasoning to tell which is which.

(a) Compare and contrast the shapes of the two shorelines.

_____

_____

_____

_____

(b) Which of these shorelines is emergent and which is submergent? Explain your reasoning.

_____

_____

_____

_____

FIGURE 14.4 Maine coastline near Boothbay Harbor.

Contour interval = 25 feet

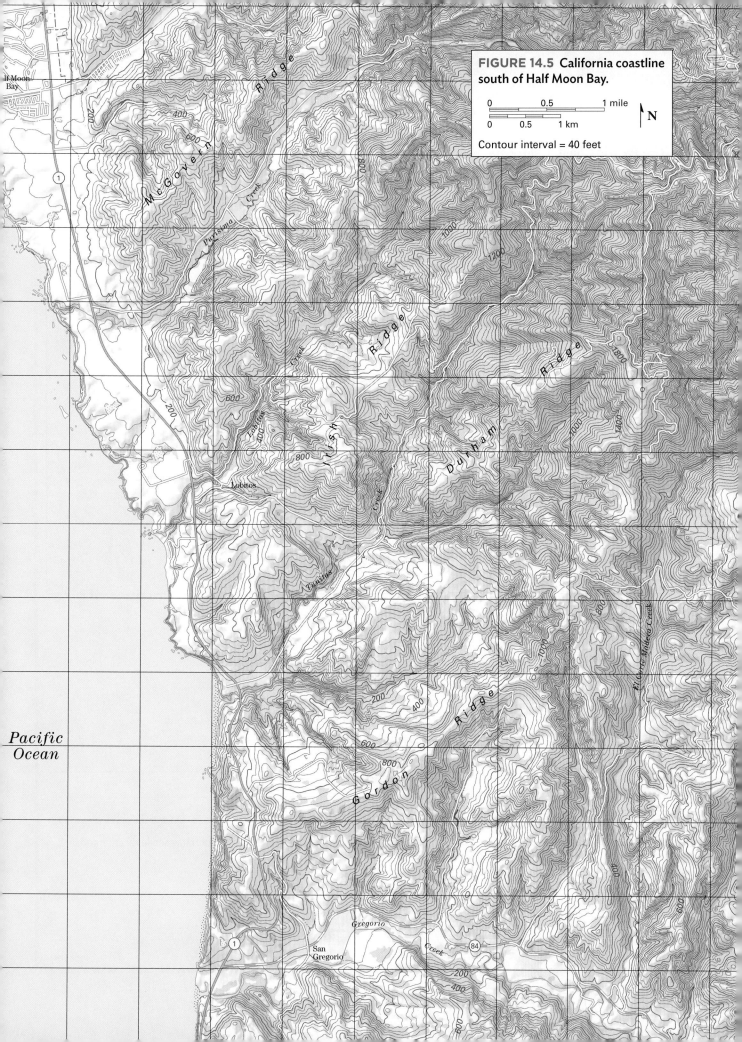

FIGURE 14.5 California coastline south of Half Moon Bay.

0   0.5   1 mile

0   0.5   1 km

Contour interval = 40 feet

N

Pacific Ocean

If Moon Bay

McGovern Ridge

Purisima Creek

Irish Ridge

Durham Ridge

Lobitos Creek

Lobitos

Tunitas Creek

Gordon Ridge

El Corte-Madero Creek

Gregorio Creek

San Gregorio

Name: _____ Section: _____
Course: _____ Date: _____

The Pacific Coast, California

The map below shows details of the California coast not far from where Figure 14.11d was photographed.

Portion of the Dos Pueblos Canyon quadrangle of California.

Contour interval = 20 feet

Santa Barbara Channel

(a) Describe the shoreline in your own words.

_____

_____

_____

(b) Sketch profiles along lines A–B and C–D on the graph paper provided at the end of this chapter.

(c) What evidence shows that a change in sea level has taken place?

_____

_____

_____

(d) Based on the map and profile, is this an emergent or submergent shoreline? By how much has sea level changed? Explain your reasoning.

_____

_____

_____

(continued)

Name: _____ Section: _____

Course: _____ Date: _____

Lake Erie, Ohio

The shorelines of many lakes that formed shortly after the retreat of Pleistocene glaciers from North America have changed markedly in the past few thousand years. Some glacial lakes have shrunk to a fraction of their former size (such as Glacial Lake Bonneville, which is now the Great Salt Lake in Utah) or disappeared entirely (Glacial Lake Hitchcock in Massachusetts). The Great Lakes have adjusted to post-glacial conditions, and their shorelines reveal those changes. **FIGURE 14.6** shows an area in Ohio just south of Lake Erie.

(e) Examine the spacing of the contour lines. What do they suggest about the evolution of Lake Erie?

_____

_____

_____

(f) Draw a profile along line A–B (using the graph paper provided at the end of this chapter) and then compare it with the profile you drew for Exercise 14.5b. What features are probably represented by the ridges? What features are probably represented by the gently sloping areas between the ridges?

_____

_____

_____

(g) Based on your answers in (f), how are Sugar, Chestnut, and Butternut ridges related to the post-glacial history of Lake Erie?

_____

_____

_____

(h) Label the previous shoreline position(s).

(i) The current elevation of Lake Erie is 174 feet. How much has lake level changed, based on the evidence on this map?

_____

(j) Was the change continuous or did it take place in sporadic episodes? Explain.

_____

_____

_____

(k) Assuming the retreat of continental glaciers took place about 10,000 years ago, calculate the rate at which lake level dropped if the change had been continuous.

_____

(l) If the change was episodic, suggest a way to estimate the relative amount of time associated with each "still-stand" of lake level. What assumptions must you make?

_____

_____

_____

FIGURE 14.6 Area just south of Lake Erie near Elyria, Ohio.

Contour interval = 5 feet

FIGURE 14.7 Mechanics of wave action offshore and near the shoreline.

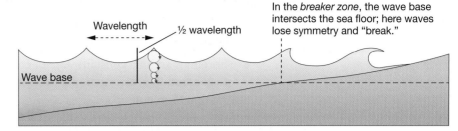

In the *breaker zone*, the wave base intersects the sea floor; here waves lose symmetry and "break."

Wavelength   ½ wavelength

Wave base

Within a deep-ocean wave, water molecules move in a circular path. The radius of the circle decreases with depth.

14.3 Shoreline Erosion and Deposition

In this section, we will learn more about how a shoreline changes. Waves are the dominant agents of shoreline erosion and deposition. They erode coastal materials when they strike the shore and deposit loose materials to form beaches and other landforms. Waves also generate currents that parallel the shoreline and build landforms offshore. Wind is not only responsible for driving the waves; it also moves sediment directly in shoreline areas, producing coastal sand dunes. We will look first at the basics of how waves operate and then at how they produce shoreline landforms.

14.3.1 How Waves Form

Waves form by friction generated when wind blows across the surface of an ocean or lake. The symmetrical shape of waves offshore shows how the kinetic energy of the wind is transferred to the water. Offshore, water molecules move in circular paths, each molecule passing some energy on to those it contacts (**FIG. 14.7**). Loss of energy in these contacts limits the depth of wave action to approximately half of the wavelength, a depth referred to as the **wave base**.

When water is shallower than the wave base, as on the right in Figure 14.7, orbiting water molecules interfere with the sea floor or lake bottom, causing waves to lose symmetry and "break." The pileup of water increases the waves' kinetic energy, enhancing their ability to erode the shoreline.

FIGURE 14.8 shows the breaker zone associated with waves in California (Figure. 14.8a) and Hawaii (Fig. 14.8b). In both cases, the wind that generated the waves blew for thousands of miles uninterrupted by land or trees, which is the most favorable condition for wave formation. The arrow in Figure 14.8b shows where the waves have cut a notch in the lava cliffs of Kauai.

FIGURE 14.8 Waves striking shorelines.

Wave-cut notch

(a) A breaker zone in California at low tide.

(b) A breaker zone in Hawaii at high tide.

14.3.2 Wave Erosion and Deposition

A few basic principles explain how waves erode and deposit materials along shore-lines and move sediment directly, forming landforms like barrier islands and spits.

- Shorelines are in a constant state of conflict between destructive wave erosion and the constructional processes of deposition, lava flows, and coral reef development.
- Waves are generated by the interaction between wind and the surface waters of oceans and large lakes.
- The kinetic energy of waves causes erosion and redeposition of *unconsolidated* sediment.
- Like streams and glaciers, waves use loose sediment to abrade the bases of solid bedrock cliffs. Waves move sediment back and forth across the tidal zone, abrading a flat wave-cut bench and carving wave-cut notches. When support of the base is undermined, the cliffs collapse by rockfall and slump.
- Waves then erode the rubble, exposing the base of the new cliffs, and the cycle repeats. In this way, shoreline cliffs gradually retreat inland. In tectonically more active coastal areas, there may be several uplifted benches; dating them enables geologists to estimate the rate of uplift.
- Wind also moves sediment by itself, forming coastal sand dunes.
- Shoreline currents redistribute sediment to produce barrier islands, spits, and other landforms.

14.3.3 Longshore Drift

In areas where there is an abundant supply of unconsolidated sediment, longshore currents move sand and silt parallel to the shoreline in a process called **longshore drift.** These currents are generated when waves strike the shoreline obliquely (i.e., not perpendicular to the shore). Each time a wave pushes onshore and then recedes, sand grains are moved onto the shore and then back toward the ocean. After many such zigzag cycles, sand grains gradually move along the beach—in what looks like a straight-line path to someone who hasn't been watching closely (**FIG. 14.9**).

FIGURE 14.9 Origin of longshore drift: Particle A eventually moves to point B by a complex zigzag path shown by the dashed lines. A-1: wave drives the grain up the beach in the direction that the wave is moving. 1-2: Water is pulled downslope to shoreline, carrying the particle with it; these processes are repeated (2-3-4-5-B) to transport the grain parallel to the shoreline.

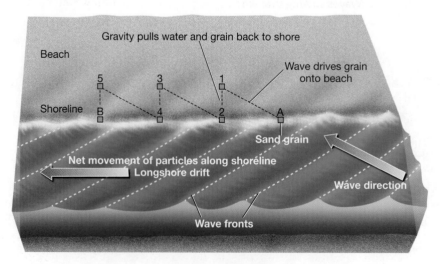

Name: _____   Section: _____
Course: _____   Date: _____

Examine the two shorelines below.

Wave-cut notch

(a) Indicate with arrows the direction in which the wind is blowing in each.

Two aspects of shoreline geometry affect the effectiveness of wave erosion: (1) Is the area exposed, projecting out into the ocean or is it an embayment, protected by flanking headlands? (2) Is the shoreline steep, allowing waves to crash onshore with full force or is it gentle, causing the waves to break offshore and lose some of their energy?

(b) Compare the shoreline geometry in the photographs above. In which would you expect wave erosion to be more effective? Explain.

_____

_____

_____

Similarly, oblique waves also generate currents just offshore that parallel the coastline. These currents are responsible for moving large volumes of sediment and for building such distinctive landforms discussed below, including familiar features like Coney Island and Cape Cod.

14.3.4 Erosional Features

Coastal erosion produces distinctive shoreline features (**FIG. 14.10**) and causes coastal landscapes to change in predictable ways. Coastal erosion is most rapid in places where the land extends out into the ocean, because this position allows waves to attack the land from nearly any direction. Conversely, coastal erosion is slowest in deep, low areas of coastal land called embayments where wave energy is diffused along a broad stretch of coastline.

Coastal erosion leads to two types of landforms to develop. The first group occurs along the shorelines: as waves drive loose sediment across shorelines underlain by bedrock, the sediment abrades a flat surface called a **wave-cut bench** and cuts into the base of bedrock cliffs to form a **wave-cut notch** (**FIG. 14.11**). Where the coastal cliffs are made of unconsolidated sediment, the waves eat into the

FIGURE 14.10 Erosional features of bedrock shorelines.

(a) A wave-cut notch along the Hawaiian coast.

(b) A wave-cut bench at the foot of the cliffs at Étretat, France.

(c) A sea arch on the coast of Hawaii.

FIGURE 14.11 Evidence for an emergent shoreline.

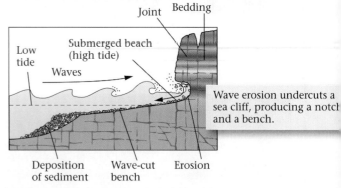

(a) Wave erosion undercuts a sea cliff, producing a notch and a bench

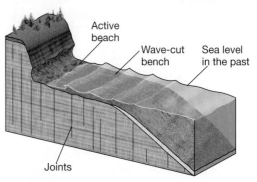

(b) Wave erosion produces a wave-cut bench along an emergent coast

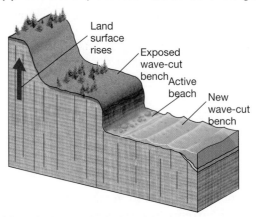

(c) As the land rises, the bench becomes a terrace, and a new wave-cut bench forms.

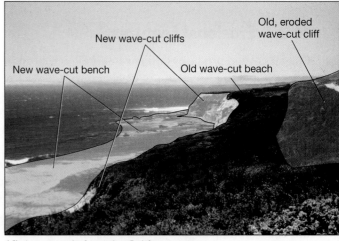

(d) An example from the California coast.

cliff and use the sediment as added abrasives. Eventually, the notch undermines the cliff, causing slumping or rockfall, and the cliff face retreats away from the shoreline. Where humans have built homes on coastal cliffs, the result can be disastrous, particularly where the cliffs are made of unconsolidated coastal plain sediments or glacial deposits as along the Atlantic Coast (**FIG. 14.12**). Over long periods of time, wave-cut benches become wider as the coastal cliffs

FIGURE 14.12 Coastal-bluff erosion caused this house in Destin, Florida, to collapse. The erosion was exacerbated by high waves during Hurricane Dennis in 2005.

FIGURE 14.13 Creation of a sea stack.

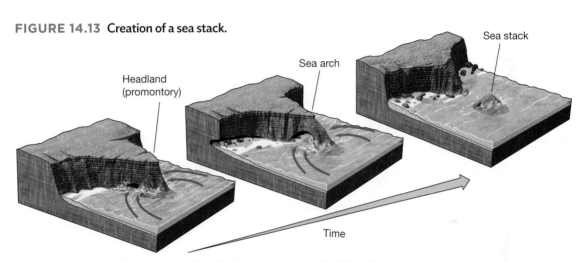

(a) Formation of sea arches and stacks by erosion at a bedrock headland.

retreat landward. In tectonically active areas, old wave-cut benches may be uplifted tens of feet above present sea level where a newer bench is being carved today. By dating the uplifted benches, geologists can estimate the amount of tectonic uplift.

The second group comprises sea arches and sea stacks: distinctive, often dramatic bedrock landforms found close to the current shoreline, a short distance from the shore. When a bedrock cliff retreats, it does not do so at the same rate in all places. Zones of weak bedrock may be eroded quickly, isolating stronger material and starting to cut into it as well. Eventually, a **sea arch**, named because of its natural arch appearance, forms. When further erosion removes the support for the arch, it will collapse and leave an isolated remnant of the bedrock called a **sea stack**

(b) Morro Rock in California, a classic sea stack.

(**FIG. 14.13a**). Morro Rock (**FIG. 14.13b**), off the coast of California, is a spectacularly beautiful sea stack. Sea stacks along a coastline mark the former position of the bedrock cliffs letting us measure the amount of cliff retreat.

Exercise 14.6 explored the factors of wave erosion and now Exercise 14.7 examines the products of wave erosion.

Name: _____   Section: _____

Course: _____   Date: _____

Refer to photos earlier in this chapter to answer the following questions.

(a) How did the large blocks in Figure 14.10a get into the surf zone? What is their eventual fate?

_____

_____

_____

(b) How did Morro Rock (Fig. 14.13b) become isolated from the bedrock shoreline?

_____

_____

_____

(c) What is the eventual fate of the sea arch in 14.10c?

_____

_____

_____

Examine the shoreline in **FIGURE 14.14**. Point Sur and False Sur are the same kind of feature and record a multistage development for this part of the California shoreline.

(d) Based on its size and steepness, is Point Sur made of a hard material like bedrock or a softer material like limestone? Explain.

_____

_____

_____

(e) Of what material or type of material is the area between Point Sur lighthouse and the California Sea Otter Game Refuge likely made of? Explain your reasoning.

_____

_____

_____

_____

(f) Suggest a sequence of events by which the Point Sur shoreline could have formed.

_____

_____

_____

_____

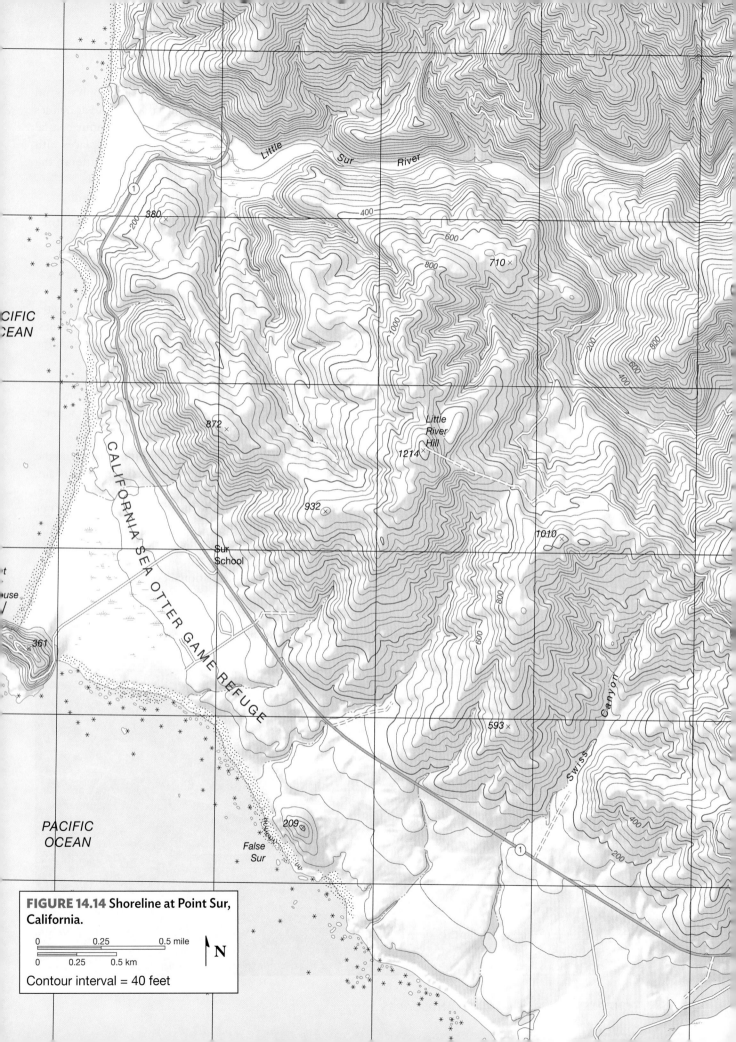

FIGURE 14.14 Shoreline at Point Sur, California.

Contour interval = 40 feet

14.3.5 Depositional Features

Prominent depositional features develop where there is an abundant supply of sand along shorelines. These range from continuous sand bars that extend for miles along the coast to small isolated beaches. The Gulf of Mexico and Atlantic coastal plains are underlain by easily eroded unconsolidated sediments, and places with familiar names display classic depositional features: Cape Cod, Cape Hatteras, the Outer Banks of North Carolina, the eastern Louisiana coast, Padre Island.

Common shoreline depositional features (**FIG. 14.15**) include the following:

- **Beaches** are the most common depositional features and consist of sand (or coarser sediment, coral and shell fragments, etc.—whatever is available).
- **Spits** are elongate sand bars attached at one end to the mainland. Some are straight, like those in Figure 14.15, but some are curved sharply and are called **hooks**.
- **Barrier islands** are elongate sand bars that lie offshore and are not connected to the mainland (e.g., the Outer Banks, Padre Island). Their name comes from the fact that they were barriers to early explorers, who had to search for inlets that would allow them to reach the shoreline.
- **Salt marshes/tidal marshes:** The area between a barrier island and the shoreline is typically a marshy wetland formed by sediment derived from the mainland. These areas are covered with vegetation tolerant to salt water; these plants are fully or partially submerged at high tide.

Sediment eroded from the mainland is deposited in bays between the shore and the barrier islands, forming marshy wetlands. These are important parts of the food chain, providing rich sources of nutrients for a wide range of aquatic organisms, and are important breeding areas for fish and the birds that feed on them. They are also part of our natural storm-protection system. If a storm surge manages to overflow the barrier island (as happened in Galveston, Texas, in 2008, for example), the wetlands act as a sponge, soaking up the water and lessening damage to the more densely inhabited mainland.

In Exercises 14.8 and 14.9, you will explore depositional evironments both in the present and the past.

FIGURE 14.15 Southwestern Long Island (NY) showing common shoreline depositional features. White areas on south shore of spits and barrier islands are *beaches.*

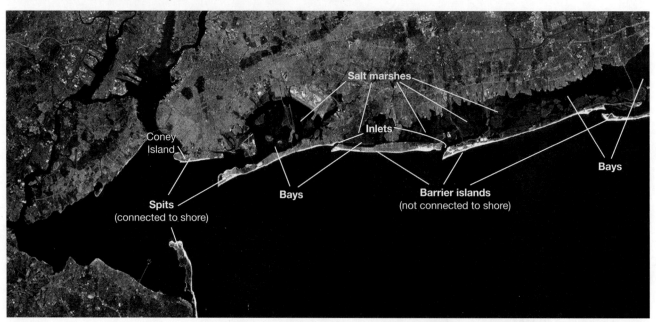

Name: _____   **Section:** _____

Course: _____   **Date:** _____

(a) Does the shoreline in **FIGURE 14.16** appear to be emergent or submergent? Explain.

_____

_____

(b) Identify and label the following depositional landforms on Figure 14.16: spit, bar, hook, barrier island, and beach.

(c) What evidence is there that sediment redistribution is taking place *landward* of the barrier island as well as on the barrier islands and spits that protect these areas?

_____

_____

(d) The rapid movement of sand by longshore drift could block access to the mainland by closing gaps in barrier islands and between spits. What steps can be taken to prevent futher erosion?

_____

_____

(e) Indicate the dominant direction of longshore drift. Explain your reasoning.

_____

_____

_____

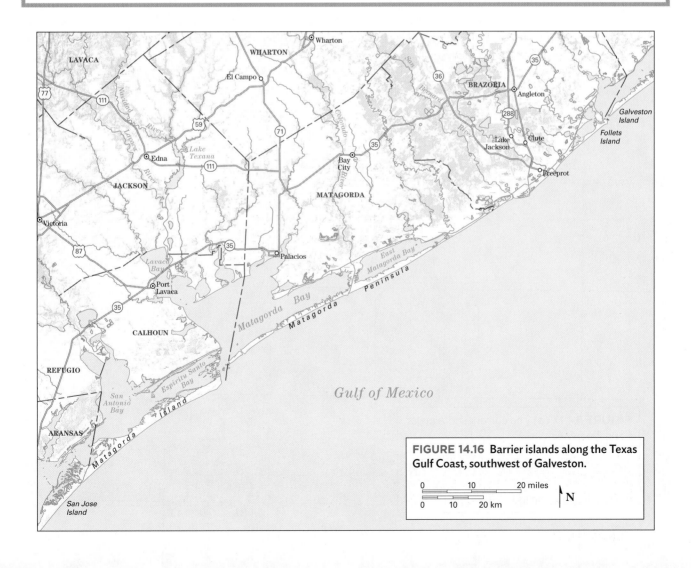

FIGURE 14.16 Barrier islands along the Texas Gulf Coast, southwest of Galveston.

ATLANTIC OCEAN

Race Point

Provincetown
Municipal
Airport

CAPE COD
NATIONAL SEASHORE

Grassy
Pond

50

50

50

Hatches
Harbor
Tidal Flat

Great
Pond

50

Pasture
Pond

6

Snake
Hills

Bennett
Pond

Duck
Pond

Clapps
Pond

6A

Provincetown

Herring Cove

Shank Painter
Pond

Provincetown
Harbor

Telegraph
Hill

Long Point

Dike

Tidal Flat

Wood
End

FIGURE 14.17 (a) The current shoreline of
Cape Cod, MA.

| 0 | | 0.5 | | 1 mile |
| 0 | | 0.5 | | 1 km |

N

Contour interval = 10 feet

CAPE COD BAY

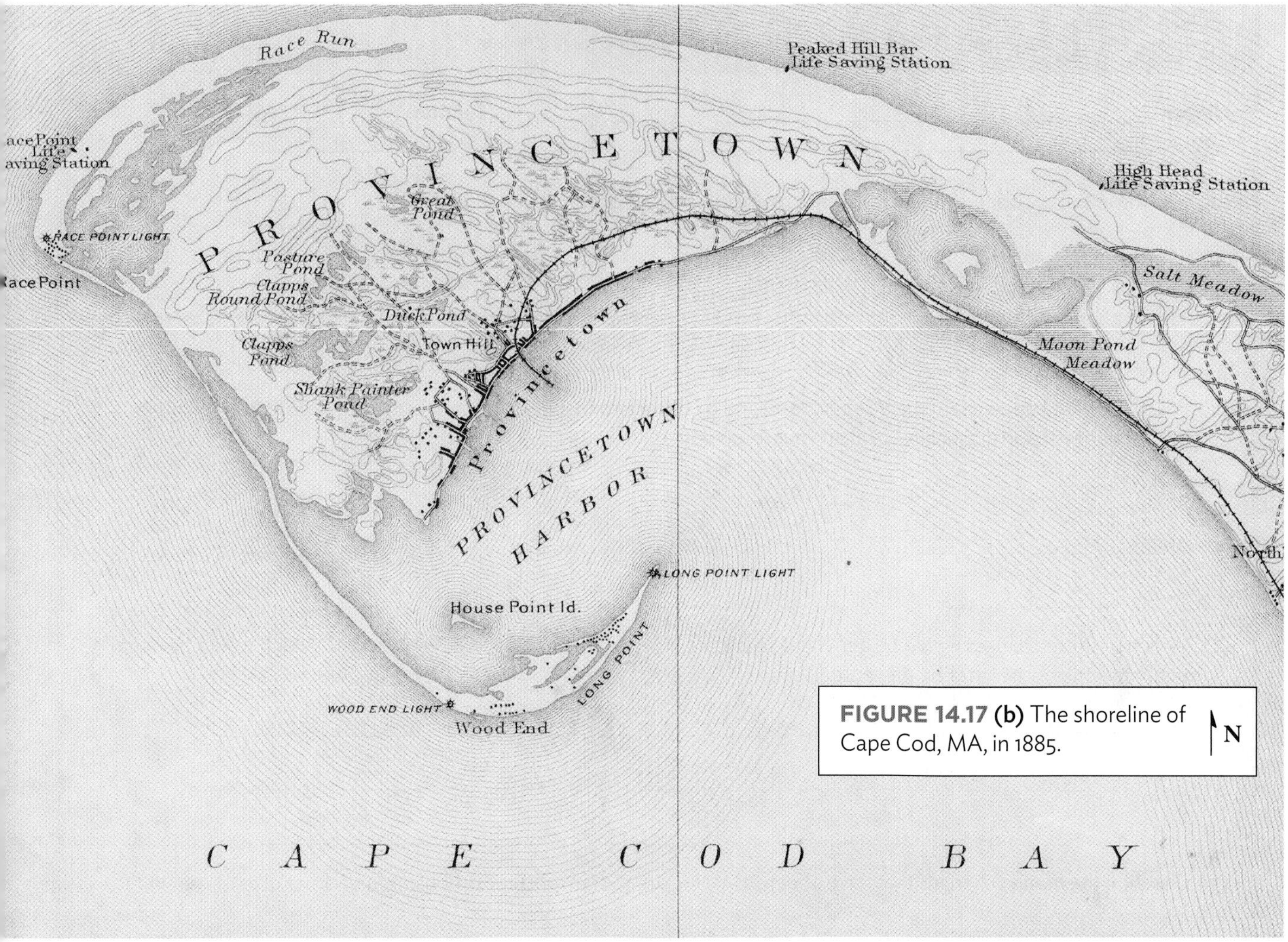

FIGURE 14.17 (b) The shoreline of Cape Cod, MA, in 1885.

Depositional Processes and Shoreline Landforms

Name: _____   Section: _____

Course: _____   Date: _____

FIGURES 14.17a–b are topographic maps of the northern end of Cape Cod made almost 100 years apart. As mentioned earlier, a terminal moraine here provides an abundant source of sand and gravel for shoreline processes. These two maps help illustrate how much change can occur along a coastline in a (geologically) short period. They also show how population pressures affect our use of limited, and therefore very valuable, shoreline space.

(a) What kind of landform is Long Point? _____

(b) Draw arrows on Figure 14.17a to indicate the direction of longshore currents in this area.

(c) Describe the *geologic* changes that have occurred in the century separating the compilation of these two maps.

_____

_____

_____

(continued)

Name: _____   Section: _____
Course: _____   Date: _____

(d) How have Cape Codders tried to prevent Provincetown Harbor from changing?

_____

_____

_____

(e) Based on what happened in the 98 years recorded by these two maps, which landforms might disappear? Which might change shape drastically? Explain your reasoning.

_____

_____

_____

(f) Which of these changes would be beneficial to people living or vacationing in Provincetown? Which would be negative? How might the latter be prevented?

_____

_____

_____

(g) Describe the *human* changes that have affected this area (e.g., transportation, housing, and other uses).

_____

_____

_____

14.4 Human Interaction and Interference with Shoreline Processes

People invest hundreds of thousands of dollars (or more) in shoreline homes and want to use those homes and the local beaches for a long time. But a storm can change a shoreline in just a few hours. The disconnect between what we want and how nature works has led us to use three expensive coastal management strategies: building seawalls to prevent wave erosion; renourishing eroded beach sand; and trapping sand moved by longshore drift with structures called *groynes*. These strategies work sometimes, but other times interfere with shoreline processes in such a way as to create new problems. We'll look first at the strategies, then at the problems.

14.4.1 Seawalls

To protect areas ravaged frequently by intense wave erosion, some communities choose to armor the shoreline with seawalls made of concrete, blocks of loose rock,

FIGURE 14.18 Examples of sea wall construction.

(a) A section of the Galveston, Texas seawall.

(b) Concrete blocks of a seawall.

or similar materials. These are designed to break the force of the waves and prevent further shoreline erosion. The 1900 hurricane that devastated Galveston, Texas, was a wake-up call for that community, and today the barrier island city is protected by an extensive seawall as shown in **FIGURE 14.18a.** Seawalls can be made with different designs and materials, like concrete as shown in **FIGURE 14.18b.**

14.4.2 Beach Nourishment

A single storm can wreak havoc along an unprotected beach, eroding vast amounts of sand as shown in **FIG-URE 14.19a.** The most common remedy is to do in a short time what it would take nature decades to do: replace the beach (renourishment) by dredging sand from offshore, pumping it onto the beach, and spreading it out with bulldozers (**FIG. 14.19b**).

FIGURE 14.19 Beach erosion and renourishment.

(a) Effect of erosion along a sandy shore.

(b) Sand pumped from offshore areas to the beach is "redeposited" by bulldozers.

FIGURE 14.20 Aerial view showing how groynes are designed to trap sediment moved by longshore drift.

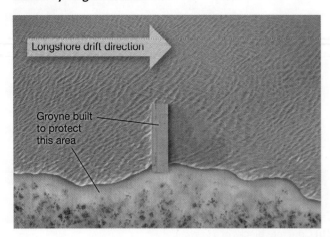

Longshore drift direction

Groyne built to protect this area

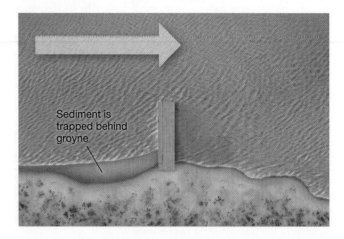

Sediment is trapped behind groyne

14.4.3 Groynes

Communities on barrier islands often use their knowledge of the longshore drift that built the islands to try to preserve their beaches. They build structures perpendicular to the shoreline in an attempt to trap moving sand and prevent its loss to downdrift areas (**FIG. 14.20**). These structures are called **groynes** (or commonly, groins).

14.4.4 Some Consequences of Human Interaction

Beach management practices sometimes backfire because the effects of building a seawall or groyne were not thought through fully. The most common problem is that while the shoreline is protected in one area, the seawall or groyne concentrates erosion in different areas, creating problems where there were none previously. For example, waves crashing against a seawall may remove the sand that accumulates naturally along the shoreline, doing exactly the opposite of what was intended. Or, as shown in **FIGURE 14.21**, sand trapped on the updrift side of a groyne is no longer available to replenish the beach naturally on the downdrift side, so the beach is preserved in one place but eroded in another. Exercise 14.10 gives you some experience recognizing these common problems.

FIGURE 14.21 Aerial view of shoreline showing potential negative effect of groyne construction.

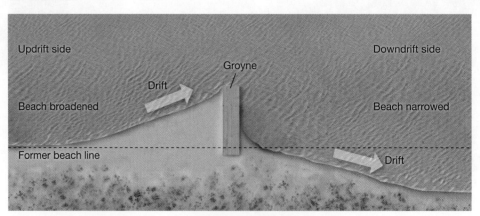

Updrift side

Downdrift side

Groyne

Drift

Beach broadened

Beach narrowed

Former beach line

Drift

Name: _____ Section: _____

Course: _____ Date: _____

Examine the two photographs directly below showing seawalls built on the Pacific and Gulf coasts. The seawalls successfully prevented further shoreline retreat in both places but also had unanticipated negative results.

(a) What negative impacts do you recognize?

_____

_____

(b) The photograph below shows a groyne field and a seawall protecting one side of the hook at North Avenue Beach, Chicago, Illinois.

  i. Draw an arrow to indicate the direction of longshore drift

  ii. Describe the width of the beach in the areas between the groynes. Suggest an explanation for this pattern.

_____

_____

_____

(continued)

Name: _____ Section: _____

Course: _____ Date: _____

iii. The following photograph of Westhampton, New York, shows the direction of longshore drift. Note the series of groynes designed to protect the narrow barrier island. What problem do you recognize beyond the groynes (closer to the foreground)?

_____

_____

14.5 When Shorelines Become Dangerous

Residents of Thailand, Indonesia, Japan, the Gulf Coast from Florida to Texas, and the east coast from Florida to New York City need no reminder of the dangers of living along a shoreline. Hurricanes (typhoons in Asia) and tsunamis strike quickly and with force that overwhelms our coastal defenses. In contrast, the gradual rise of sea level is almost imperceptible but will eventually disrupt society more than any single catastrophic event. This section explores three ways in which shorelines become potential natural disasters.

14.5.1 Sea-Level Change

The slow rise of sea level is already causing problems worldwide that will worsen in the future. By 2100, moderate estimates project a further rise of about 6.5 feet (2 m), and in a few hundred years a rise of 26 feet. The highest point in the island nation of Maldives in the Indian Ocean is 8 feet above sea level, so the 350,000 Maldives citizens will be without a country within your lifetime. Exercise 14.12 will examine problems caused by sea-level rise.

14.5.2 Coastal Storms

When the normal force of wind is multiplied fivefold in winter "nor'easters" in New England or tenfold in hurricanes, wave-related processes are amplified catastrophically. Typical coastal storms come from a single direction and are easier to prepare for than hurricanes, whose wind velocities can be above 150 miles per hour and whose direction can shift rapidly.

Hurricanes and typhoons, their Pacific Ocean relatives, are enormous storm systems that develop when an atmospheric disturbance passes over warm ocean water (more than 80°F). The storm absorbs energy from the ocean water and forms a low-pressure system as hot air rises. Low-level winds flow toward the center of the system, with a swirling counterclockwise pattern caused by the Coriolis effect (**FIG. 14.22**).

Heat from the ocean adds energy and moisture, and water vapor condensing at high altitudes adds even more energy. If winds in the upper atmosphere are weak, they cannot prevent the storm from intensifying to hurricane levels (wind velocities greater than 74 miles per hour). While most wind associated with hurricanes is horizontal, winds near the center are a vertical downdraft of warm air that surrounds the eye, a generally clear and ominously calm sector of the storm. Most hurricanes are around 300 miles wide, including the eye, which is typically 20 to 40 miles across.

Hurricanes are pushed slowly by *steering currents* in the atmosphere, generally at 10 to 15 miles per hour. Winds throughout the lower atmosphere, ocean temperature, and interaction with landmasses can make it difficult to predict the path of a hurricane. Hurricanes weaken after making landfall because they are no longer nourished by the warm ocean water, but they may strengthen if they cross over water again. This happened to Katrina, which was only a category 1 hurricane when it hit Florida but grew to category 5 as it headed across the Gulf of Mexico toward New Orleans.

The hazards posed by hurricanes were demonstrated all too well by the effects of Katrina on New Orleans in 2005 and Sandy on New York City in 2013. Both cities were flooded and damaged by strong storm surge (**FIG. 14.23a, b**), but New York's vast underground infrastructure proved vulnerable in ways that New Orleans escaped.

A common misconception about coastal storms is that wind is the major hazard. Winds of 100 to 175 miles per hour are truly dangerous, but the storm surge, a wall of water driven onshore by the hurricane, is much more hazardous. Katrina's storm surge was estimated at 30 feet above normal sea level; it carried boats, houses, and other debris inland, causing widespread destruction and wiping out entire communities.

In addition to storm surge, coastal flooding typically results from heavy rain that accompanies hurricanes, sometimes as much as 15 inches. Rain from Katrina raised the level of Lake Pontchartrain, which breached the levees separating it from New Orleans (**FIG. 14.23c**), and storm surge inundated New York's network of subway and commuter tunnels (**FIG. 14.23d**). Salt water and electrical utilities should never mix—**FIGURE 14.23e** shows the result in the barrier island Breezy Point neighborhood of New York.

Problems related to a major storm last much longer than the storm itself. Floodwaters damage buildings, which must be inspected to guarantee their safety. Saltwater damage must be repaired, bacteria and mold disinfected, and sand deposited by storm surge removed. Downed trees block traffic, making it difficult to rescue isolated families and deliver emergency food and medicine. Water treatment plants overwhelmed by storm surge dump millions of gallons of sewage into the flood areas. Exposure to toxic materials in floodwaters, like gasoline from damaged gas stations and cars, is a long-term health problem. Exercise 14.11 examines how different communities might anticipate the strength and paths of hurricanes and some of the problems they cause.

FIGURE 14.22 Comparison of Hurricanes Katrina (a) and Sandy (b).

(a) Note the spiral form with a well-developed eye and swirling rain bands outlining the counterclockwise wind circulation pattern.

(b) Hurricane Sandy displayed the same spiral, counterclockwise wind circulation as Katrina, but its centre was less defined and, overall, had a larger ring of rain bands.

FIGURE 14.23 Damage caused by Hurricanes Katrina and Sandy.

(a) A neighborhood complex flattened by the storm surge in Biloxi, Mississippi.

(b) Storm surge hitting Manhattan.

(c) Approximately 80% of New Orleans was flooded when levees that protected the city from Lake Pontchartrain failed.

(d) New York's subway and highway tunnels were flooded, stranding many commuters. This station was lucky–some were completely filled, almost to street level.

(e) Contact between storm-surge saltwater and live electrical lines started fires that destroyed a large part of Breezy Point, New York.

Name: _____ Section: _____

Course: _____ Date: _____

Emergency planners must consider all the factors that determine how dangerous a storm will be: shoreline topography and composition, population density, and type of building construction as well as the strength and path of the storm. These conditions can vary widely over short distances and change rapidly if a storm changes direction.

Storm Path

The path of a hurricane determines whether an area is spared or severely damaged, as the path determines effective wind velocity at any point along the coast. The effective velocity of hurricane winds is a combination of the wind speed in the hurricane and that of the steering winds. In the figure below, Point A, in the direct path of the hurricane, will receive winds of 125 mph because the velocity of the steering winds adds to that of the hurricane itself.

(a) What will be the effective wind velocity of the hurricane at Point B? Explain.

_____

_____

_____

In general, the right side of a hurricane (the side that would be on your right if you were standing directly behind the hurricane) is the most dangerous because the full brunt of steering and wind hurricane velocities are there. The direction from which hurricane winds strike an area also depends on the precise storm path. Remembering that hurricane winds flow in a counterclockwise direction, answer the following questions referring to the figure below.

(b) As a hurricane passes, the direction of its winds shifts. Explain how a single storm could cause winds from opposite directions to affect an area.

_____

_____

_____

(c) The deep estuaries indicated by the arrows in the figure below are highly vulnerable to storm surge because their funnel shapes concentrate water to heights well above those of typical storm surge. Sketch the path that would cause the greatest storm surge into each estuary—the worst-case scenario—on the map below. Remembering Hurricane Katrina, makes the task easy for A, as it followed almost exactly the worst-case scenario.

Effect of storm path on effective wind velocity.

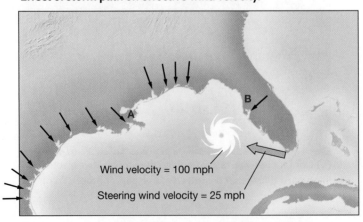

Wind velocity = 100 mph

Steering wind velocity = 25 mph

(continued)

Name: _____ Section: _____
Course: _____ Date: _____

(d) The following figure shows a hurricane in the area of the Bahama Islands. Using a colored pencil, sketch a hurricane path on the map that would cause the first winds from the hurricane to strike each of the following locations from the direction indicated:

- Location A from the south
- Location B from the north
- Location C from the east
- Location D from the north
- Location E from the west
- Location F from the east

Relationship between wind direction and hurricane path.

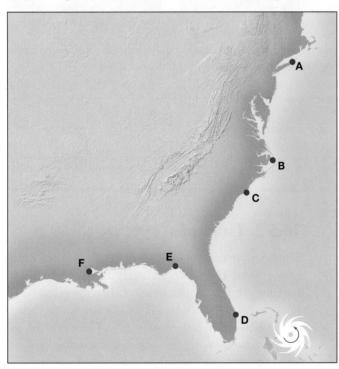

Landforms

Barrier islands and the wetlands behind them along shorelines are a natural line of defense against hurricane damage. The barrier islands break the force of the storm surge and the wetlands absorb water like a sponge, lessening inland flooding. Floridians are well aware of these phenomena.

(e) Compare the pictures of the east and west coasts of Florida on the following page. Which shoreline is more protected from a storm, Naples or Port. St. Lucie? Explain.

_____

_____

_____

(continued)

A tale of two coasts.

Naples on Florida's west coast.

Port St. Lucie on Florida's east coast.

(f) Developers often request rezoning so they can build highly profitable housing on barrier islands and fill in wetlands for shopping centers serving the new inhabitants. Why is this a problem?

_____

_____

_____

14.5.3 Tsunamis

Tsunamis are enormous waves generated by earthquakes or volcanic explosions in the ocean basins. Harmless and barely detectable in mid-ocean, they become walls of water tens of feet high when funneled into narrow coastal embayments. Tsunamis are discussed in more detail in Chapter 16; this section shows their effects on a single coastal city.

Earthquakes in Indonesia in 2004, Chile in 2010, and Japan in 2011 generated devastating tsunamis, waves that traveled across the Pacific Ocean at the speed of a jet plane and slammed into low-lying shorelines. Tsunami waves as high as 25–30 feet were reported from coastal cities in Japan. Television reports captured the awesome power of tsunamis, showing cars, large boats, and even small buildings being carried or smashed into pieces. **FIGURE 14.24** captures some of that power, even without the live action.

Eight hours later and thousands of miles away, a tsunami generated by the Sendai earthquake hit Hawaii and, later still, the northwest coast of the United States. The waves were much lower, and damage was nothing compared with that in Japan.

Tsunamis travel at approximately 500 mph. At that rate, Japanese coastal cities 60–100 miles from the earthquake epicenter had only a few minutes warning—not nearly enough to evacuate to higher ground. That is why the damage was so severe and the loss of life so great, even though Japan is the world's leader in earthquake readiness.

Nations surrounding the Pacific Ocean's "Ring of Fire" have cooperated in forming a tsunami warning network so that some warning can be given to island and continental coastlines throughout the ocean. Sensors on strategically placed buoys can track the passage of a tsunami and relay that information to civil defense workers. **FIGURE 14.25** shows estimated travel times for a tsunami generated by the Sendai earthquake.

FIGURE 14.24 Damage from tsunamis generated by the Sendai, Japan, earthquake (March 10 and 11, 2011).

(a) Tsunami slamming into Japanese coastline.

(b) Sendai, Japan, airport inundated by tsunami water and debris.

(c) Debris, including houses and a large ship, carried by the tsunami into the city center of Kesunnuma in northeastern Japan.

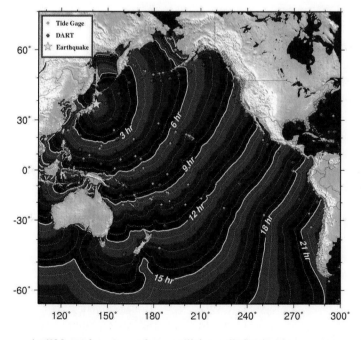

FIGURE 14.25 Tsunami tracking stations and estimated arrival times from the Sendai, Japan, earthquake. (DART = Deep-ocean Assessment and Reporting of Tsunamis sensors operated by NOAA.)

At 500 mph, some places will have little time to prepare, while others may have many hours. Note that the west coast of South America had nearly a full day to evacuate coastal areas, California and Oregon about 10 hours, and Honolulu about 8 hours. In Exercise 14.12, you will consider the risks of damage to different shorelines of the United States and make recommendations on how to reduce those risks.

Name: _____   Section: _____
Course: _____   Date: _____

? **What Do You Think**   Half of the population of the United
States lives within 50 miles (80 km) of a shoreline. Disasters caused by
recent tsunamis, hurricanes, and typhoons are making insurance companies
reexamine risks and reevaluate premiums for insurance along shorelines—or
consider if they should even offer insurance in some areas. They rely on geologists' expertise, and you have been contacted by a company for your advice
about whether there should be different policies and/or rates for the east and
gulf coasts of North America as compared with the west coast. *Your job is to
outline the factors that control potential damage to coastal properties in these two
regions.* Questions to consider:

- Do the regions have the same risk of damage by tsunamis and hurricanes?
- What factors determine the amount of potential damage in each of the regions?
- How would continued global climate change affect risk? How far inland would a 5-foot
  sea-level rise shift the shoreline? A 25-foot rise?

To help, six shoreline profiles have been provided, each ~5 miles long from either the Gulf of Mexico, or the Pacific or Atlantic coasts of the United States. On a separate sheet of paper, provide your
report and recommendations for each coast.

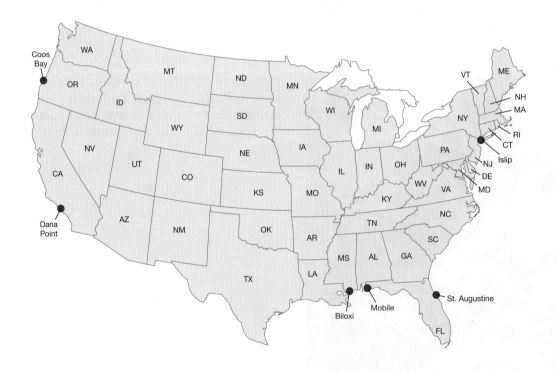

(continued)

Name: _____   Section: _____

Course: _____   Date: _____

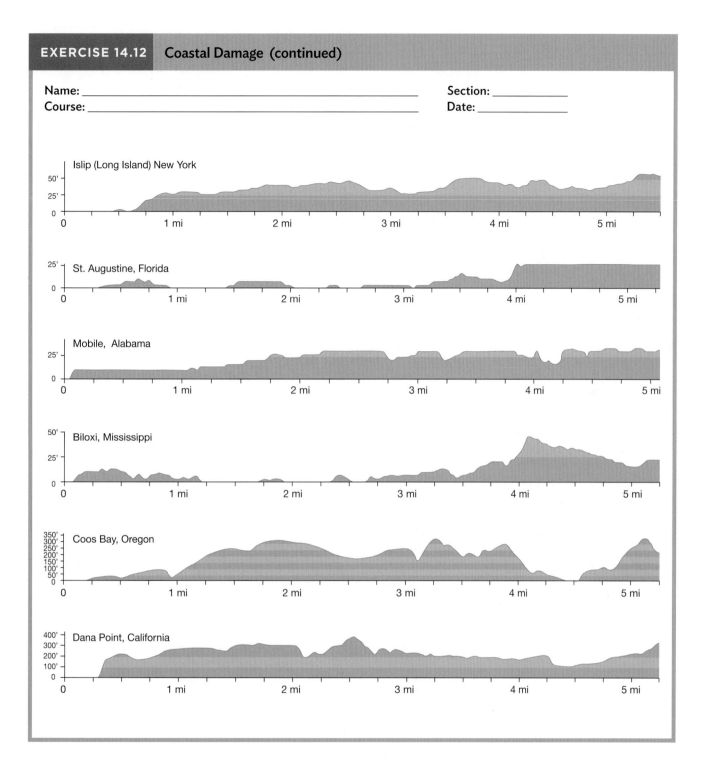

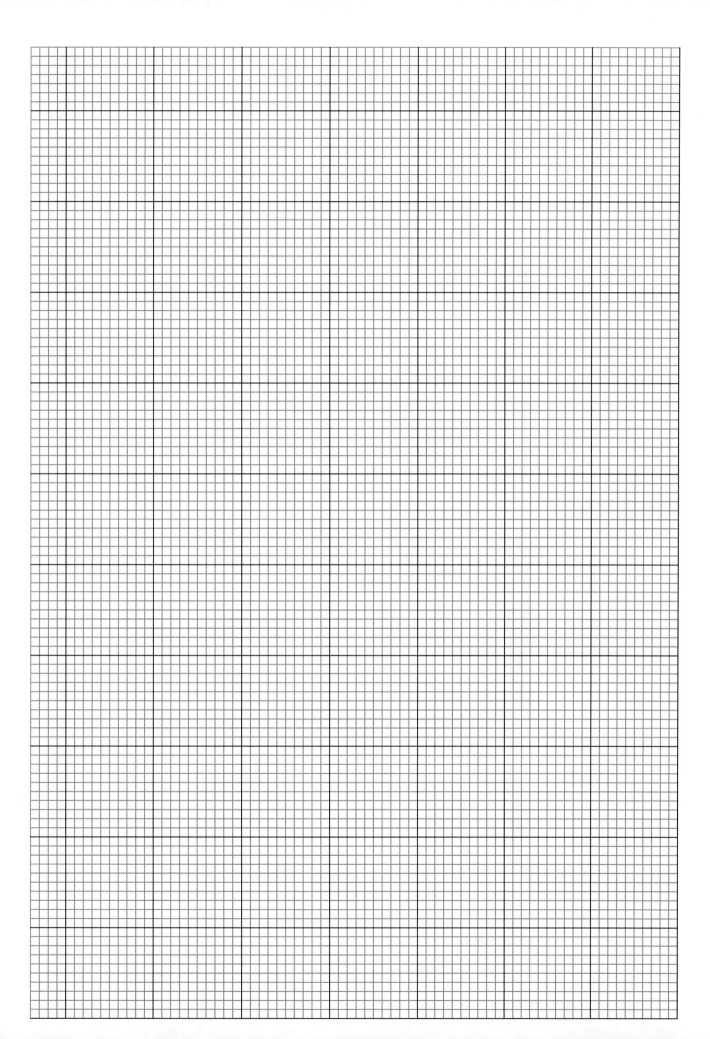

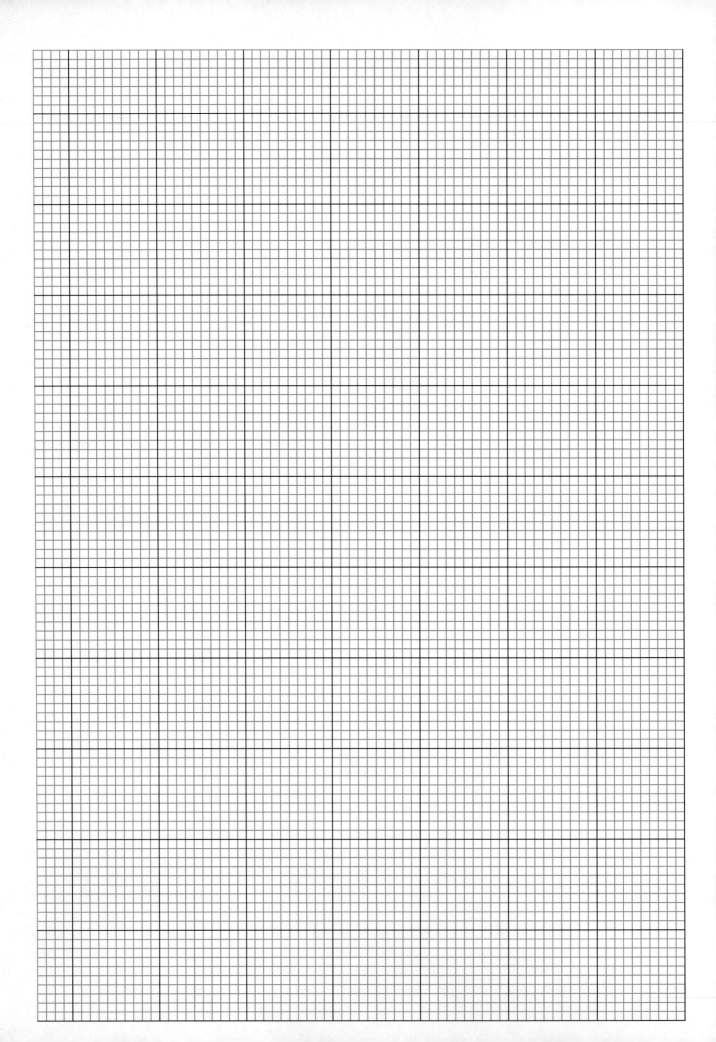

15

Interpreting Geologic Structures on Block Diagrams, Geologic Maps, and Cross Sections

Stress applied to rock produces deformation, such as the recumbent folds and thrust fault seen in this image of a coastal cliff in Cornwall, United Kingdom.

LEARNING OBJECTIVES

- Become familiar with common geologic structures such as folds and faults
- Visualize structures in three dimensions using block diagrams, maps, and cross sections
- Recognize the presence of folded and faulted rock from landscape features
- Interpret the geologic structure of an area from a geologic map
- Learn how to read a geologic map of a region

MATERIALS NEEDED

- Colored pencils
- A fine-tipped black pen
- Tracing paper
- A pair of scissors and tape
- A protractor
- A straightedge

15.1 Introduction

The Earth is a dynamic place! Over time, lithosphere plates move relative to one another: at convergent boundaries, one plate sinks into the mantle beneath another; at rifts, a continental plate stretches and may break apart; at a mid-ocean ridge, two oceanic plates move away from each other; at a collision zone, continents press together; and at a transform boundary, two plates slip sideways past each other. All these processes generate stress that acts on the rocks in the crust. In familiar terms, *stress* refers to any of the following (**FIG. 15.1**): **pressure**, which is equal squeezing from all sides (Fig. 15.1a); **compression**, which is squeezing or squashing in a specific direction (indicated by the inward-pointing arrows in Fig. 15.1b); **tension**, which is stretching or pulling apart (indicated by the outward-pointing arrows in Fig. 15.1c); **shear**, which happens when one part of a material moves relative to another part in a direction parallel to the boundary between the parts (indicated by adjacent arrows pointing in opposite directions in Fig. 15.1d). Exercise 15.1 relates different types of stress to everyday processes.

The application of stress to rock produces **deformation**, which includes many phenomena, such as the displacement of rocks on sliding surfaces called **faults**, the bending or warping of layers to produce arch-like or trough-like shapes called **folds**, or the overall change in the shape of a rock body by thickening or thinning. Under certain conditions, a change in the shape of a rock body produces **foliation**, a fabric caused by the alignment of platy or elongate minerals.

The products of deformation, such as faults, folds, and foliations, are called **geologic structures**. Some geologic structures are very small and can be seen in their entirety within a single hand specimen. Typically, however, geologic structures in the Earth's crust are large enough that they affect the orientation and geometry of rock layers, which in turn may control the pattern of erosion and, therefore, the shape of the land surface.

Geologists represent the shapes and configurations of geologic structures in the crust with the aid of three kinds of diagrams. A **block diagram** is a three-dimensional representation of a region of the crust that depicts the configuration of structures on the ground (the map surface) as well as on one or two vertical slices

FIGURE 15.1 Kinds of stress.

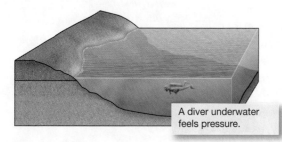

(a) Pressure.

A diver underwater feels pressure.

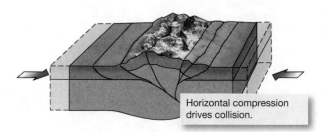

(b) Horizontal compression.

Horizontal compression drives collision.

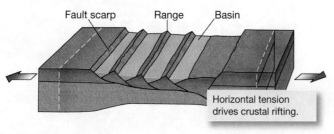

Fault scarp   Range   Basin

Horizontal tension drives crustal rifting.

(c) Horizontal tension.

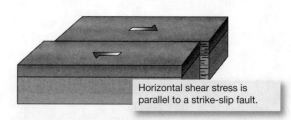

Horizontal shear stress is parallel to a strike-slip fault.

(d) Horizontal shear.

into the ground (cross-section surfaces). A **geologic map** represents the Earth's surface as it would appear looking straight down from above, showing the boundaries between rock units and where structures intersect the Earth's surface. A **cross section** represents the configuration of structures as seen in a vertical slice through the Earth. **FIGURE 15.2** shows how these different representations depict Sheep Mountain in Wyoming.

EXERCISE 15.1   **Picturing Stress**

Name: _____   Section: _____

Course: _____   Date: _____

For each of the phenomena described below, name the stress state involved.

(a) You spread frosting on a cake with a knife. The frosting starts out as a thick wad, then smears into a thin sheet.

_____

(b) You step on a filled balloon until the balloon flattens into a disk shape.

_____

(c) You pull a big rubber band between your fingers so that it becomes twice its original length.

_____

(d) A diver takes an empty plastic milk jug, with the lid screwed on tightly, down to the bottom of a lake. The jug collapses inward from all sides.

_____

The purpose of this chapter is to help you understand the various geometries of geologic structures and develop the ability to visualize structures and other geologic features by examining block diagrams, geologic maps, and cross sections. In addition, this chapter will help you to see how the distribution of rock units, as controlled by geologic structures, influences topography, as depicted on topographic maps and digital elevation maps (DEMs). Geologic structures can be very complex, and in this chapter we can only work with the simplest examples. Again, our main goal here is to help you develop the skill of visualizing geologic features in three dimensions.

15.2 Beginning with the Basics: Contacts and Attitude

15.2.1 Geologic Contacts and Geologic Formations

When you looked at Figure 15.2d, you saw patterns of lines. What do these lines represent? Each line is the trace of a **contact**, the boundary between two geologic units. In this context, a **trace** is simply the line representing the intersection of a planar feature with the plane of a map or cross section; a *unit* may be either a **stratigraphic**

FIGURE 15.2 Geology of Sheep Mountain in Wyoming.

(a) Oblique air photo.

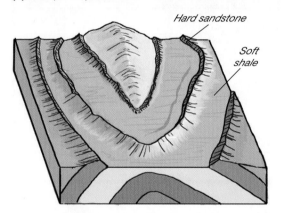

Hard sandstone

Soft shale

(b) Block diagram.

(c) Cross section.

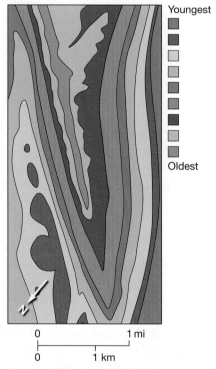

Youngest

Oldest

(d) Geologic map. The color bands represent rock units. The map is oriented to correspond with the photo in part (a).

0   1 mi

0   1 km

formation (a sequence of sedimentary and/or volcanic layers that has a definable age and can be identified over a broad region), an igneous intrusion, or an interval of a specified type of metamorphic rock. Geologists recognize several types of contacts: (1) an **intrusive contact** is the boundary surface of an intrusive igneous body; (2) a **conformable contact** is the boundary between successive beds, sedimentary formations, or volcanic extrusions in a continuous stratigraphic sequence; (3) an **unconformable contact** (or **unconformity**) occurs where a period of erosion and/or deposition has interrupted deposition; and (4) a **fault contact** is where two units are juxtaposed across a fault. Throughout this chapter, you'll gain experience interpreting contacts, but to be sure you understand the definitions from the start, complete Exercise 15.2.

Name: _____   Section: _____
Course: _____   Date: _____

In the figure below, each arrow points to one of the four basic kinds of contacts. Add the labels.

Contacts.

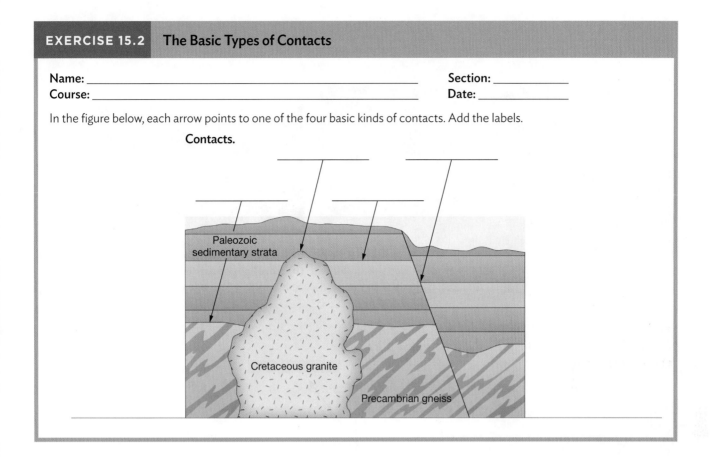

15.2.2 Describing the Orientation of Layers: Strike and Dip

You can efficiently convey information about the orientation, or **attitude**, of any planar geologic feature, such as a bed or a contact, by providing two numbers. The first number, called the **strike**, is the compass direction of a horizontal line drawn on the surface of the feature (**FIG. 15.3a**). You can think of the strike as the intersection between a horizontal surface (e.g., the flat surface of a lake) and the surface of the feature. We can give an approximate indication of strike by saying "the bed strikes northeast" or we can be very exact by saying "the bed has a strike of N 45° E," meaning that there is a 45° angle between the strike line and due north, as measured in a horizontal plane. The second number, called the **dip**, is the angle of tilt or the angle of slope of the bed, measured relative to a horizontal surface. A horizontal bed has a dip of 00°, and a vertical bed has a dip of 90°. A bed dipping 15° has a gentle dip, and a bed dipping 60° has a steep dip. The direction of dip is perpendicular to the direction of strike (**FIG. 15.3b, c**).

Because strike represents a line with two ends, a strike line actually trends in two directions: a strike line that trends north must also trend south; one that trends northwest must also trend southeast. How do we pick which direction a strike line trends? By convention, strike is read relative to north, so you will generally only see strikes described as angles east or west of north, or due north, east, or west (**FIG. 15.3d**). Thus, the beds on the left side of Figure 15.3A strike N 38°E or 38° east of north. Also by convention, beds that dip directly to the north are considered to strike west, while those that dip directly to the south strike east.

FIGURE 15.3 Strike and dip show the orientation of planar structures.

(a) A strike is the intersection of a horizontal plane with the bed surface. Strike lines for two sets of beds oriented in opposite directions are shown here.

(b) Strike line and dip direction shown for sloping beds at Turner Falls in Massachusetts.

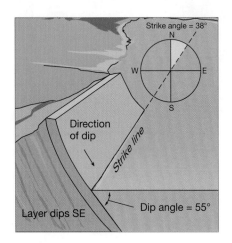

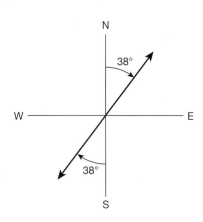

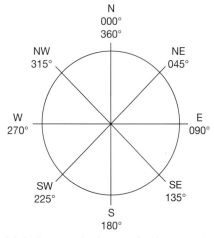

(c) This detail of the beds on the left side of (a) shows their strike and dip, which are used to describe the orientation, or attitude, of these beds.

(d) The strike line in (c) can be described as trending 38° northeast or 38°southwest. By convention, geologists will say the strike is N38°E.

(e) Strikes can also be specified by azimuths (compass angles between 000° and 360°). In azimuth notation, the strike line in (c) of N38°E is written as 038°.

When a strike is written as some angle relative to north, such as N 30° E, this is referred to as **quadrant notation**. However, geologists also use a shorthand notation for giving the strike and dip of a bed. We can also write the strike as a three-digit number, for we divide the compass dial into 360 degrees, or azimuths (**FIG. 15.3e**). A strike of 000° (or 360°) means the bed strikes due north; a strike of 045° means that the strike line trends 45° east of north (i.e., northeast); a strike of 090° is 90° east of north (i.e., due east); and a strike of 320° is 60° west of north. Writing strike as a three-digit number is referred to as **azimuth notation**. Because strike is measured relative to north, the allowable azimuth values are 000° to 090° for a northeast to east-trending strike line and 270° to 360° for a northwest to west-trending strike line. We write the dip as a two-digit number (an angle between 00° and 90°) followed by a general direction. Let's consider an example: if a bed has an attitude of 045°/60° NW, we mean that it strikes northeast and dips steeply northwest. A bed with an attitude of 053°/72° SE strikes *approximately* northeast and dips steeply to the southeast. Exercise 15.3 will give you some practice measuring strike and dip.

Name: _____ Section: _____
Course: _____ Date: _____

(a) Use a protractor to draw the indicated strike lines on each compass. Translate each azimuth into a direction (e.g., northeast).

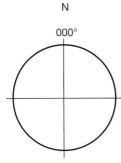

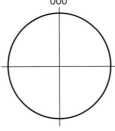

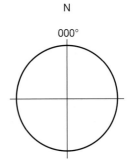

Strike: 060°   Strike: 340°   Strike: 090°

Direction: _____   Direction: _____   Direction: _____

(b) Use a protractor to measure the strikes below. Give the strike direction in both azimuth and quadrant notations, and describe the direction (e.g., northeast).

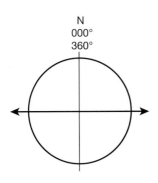

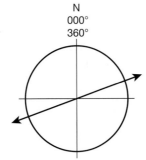

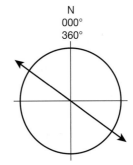

(Here the bed dips north.)

Azimuth: _____   Azimuth: _____   Azimuth: _____

Quadrant: _____   Quadrant: _____   Quadrant: _____

Direction: _____   Direction: _____   Direction: _____

(c) Use a protractor to measure the angle of dip for the two beds on the next page. The dashed lines indicate the level of the horizontal. The shaded surfaces are the surfaces of the beds. Using the strikes given and diagrams for each bed, indicate the rough direction of dip (e.g., west, northwest) for each bed. Remember, the direction of dip is perpendicular to the direction of strike.

(continued)

Name: _____   Section: _____

Course: _____   Date: _____

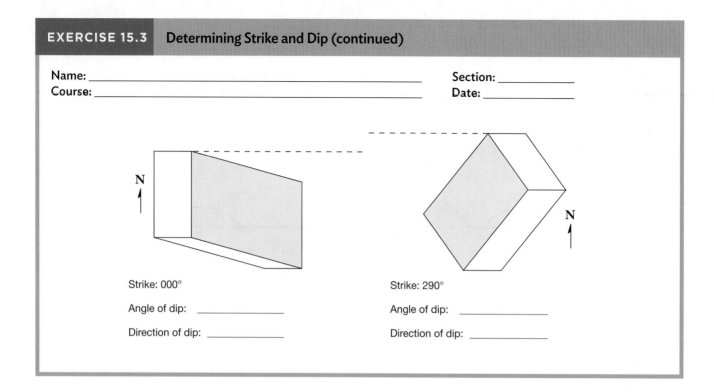

Strike: 000°

Angle of dip: _____

Direction of dip: _____

Strike: 290°

Angle of dip: _____

Direction of dip: _____

15.3 Working with Block Diagrams

We start our consideration of how to depict geologic features on a sheet of paper by considering block diagrams, which represent a three-dimensional chunk of Earth's crust and utilize the artist's concept of perspective (**FIG. 15.4a**). Typically, geologists draw blocks so that the top surface and two side surfaces are visible. The top is called the **map view**, and the side is a **cross-section view**. In the real world, the map-view surface would display the topography of the land surface, but for the sake of simplification, our drawings portray the top surface as a flat plane. In the following subsections, we introduce a variety of structures as they appear on block diagrams.

15.3.1 Block Diagrams of Flat-Lying and Dipping Strata

The magic of a block diagram is that it allows you to visualize rock units underground as well as at the surface. For example, **FIGURE 15.4b** shows three horizontal layers of strata. If the surface of the block is smooth and parallel to the layers, you can see only the top layer in the map view; the layers underground are visible only in the cross-section views. But if a canyon erodes into the strata, you can see the strata on the walls of the canyon, too (**FIG. 15.4c**).

Now, imagine what happens if the layers are tilted during deformation so that they have a dip. **FIGURE 15.4d** shows the result if the layers dip to the east. In the front cross-section face, we can see the dip. Because of the dip, the layers intersect the map-view surface, so the contacts between layers now appear as lines (the traces of the contacts) on the map-view surface. Note that, in this example, the beds strike due north, so their traces on the map surface trend due north. Also note that, in the case of tilted strata, the true dip angle appears in a cross-section face only if the face is oriented perpendicular to the strike. On the right-side face in Figure 15.4d, the beds look horizontal because the face is parallel to the strike. (On a randomly oriented cross-section face, the beds have a tilt somewhere between 0° and the true dip.) Practice drawing tilted strata in Exercise 15.4.

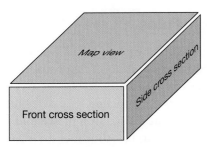

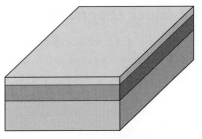

(a) Construction of a block diagram.

(b) A block diagram of three horizontal strata represented by different colors.

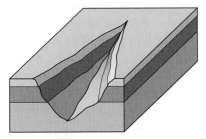

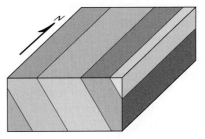

(c) A canyon cut into horizontal strata.

(d) A block diagram of east-dipping strata.

FIGURE 15.4 Block diagrams.

EXERCISE 15.4	**Portraying Tilted Strata on a Block Diagram**

Name: _____ **Section:** _____

Course: _____ **Date:** _____

(a) On the block diagram template below, sketch what a sequence of three layers would look like if their contacts had traces that trended north–south on the map view and dipped to the west at about 45°.

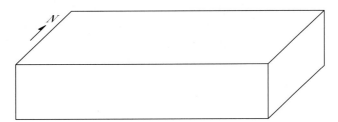

(b) On the block diagram template below, sketch what a sequence of three layers would look like if the traces had an east–west trend and dipped south at about 45°.

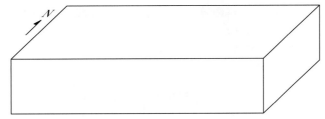

(c) On the block diagram template below, sketch what three layers would look like if the traces had a northeast–southwest trend and the layers dipped to the southeast at about 45°. (*Hint:* This is a bit trickier, because tilt appears in both cross-section faces.)

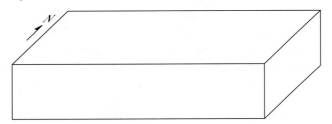

15.3.2 Block Diagrams of Simple Folds

When rocks are deformed and structures develop, the geometry of layers depicted on a block diagram become more complicated (**FIG. 15.5**). If deformation causes rock layers to bend and have a curve, we say that a **fold** has developed. Geologists distinguish between two general shapes of folds: an **anticline** is an arch-like fold whose layers dip away from the crest, whereas a **syncline** is a trough-like fold whose layers dip toward the base of the trough (Fig. 15.5a, b). Anticlines arch layers of rocks upward; synclines do the opposite—their strata bow *downward*. For the sake of discussion, the side of a fold is a **fold limb**, and the line that separates the two limbs (i.e., the line along which curvature is greatest) is the **fold hinge**. We can represent the hinge with a line and associated arrows on the map—the arrows point outward on an anticline and inward on a syncline. On a block diagram of the folds, we see several layers exposed (Fig. 15.5c). Note that the same set of layers appears on both sides of the hinge. (Exercise 15.5 allows you to discover that the age relations of layers, as seen on the map view, indicate whether a given fold is an anticline or syncline.) If the folded strata include a bed that is resistant to erosion, the bed may form topographic ridges at the ground surface (Fig. 15.5d). Note that the layers are repeated *symmetrically*, in mirror image across the fold hinges.

The hinge of a fold may be horizontal, producing a **nonplunging fold** (**FIG. 15.6a**), or it may have a tilt, or "plunge," producing a **plunging fold** (**FIG. 15.6b**); an arrowhead on the hinge line in Fig. 15.6 indicates the direction of plunge. Note that if the fold is nonplunging, the contacts are parallel to the hinge trace, whereas if the fold is plunging, the contacts curve around the hinge—this portion of a fold on the map surface is informally called the fold "nose." Note that anticlines plunge *toward* their noses, synclines *away from* their noses. Curving ridges may form if one or more of the beds that occur in the folded sequence are resistant to erosion.

In some situations, the hinge of a fold is itself curved, so that the plunge direction of a fold changes along its length. In the extreme, a fold can be as wide as it is long. In the case of down-warped beds, the result is a **basin**, a bowl-shaped structure; and in the case of up-warped beds, the result is a **dome**, shaped like an overturned bowl. Try Exercise 15.6 to see the differences between basins and domes.

FIGURE 15.5 The basic types of folds.

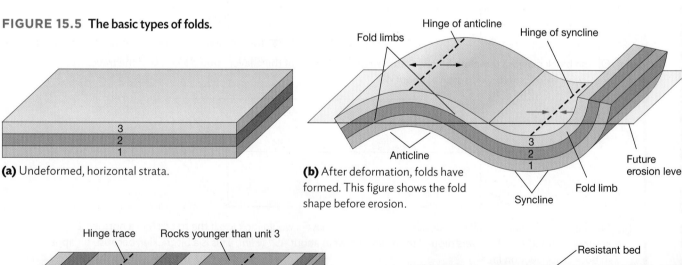

(a) Undeformed, horizontal strata.

(b) After deformation, folds have formed. This figure shows the fold shape before erosion.

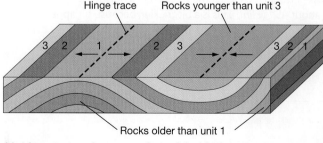

(c) After erosion, the map surface of the block exposes several different layers of strata.

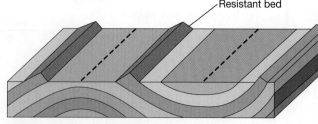

(d) Topographic ridges may form if one of the beds of the folded strata is resistant to erosion.

FIGURE 15.6 **Block diagrams showing the contrast between nonplunging and plunging folds.**

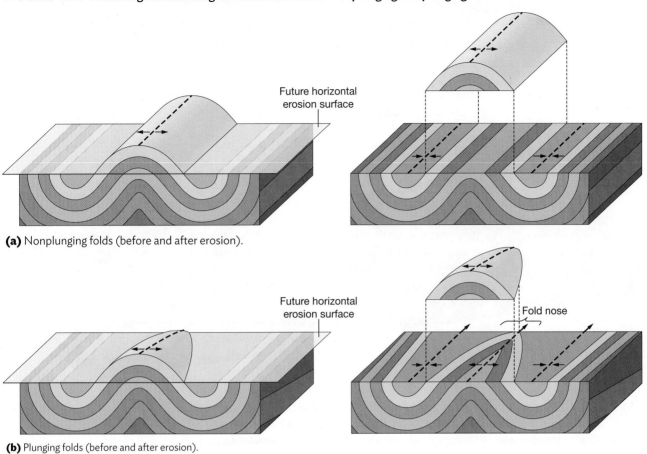

(a) Nonplunging folds (before and after erosion).

(b) Plunging folds (before and after erosion).

EXERCISE 15.5 | **Age Relations of Folded Strata**

Name: _____ Section: _____

Course: _____ Date: _____

Refer to the block diagram below. When erosion bevels the land surface, the map surface is like a horizontal slice through the fold. Keeping in mind that anticlines bow strata up and synclines bow them down, answer the following:

If resistant beds occur in the folded sequence, they form curving ridges.

(a) Are the strata along the hinge of the anticline younger or older than the strata on the exposed part of the limbs, as seen in the map-view surface? _____

(b) Are the strata along the hinge of the syncline younger or older than the strata on the exposed part of the limbs, as seen in the map-view surface? _____

Name: _____   Section: _____

Course: _____   Date: _____

(a) In the figure below, which of the following block diagrams illustrates a basin and which illustrates a dome? Add labels to the figure.

The difference between a basin and a dome.

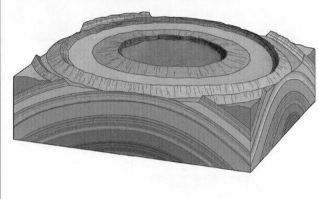
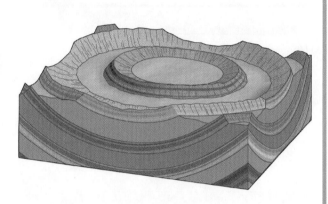

(b) Look at the distribution of strata in each of the blocks and circle the correct term in each statement.
  i. The center of a dome exposes (older/younger) strata relative to its outer edge.
  ii. The center of a basin exposes (older/younger) strata relative to its outer edge.

15.3.3 Block Diagrams of Faults

As we noted earlier, a fault is a surface on which one body of rock slides past another by an amount called the **fault displacement** (**FIG. 15.7**). Faults come in all sizes—some have displacements of millimeters or centimeters and are contained within a single layer of rock; others are larger and offset contacts between layers or between formations by many miles. Not all faults have the same dip—some faults are nearly vertical, whereas others dip at moderate or shallow angles. If the fault is not vertical, rock above the fault surface is the **hanging wall**, and rock below is the **footwall** (Fig. 15.7a).

Geologists distinguish among different kinds of faults based on the direction of displacement. **Strike-slip faults** tend to be nearly vertical, and the displacement on them is horizontal, parallel to the *strike* of the fault (Fig. 15.7b). On **dip-slip faults**, the displacement is parallel to the dip direction on the fault; if the hanging wall block moves up dip, it's a **reverse fault** (Fig. 15.7d), and if it moves down dip, it's a **normal fault** (Fig. 15.7c). If a reverse fault has a gentle dip (less than about 30°) or curves at depth to attain a gentle dip, then geologists generally refer to it as a **thrust fault** (Fig. 15.7e). Reverse and thrust faults form in response to compression, and normal faults form in response to tension.

You can recognize faulting, even if the fault surface itself is not visible (due to cover by soil or vegetation), if you find a boundary along which contacts terminate abruptly (**FIG. 15.8**). The configuration that you find depends on both the attitude of the fault and the attitude of the layers, as you will see in Exercise 15.7.

FIGURE 15.7 Hanging wall, footwall, and the classification of faults.

If you look across a strike-slip fault and the opposite side moved to your right, it's a right-lateral fault. If the opposite side moved to your left, it's a left-lateral fault.

This fault is left-lateral.

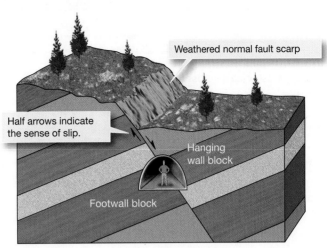

Weathered normal fault scarp

Half arrows indicate the sense of slip.

Hanging wall block

Footwall block

(a) The hanging wall is above the fault surface; the footwall is below.

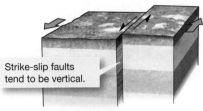

Strike-slip faults tend to be vertical.

(b) On a strike-slip fault, one block slides laterally past another, so no vertical displacement takes place.

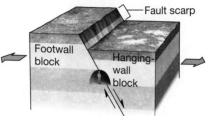

Fault scarp

Footwall block

Hanging-wall block

(c) Normal faults form during extension of the crust. The hanging wall moves down.

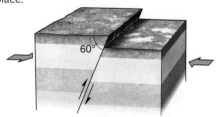

60°

(d) Reverse faults form during shortening of the crust. The hanging wall moves up and the fault is steep.

30°

(e) Thrust faults also form during shortening. The fault's slope is gentle (less than 30°).

EXERCISE 15.7 Faulted Strata on a Block Diagram

Name: _____   Section: _____
Course: _____   Date: _____

The following questions refer to the figures on the next page.

(a) The block in (a) shows a vertical fault cutting across a nonplunging syncline. Complete the block diagram by adding arrows to show the sense of slip across the fault and by adding colored bands for the appropriate stratigraphic units in the blank areas. What type of fault is it? _____

(b) The block in (b) shows a dip-slip fault. Is this a normal or reverse fault? _____

(continued)

Name: _____   Section: _____
Course: _____   Date: _____

(c) As you walk from west to east across the map surface of (b), you cross layer 3 more than once. Explain how the faulting caused this.

_____

_____

(d) The red line on the front cross-section face of (b) represents a drill hole. Does the drill hole cut through the complete stratigraphic section, or do you see repetition or loss of section?

_____

_____

(a) Vertical fault.

(b) Dip-slip fault.

FIGURE 15.8 **Examples of the consequences of faulting on strata.**

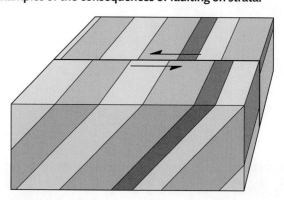

In this case, displacement on a strike-slip fault causes layers to terminate abruptly.

15.3.4 Block Diagrams of Unconformities

An **unconformity** is a contact that represents a period of nondeposition and/or erosion, as we noted earlier. Geologists recognize three different kinds: (1) at a **disconformity**, bedding above and below the unconformity is parallel, but there is a significant time gap between the age of the strata below and the age of the strata above; (2) at a **nonconformity**, strata are deposited on a "basement" of intrusive igneous and/or metamorphic rock; and (3) at an **angular unconformity**, the orientation of the beds above the unconformity is not the same as that below. Exercise 15.8 gives you a chance to distinguish among these three types. Unconformities will be examined in more detail in Chapter 17.

15.3.5 Block Diagrams of Igneous Intrusions

An igneous intrusion forms when molten rock (magma) pushes into or "intrudes" preexisting rock. Geologists distinguish between two general types of igneous intrusions. (1) **Tabular intrusions** have roughly parallel margins; these include wall-like intrusions called **dikes**, which cut across preexisting layering, and sheet-like intrusions called **sills**, which are parallel to layering. (2) **Plutons** are irregularly shaped, blob-shaped, or bulb-shaped intrusions. On a block diagram, you can generally distinguish among different types of intrusions based on their relationship with adjacent layering. To see how, try Exercise 15.9.

| EXERCISE 15.8 | Interpreting Unconformities On a Block Diagram |

Name: _____   Section: _____
Course: _____   Date: _____

(a) In the space provided below each block in the figure that follows, indicate what type of unconformity is shown.

Block diagrams of unconformities.

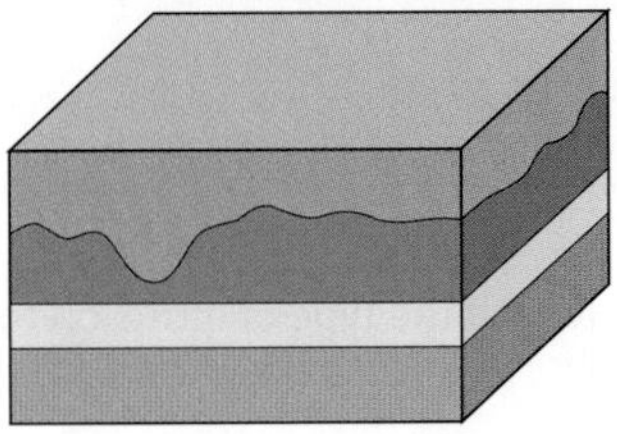

Block 1: 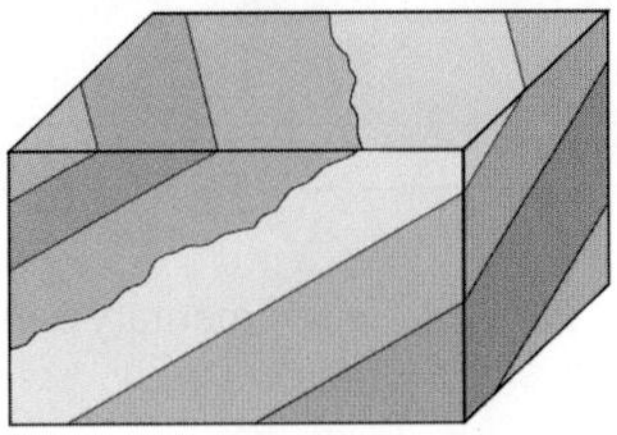 _____   Block 2: _____

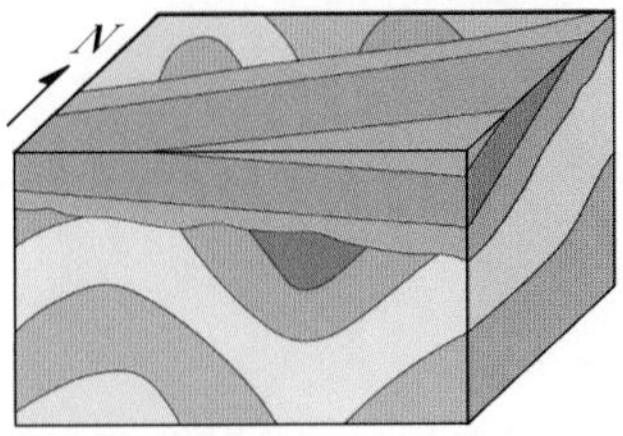

Block 3: 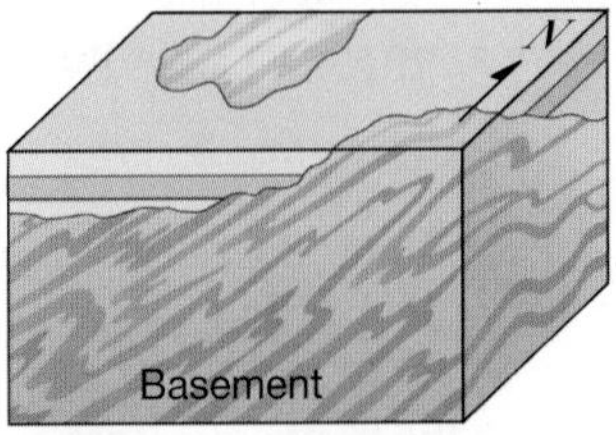 _____   Block 4: _____

(b) The sedimentary rocks in Block 2 were deposited as horizontal layers. What has happened to the layers? Was the unconformity originally horizontal or tilted? Explain.

_____

_____

(continued)

Name: _____ Section: _____

Course: _____ Date: _____

(c) In Block 3, in which direction are the post-unconformity strata dipping?

_____

(d) In Block 4, a gray area at the north edge of the block appears on the map surface. What geologic observation(s) could prove that the contact between the gray area and the sedimentary beds is an unconformity and not an intrusive contact?

_____

_____

_____

_____

EXERCISE 15.9 Interpreting Intrusions on a Block Diagram

Name: _____ Section: _____

Course: _____ Date: _____

(a) The following two blocks (on the next page) show sedimentary beds and intrusions. Match the type of intrusion to the appropriate letter on the block.

Pluton   Block 1: _____   Block 2: _____

Dike   Block 1: _____   Block 2: _____

Sill   Block 1: _____   Block 2: _____

(b) Using common sense to interpret cross-cutting relationships, list the sequence of intrusions for each block. If you can't determine an answer from the information shown, indicate so.

	Oldest	Middle	Youngest
Block 1:	_____	_____	_____
Block 2:	_____	_____	_____

(c) Analyses indicate the unlabeled intrusion in the front cross-section face of Block 2 is part of the same body as Intrusion A exposed on the top surface. Explain why you can't *see* the connection between the map-view exposure and the subsurface cross section on this block.

_____

_____

_____

(continued)

Name: _____ Section: _____

Course: _____ Date: _____

Blocks of igneous intrusions.

Block 1   Block 2

EXERCISE 15.10   **Completing Block Diagrams**

Name: _____ Section: _____

Course: _____ Date: _____

Four cutout block diagrams are provided at the end of the book for additional practice and to help visualize structures in three dimensions. Cut and fold the diagrams as indicated, and use tape to hold the tabs together to make three-dimensional block diagrams.

(a) Complete the blank cross-section panels for Block 1, and describe the structure present. Does the block show horizontal or tilted strata? Folds? Faults?

_____

_____

_____

_____

(b) Complete the map view and blank cross-section views for Blocks 2 and 3. Compare and contrast the structures in these two blocks.

_____

_____

_____

_____

_____

(continued)

Name: _____ Section: _____
Course: _____ Date: _____

(c) Complete the map view and cross-section panels for Block 4. Be sure to add arrows showing the direction of slip. Describe the nature of the faulting. Is it dip-slip? If so, is it normal or reverse? Is it strike-slip? If so, is it left-lateral or right-lateral? Explain your reasoning.

_____

_____

_____

_____

15.4 Geologic Maps

15.4.1 Introducing Geologic Maps and Map Symbols

Now that you've become comfortable reading and interpreting block diagrams, we can focus more closely on how to interpret geology on the map-view surface. A map that shows the positions of contacts, the distribution of rock units, the orientation of layers, the position of faults and folds, and other geologic data is called a **geologic map** (**FIG. 15.9a**). Contacts between rock units are shown by lines (traces), and the units themselves are highlighted by patterns and/or colors and symbols that indicate their ages. The orientation of beds, faults, and foliations, as well as the position of fold hinges, can be represented by strike and dip symbols. The map's **explanation** (or *legend*) defines all the symbols, abbreviations, and colors on the map. **FIGURE 15.9b**, a geologic map of the Bull Creek quadrangle in Wyoming, illustrates the components of a geologic map. Note that all maps should have a scale, north arrow, and explanation.

FIGURE 15.9 Geologic maps.

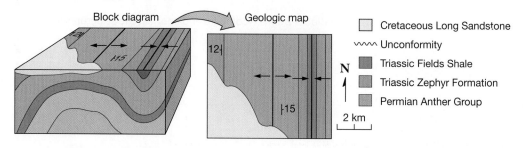

(a) A geologic map is the top surface of a block diagram. It shows the pattern of geologic units and structures as you would see them by looking straight down from above.

FIGURE 15.9 Geologic maps (continued)

M.L. Schroeder, 1976

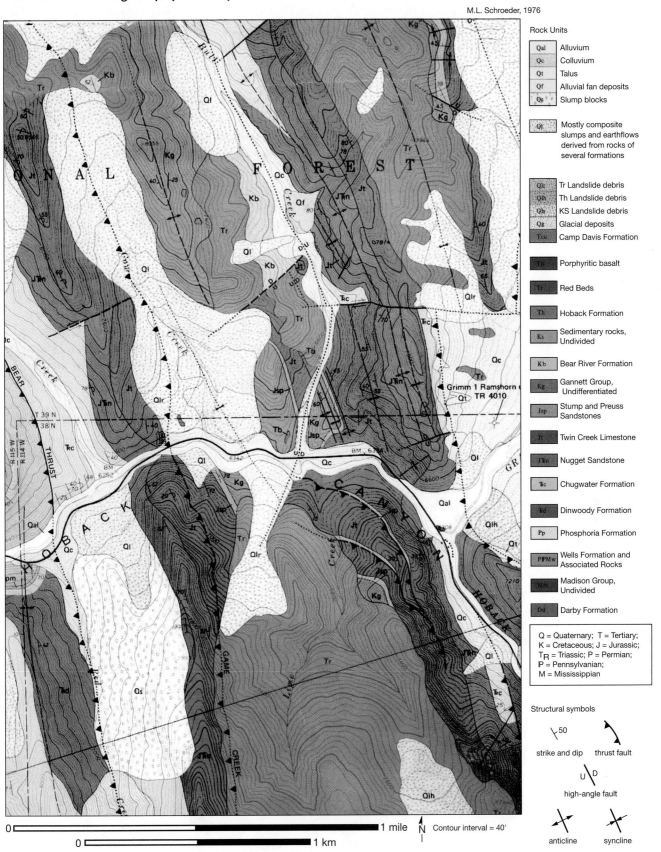

Rock Units

Qal	Alluvium
Qc	Colluvium
Qt	Talus
Qf	Alluvial fan deposits
Qs	Slump blocks
Ql	Mostly composite slumps and earthflows derived from rocks of several formations
Qlr	Tr Landslide debris
Qlh	Th Landslide debris
Qls	KS Landslide debris
Qg	Glacial deposits
Tcu	Camp Davis Formation
Tb	Porphyritic basalt
Tr	Red Beds
Th	Hoback Formation
Ks	Sedimentary rocks, Undivided
Kb	Bear River Formation
Kg	Gannett Group, Undifferentiated
Jsp	Stump and Preuss Sandstones
Jt	Twin Creek Limestone
JTRn	Nugget Sandstone
TRc	Chugwater Formation
TRd	Dinwoody Formation
Pp	Phosphoria Formation
PIPMw	Wells Formation and Associated Rocks
Mm	Madison Group, Undivided
Dd	Darby Formation

Q = Quaternary; T = Tertiary;
K = Cretaceous; J = Jurassic;
T$_R$ = Triassic; P = Permian;
IP = Pennsylvanian;
M = Mississippian

Structural symbols

├50
strike and dip

thrust fault

U|D
high-angle fault

anticline   syncline

0 ——————————————— 1 mile
0 ——————————————— 1 km

N

Contour interval = 40'

(b) Geologic map of the Bull Creek quadrangle, Teton and Sublette Counties, in Wyoming.

Let's begin our discussion of geologic maps by considering the various features that can be portrayed on these maps. (You will practice mapping features in Exercise 15.11.)

■ *Rock units*: Geologic maps show the different rock units in an area. These units may be bodies of intrusive igneous rock, layers of volcanic rock, sequences of sedimentary rocks, or complexes of metamorphic rock. The most common unit of sedimentary and/or volcanic rock is a stratigraphic formation, as noted earlier. These are commonly named for a place where it is well exposed. A formation may consist entirely of beds of a single rock type (e.g., the Bright Angel Shale consists only of shale) or it may contain beds of several different rock types (e.g., the Bowers Mountain Formation contains shale, sandstone, and rhyolite).

Typically, maps use patterns, shadings of gray, or colors to indicate the area in which a given unit occurs. On geologic maps produced in North America, an abbreviation for the map unit may also appear within the area occupied by the formation. This abbreviation generally has two parts: the first part represents the formation's age, in capital letters; the second part, in lowercase letters, represents the formation's name. For example, O indicates rocks of Ordovician age (Oce = Cape Elizabeth Formation, Osp = Spring Point Formation), and SO indicates Silurian or Ordovician age (SOb = Berwick Formation, SOe = Eliot Formation).

■ *Contacts:* Different kinds of contacts are generally shown with different types of lines. For example, a conformable or intrusive contact is a thin line, a fault contact is a thicker line, and an unconformity may be a slightly jagged or wavy line. In general, a visible contact is a solid line, whereas a covered contact (buried by sediment or vegetation) is a dashed line.

■ *Strike and dip:* On maps produced in North America, geologists use a symbol to represent the strike and dip of a layer. The symbol consists of a line segment drawn exactly parallel to the direction of strike and a short tick mark drawn perpendicular to the strike and pointing in the direction of dip (**FIG. 15.10a, b**). A number written next to the tick mark indicates the angle of dip. (It is not necessary to write a number indicating the strike angle because that is automatically represented by the map trend of the strike line.) Different symbols are used to represent bedding and foliation; in this book, we use only bedding symbols.

■ *Other structural symbols:* The explanation also includes symbols representing the traces of folds and faults. **FIGURE 15.10c** illustrates some of these symbols.

FIGURE 15.10 Indicating features on geologic maps.

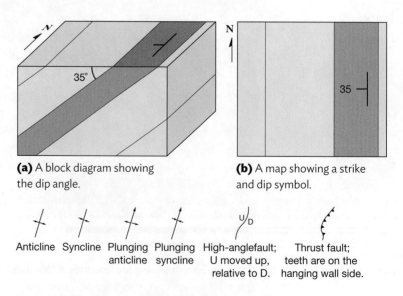

(a) A block diagram showing the dip angle.

(b) A map showing a strike and dip symbol.

Anticline Syncline Plunging anticline Plunging syncline High-anglefault; U moved up, relative to D. Thrust fault; teeth are on the hanging wall side.

(c) Basic structural symbols used on geologic maps.

Name: _____   Section: _____

Course: _____   Date: _____

(a) On the blank map below of a region with no hills or valleys, draw the appropriate strike and dip symbol next to the appropriate point. To do this, you must use a protractor and measure the angle between the north direction (the side edge of the map) and the strike angle. Then, look at the direction of dip so that you put the dip tick on the correct side of the strike symbol. These points are on a contact between two formations:

     A: 045°/30° SE

     B: 280°/10° SW

     C: 350°/25° W

(b) Based on the strike and dip symbols you show, draw a line representing the contact that passes through these points. Remember, the line needs to be parallel to the strike symbol.

_____

_____

(c) What is the structure shown by the structure symbols?

_____

_____

Blank map.

15.4.2 Constructing Cross Sections

We've seen that a cross section represents a vertical slice through the crust of the Earth. Thus, the sides of a block diagram are cross sections. If you start with a block diagram, you can construct the structure in the cross-section planes simply by drawing lines representing the contacts so that they connect to the contact traces in the map plane—the strike and dip data on the map tell you what angle the contact makes, relative to horizontal, and you use a protractor to draw the correct angle. If a fold occurs on the map surface, it generally also appears in the cross section.

So far, we've worked with data depicted on block diagrams. Now let's consider the more common challenge of producing a cross section from a geologic map. This takes a couple of extra steps—to see how to do it, refer to **FIGURE 15.11**.

On the left side of Figure 15.11a, you see a simple geologic map. The **line of section** (XX′) is the line on the map view along which you want to produce the cross section. The cross section is a vertical plane inserted into the ground along the line

FIGURE 15.11 **Constructing a cross section.**

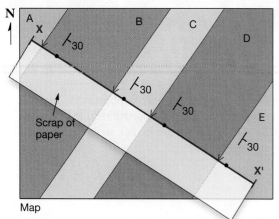

Scrap of
paper

Map

Step 1: Mark data locations on the cross-section paper.

(a) Example with gently dipping beds.

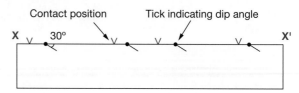

Contact position   Tick indicating dip angle

Step 2: Identify contact positions. Add dip marks at correct angles.

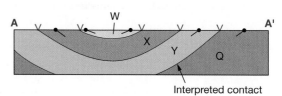

Step 3: Draw contacts so they obey location and dip data.

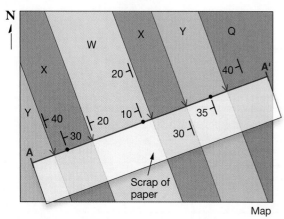

Scrap of
paper

Map

Step 1: Mark data locations on the cross-section paper.

(b) Example with folded beds (syncline).

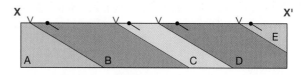

Contact position   Tick indicating dip angle

Step 2: Identify contact positions. Add dip marks at correct angles.

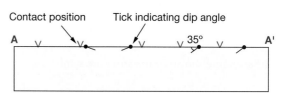

Interpreted contact

Step 3: Draw contacts so they obey location and dip data.

of section. Now, take a scrap of paper and align it with your line of section. Mark off the points where contacts cross the scrap of paper. Transfer these points to the cross-section frame on the right. Using a protractor, make a little tick mark indicating the dip of the contact; use the strike and dip symbol closest to each contact to provide this angle. Next, in the subsurface, sketch in lines that conform to the positions of the contacts and the dip angles (the contacts dip gently to the southeast). In Figure 15.11b, the contacts curve underground to define a syncline. Unless there is a reason to think otherwise, the layers should have constant thickness. Note that, because of this constraint, Layer Q in the second example appears in the lower left corner of the cross section; it would come to the surface to the west of the map area. Contrast your own cross sections in Exercise 15.12.

15.4.3 Basic Geologic Map Patterns

Geologic maps can get pretty complex, especially where structures are complicated or where topography is rugged. But, by applying what you have learned so far about block diagrams, you can start to interpret them. To make things simple, we begin with some very easy maps of areas that have no topography (i.e., the ground surface is flat), as in the block diagrams that you've worked with. Exercise 15.13 challenges you to look at a map and imagine the three-dimensional structure it represents. Also keep in mind that sedimentary and *extrusive* igneous rocks are commonly deposited in horizontal layers with the youngest layer at the top of the pile and the oldest at the bottom.

Name: _____   Section: _____

Course: _____   Date: _____

(a) The map surface of the block diagram in the figure below provides strikes and dips of the layers shown. From this information, show the layers with their proper angles in the front and side cross-section faces. Note that the strike of the layers is perpendicular to the front face of the block.

Block diagram with strikes and dips of the layers.

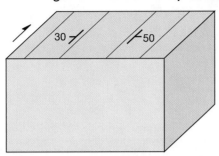

What kind of structure is shown? _____

(b) Complete the map view and cross-section views of the block diagrams below by showing a sequence of sedimentary rocks with the indicated orientations. Show at least three layers in each block, and plot the strike and dip symbol on the top surface.

 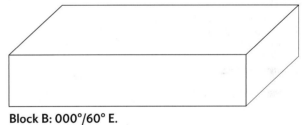

Block A: 090°/40° S.   **Block B: 000°/60° E.**

(c) Complete the cross-section views of the block below.

Block diagram for creating cross-section views.

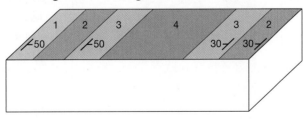

Earlier, we distinguished between nonplunging folds and plunging folds, in the context of discussing block diagrams. We can recognize these folds on geologic maps simply by the pattern of color bands representing formations—on a map, formation contacts of nonplunging folds trend parallel to the hinge trace, whereas those of plunging folds curve around the hinge trace so we can see the fold nose. Furthermore, we can distinguish between anticlines and synclines by the age relationships of the color bands—strata get progressively younger away from the hinge of an anticline and progressively older away from the hinge of a syncline. If the hinge isn't shown on the map, you can draw it in where the reversal of age takes place. See **FIGURE 15.12** for an example of a map and cross section of plunging folds.

FIGURE 15.12 Patterns of plunging folds.

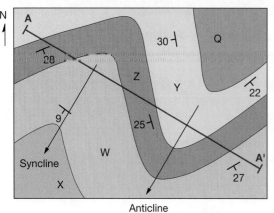

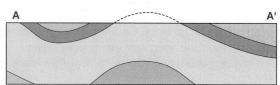

(b) Cross section of folds along the A–A' line.

(a) Map view of plunging anticline and syncline. Hinge traces of the folds are indicated by labeled lines, with arrows indicating the direction of plunge.

EXERCISE 15.13	Interpreting Simple Geologic Maps

Name: _____   **Section:** _____

Course: _____   **Date:** _____

For each of the following maps, identify the structure or geologic features portrayed. Does the map show a fault, fold, tilted strata, dike, pluton, unconformity, or some combination? To answer these questions, you may need to refer to the block diagrams presented earlier in the chapter. Remember—think in three dimensions! With a little practice, geologists learn to recognize the basic patterns quickly. *Note:* the geologic periods in the maps below in order from oldest to youngest are the Ordovician, Silurian, Devonian, Triassic, Jurassic, Cretaceous, and Tertiary.

Geologic maps.

Map A

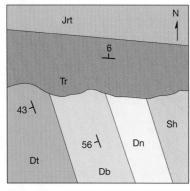

Map B

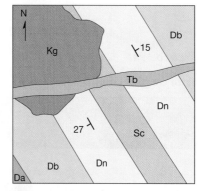

Map C

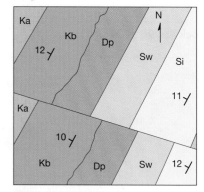

Jrt	Jurassic Tinta Fm.
Tr	Triassic Jones Sh.
Dt	Devonian Tella Fm.
Db	Devonian Bouser Fm.
Dn	Devonian Norfolk Sh.
Sh	Silurian Hallo Fm.

Tb	Tertiary basalt
Kg	Cretaceous granite
Da	Devonian Alsen Fm.
Db	Devonian Becraft Ls.
Dn	Devonian Norfolk Sh.
Sc	Silurian Cligfell Fm.

Ka	Cretaceous Altoona Fm.
Kb	Cretaceous Barrell Fm.
Dp	Devonian Potomoo Ls.
Sw	Silurian Wala Sh.
Si	Silurian Jack Fm.
Ot	Ordovician Trent Fm.

(a) Describe the geologic features of Map A.

i. _____-aged strata are overlain at an unconformity by _____-aged strata. What type of unconformity is it? _____

ii. The Triassic and Jurassic strata dip in a _____ direction at a _____ angle.

iii. The Silurian and Devonian strata dip in a _____ direction at a _____ angle.

(continued)

Name: _____   Section: _____
Course: _____   Date: _____

(b) Describe the geologic features of Map B.

    i. _____-aged strata are folded into a(n)_____. The hinge of the fold trends _____. Do the folded layers plunge? _____ Are the folded layers symmetrical? _____ If not, which side is steeper? _____

    ii. A _____ intrudes the western portion of the folded layers (give age, rock, and type of the intrusion).

    iii. A _____ intrudes both the folded layers and the intrusion described in (ii) (give age, rock, and type of the intrusion).

(c) Describe the geologic features of Map C.

    i. A _____ trending fault cuts strata that strike in a _____ direction and dip to the _____.

    ii. The fault is either a _____ fault or a dip-slip fault in which the _____ (N,S,E,W) block moved _____ relative to the _____ (N,S,E,W) block. Without seeing a cross section of the fault, it is not possible to determine if the fault is normal or _____.

    iii. _____-aged strata are overlain at an unconformity by _____-aged strata. What type of unconformity is it?_____

(d) If you walk from left to right (west to east) along the southern edge of Map A, are you walking "up section" (i.e., into rocks of progressively younger age) or "down section" (i.e., into rocks of progressively older age)? _____

15.4.4 Geologic Maps with Contour Lines

When the map surface is not flat, geologic maps become even more challenging to interpret. That's because the trace of a contact that appears on a geologic map depends on both the slope angle and the slope direction of the land surface as well as the strike and dip of beds. We introduce only two simple situations here—the pattern of horizontal contacts and the pattern of vertical contacts. You'll address more complex situations in other geology courses. Work through Exercise 15.14 to see how these patterns appear on a map.

Name: _____   Section: _____
Course: _____   Date: _____

(a) Look at the block diagram and the geologic map on the next page. What is the relationship between a horizontal contact and a contour line—parallel, perpendicular, or oblique? Keeping in mind the definition of a contour line, explain why. Geologists refer to the arrangement of valleys on this map as a *dendritic pattern,* because it resembles the veins of a leaf. Dendritic drainage patterns tend to occur where the strata are horizontal.

_____

_____

_____

(continued)

Name: _____   Section: _____
Course: _____   Date: _____

(b) The gray stripe is a vertical basalt dike. Remember that when you are looking at a map, you are looking straight down from the sky. With this in mind, explain why the dike appears as a straight line on the map.

_____

_____

Map pattern of horizontal strata in a valley.

- ▬ Dike
- — Contour
- — Contact
- ▬ Stream

15.5 Structures Revealed in Landscapes

Unless you live in the Great Plains or along the Gulf Coast of the United States, you know that landscapes tend not to be as flat as the tops of the idealized block diagrams that we've worked with so far. In many cases, the distribution of rock units controls the details of the landscape, so erosion may cause structures to stand out in the landscape, especially in drier climates. For example, in regions of flat-lying strata, resistant rocks form cliffs, whereas nonresistant rocks form gentler slopes. Thus, a cliff exposing alternating resistant and nonresistant rocks develops a **stair-step profile**. Where strata are tilted, resistant rocks form topographic ridges, whereas nonresistant rock types tend to form valleys (**FIG. 15.13a**). Generally, the ridges are asymmetric—a dip slope parallel to the bedding forms on one side, and a scarp cutting across the bedding forms on the other. In the Appalachian Mountains of Pennsylvania, ridges trace out the shape of plunging folds (**FIG. 15.13b**). A region in which the structure of bedrock strongly influences topography is called a **structurally controlled landscape**, and you will see how this is manifested in Exercise 15.15.

FIGURE 15.13 **Structural control of topography.**

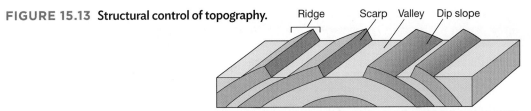

(a) A block diagram showing how resistant layers hold up ridges.

(b) A satellite photo of the Valley and Ridge Province of Pennsylvania (from *Google Earth*™).

EXERCISE 15.15 **Interpreting Structurally Controlled Landscapes**

Name: _____   Section: _____
Course: _____   Date: _____

The figure on the next page shows a region of central Pennsylvania (a) that includes the boundary between two different structural provinces (the Valley and Ridge Province to the southeast and the Plateau Province to the northwest).
Enlargements of the two provinces are shown in parts (b) and (c) of the figures. Based on the general shape of the land
surface, as indicated by the maps, answer the following:

(a) In which of the two provinces is the landscape structurally controlled?_____

(b) Compare the pattern of stream valleys in the Plateau Province with that depicted in the topography of the
Valley and Ridge Province. Approximately what is the dip of the beds beneath the Plateau
Province?_____

(c) In the Valley and Ridge Province, are the folds plunging or nonplunging? What is
your evidence?_____

(continued)

Name: _____   Section: _____
Course: _____   Date: _____

(d) What is the overall trend of fold hinges in the portion of the Valley and Ridge Province depicted in the lower map?_____

(e) If fold hinges trend roughly perpendicular to the direction of compression, in what direction was the compressive "push" during the development of the folds in the Valley and Ridge Province? (Geologists have determined that these folds formed when Africa collided with eastern North America at the end of the Paleozoic Era.)_____

(a) Shaded relief of central Pennsylvania.

10 km   5 mi

Location

(b) Pennsylvania Valley and Ridge Province; the ridges are underlain by resistant sandstone layers.

5 km   2 mi

(c) Pennsylvania Plateau Province; the pattern of rivers and tributaries is called a dendritic drainage pattern.

5 km   2 mi

15.6 Reading Real Geologic Maps

15.6.1 Geologic Maps of Local Areas

You are now ready to apply what you've learned to interpreting the structure of selected areas of North America using excerpts from published geologic maps. Exercises 15.16 and 15.17 give you a sense of how to see "clues" in a map that help you to picture the three-dimensional configuration of rocks underground.

We finish this chapter by talking about how geologists make geologic maps in the first place. It isn't easy! Students who want to learn the skill generally attend a summer geology field camp, where they practice the art of mapping for several weeks and gradually develop an eye for identifying rock types, contacts, folds, and faults. Typically, in arid regions, not much soil forms, so bedrock may be abundantly exposed; in such areas, geologists may actually see contacts and can walk out the traces of contacts. Commonly, however, soil and vegetation cover much of the rocks, so outcrops are discontinuous and separated from one another by "covered intervals." In such cases, geologists must extrapolate contacts, using common sense and an understanding of geologic structures. Exercise 15.18 provides the opportunity for you to construct a map in an area where limited outcrop data are available. The map shown is called an outcrop map because individual outcrops of rock are outlined. To complete the map, you need to extrapolate contacts. In Exercise 15.19, you will use all of your skills to interpret the structural geology of a region and present your findings.

EXERCISE 15.16   **The Observation Peak Quadrangle of Wyoming**

Name: _____   Section: _____
Course: _____   Date: _____

Examine the geologic map of a portion of the Observation Peak quadrangle in Wyoming in **FIGURE 15.14**. This map area contains a number of interesting geologic features, some of which you can understand based on the work you have done earlier in this chapter. Each of the questions below refers to a specific feature on the map.

(a) In the northern part of the map, there is a large area of yellow (i.e., Quaternary deposits). What kind of contact forms the boundary between these deposits and older bedrock?_____

(b) The bright-red color represents igneous intrusions.

  • What kind of intrusion is the larger round area?_____
  • What kind of intrusions are the narrow bands? _____

(c) A thrust fault (a gently to moderately dipping reverse fault) is exposed in the southwest quarter of the map. The "teeth" of the thrust fault symbol lie on the hanging wall. In this locality, the fault dips about 40° in a westerly direction.

  • Where Indian Creek crosses the fault, what rock unit is in the hanging wall, and what rock unit is in the footwall?_____
  • Is the older rock in the hanging wall or in the footwall?_____
  • Thinking about the movement direction on a thrust fault, does this make sense? (Explain your answer.)_____

(d) In the southeastern quarter of the map, you can see the trace of a fold.

  • What type of fold is it?_____
  • In which direction does it plunge?_____
  • In the Ankareh Formation, what is the strike and dip of the strata on the western limb of the fold?_____

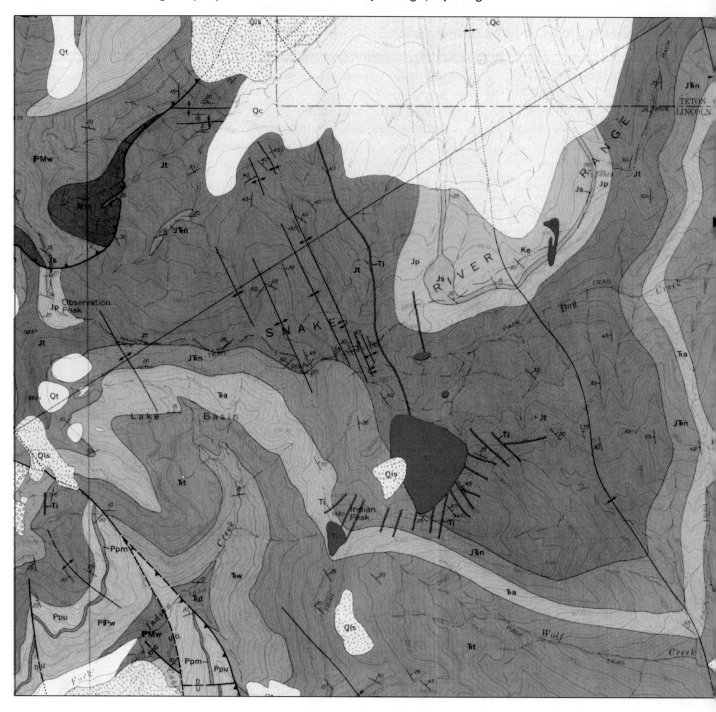

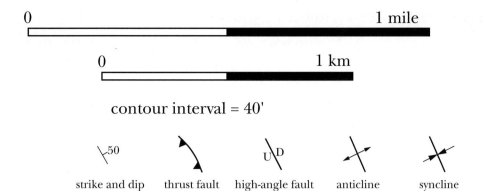

0   1 mile

0   1 km

contour interval = 40'

strike and dip   thrust fault   high-angle fault   anticline   syncline

Geologic Map of the
Observation Peak Quadrangle, Wyoming
by
Howard F. Albee
1973

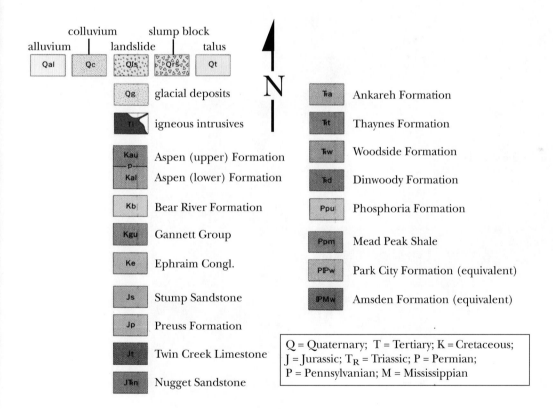

alluvium   colluvium   landslide   slump block   talus

Qal   Qc   Qls   Qrs   Qt

N

Qg — glacial deposits

Ti — igneous intrusives

Kau — Aspen (upper) Formation

Kal — Aspen (lower) Formation

Kb — Bear River Formation

Kgu — Gannett Group

Ke — Ephraim Congl.

Js — Stump Sandstone

Jp — Preuss Formation

Jt — Twin Creek Limestone

JŦn — Nugget Sandstone

Ŧa — Ankareh Formation

Ŧt — Thaynes Formation

Ŧw — Woodside Formation

Ŧd — Dinwoody Formation

Ppu — Phosphoria Formation

Ppm — Mead Peak Shale

PPw — Park City Formation (equivalent)

PMw — Amsden Formation (equivalent)

Q = Quaternary; T = Tertiary; K = Cretaceous;
J = Jurassic; T_R = Triassic; P = Permian;
P = Pennsylvanian; M = Mississippian

Name: _____   Section: _____

Course: _____   Date: _____

Examine the geologic map of part of the Grand Canyon area in Arizona (below). There is no legend because you don't have to know the specific ages of the units to answer the question.

In the lower right part of the map, what is the relative age of the bright-red rock layer compared to the pale-green layer? Note that the thin blue line running through the red layer represents the Colorado River, which flows through the bottom of the Grand Canyon. _____

Geologic map of part of the Grand Canyon.

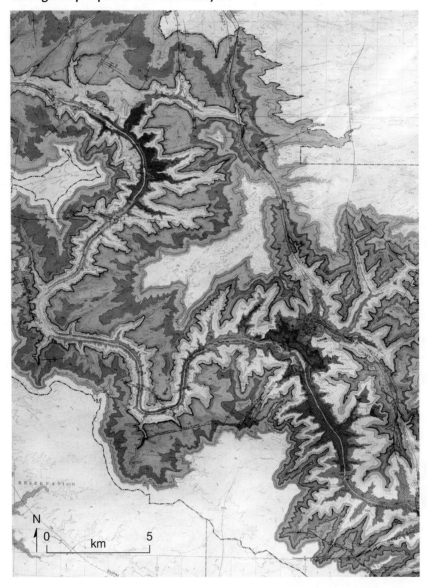

EXERCISE 15.18 Making a Geologic Map from Outcrop Data

Name: ______________________________ **Section:** ____________

Course: ______________________________ **Date:** ____________

The figure below is an outcrop map of an area showing several different rock layers and their attitudes. Using the structural information available on the map and the stratigraphic column, draw a geologic map that shows the contacts between the rock units present. Note that in a few outcrop areas, a contact is visible.

Completing an Outcrop Map

A geologist has just finished mapping a few square kilometers. She has drawn the approximate shapes of outcrops, has plotted strike and dip measurements, and has located contacts where they were exposed. Complete the map by extrapolating contacts across the covered areas and by adding appropriate symbols for folds, if they are present. When you have completed the map, describe the basic structure of the map area in words using the space provided here.

Structural Description: ____

____

____

____

An outcrop map for structural interpretation.

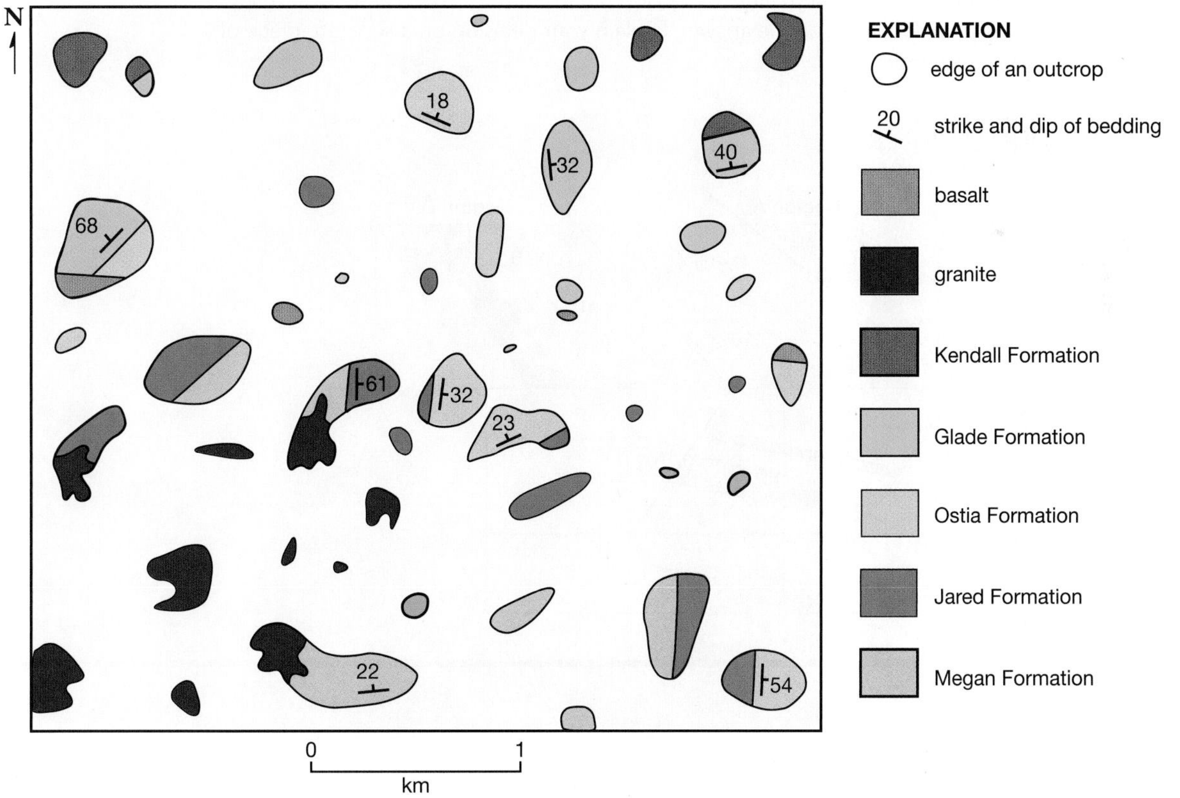

Name: _____   Section: _____

Course: _____   Date: _____

❓What Do You Think Knowledge of an area's deformation history helps avoid potentially calamitous situations, such as building a school directly on an active fault. Structural information can also pay off—big time—in our search for energy and mineral resources. For example, geologists have learned that oil and natural gas are often trapped in the crests of anticlines. Because it costs millions of dollars to drill an exploratory well, knowing where the anticlines are (or aren't) can mean the difference between a fortune and bankruptcy.

Imagine that a company has dug a shaft for an underground mine into a coal bed (gray layer in the figure below). The coal bed and all beds above and below it dip 10° W. A drill hole through the coal layer has provided the stratigraphic sequence and the thickness of the layers, as represented in the column on the left (drawn to scale). Mapping at the ground surface reveals two faults and the map traces of two distinctive beds. The miners have found that the coal layer terminates at the faults, as shown in the cross-section face. It is not economical to mine at a depth greater than the red line. Based on the data available, would you recommend the mining company buy the mineral rights beneath Region A (west of the western fault) or Region B (east of the eastern fault)? Complete the cross-section face to help you visualize the answer. Explain your decision on a separate piece of paper, or as a brief slide presentation.

Practical application problem.

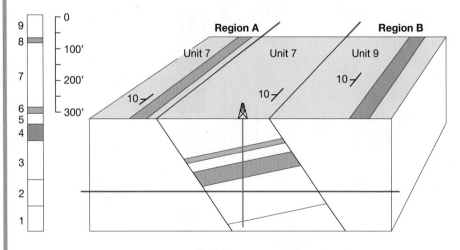

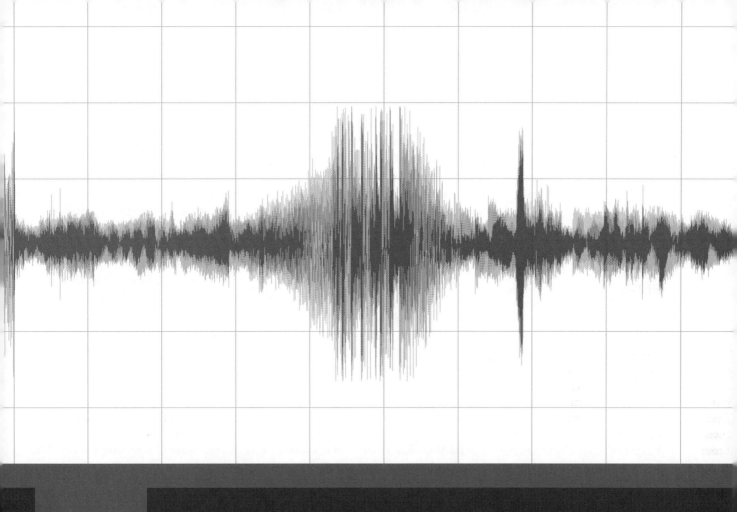

Earthquakes and Seismology

Earthquakes create different kinds of seismic waves that can be measured by a seismometer and that describe its strength and location.

LEARNING
OBJECTIVES

- Understand how faulting
  causes the ground to move
  during earthquakes

- Recognize how earthquakes
  cause damage to buildings
  and other structures

- Learn how to locate an
  earthquake epicenter and
  determine its magnitude and
  when it occurred

MATERIALS
NEEDED

- Sharp pencil

- Ruler with divisions in
  millimeters

- Architect's compass
  (or piece of string)

16.1 Introduction

Few things are as fearsome as a major earthquake. Unpredictable and enormously powerful, a great earthquake destroys more than buildings and other structures. It shakes our sense of safety and stability as it shakes the solid rock beneath our feet. But it is not just the shaking that is dangerous. Devastating tsunamis in 2004 and 2011 reminded us that oceanic earthquakes can ravage coastlines thousands of miles from the earthquake origin; landslides and mudslides triggered by earthquakes can engulf towns and villages; and the loss of water when rigid pipes break beneath city streets can cause health problems and make it difficult to fight fires.

In our attempt to understand earthquakes, we have developed tools that reveal Earth's internal structure, define the boundaries between tectonic plates, prove that the asthenosphere exists, and track the movement of plates as they are subducted into the mantle. In this chapter, we look at what an earthquake is and why it causes so much damage and learn how seismologists locate earthquakes and estimate the amount of energy they release. You will learn to read a seismogram and use it to locate an earthquake, determine when it happened, and measure its strength. First, let's review some basic facts about the causes and nature of earthquakes that are discussed in detail in your textbook.

16.2 Causes of Earthquakes: Seismic Waves

- Earthquakes occur when rocks in a fault zone break, releasing energy. The energy is brought to the surface by two kinds of seismic waves called **body waves** because they travel through the body of the Earth. It is this energy that causes the ground to shake initially.

- The point beneath the surface where the energy is released is called the **focus** (or hypocenter) of the earthquake. The point on the surface directly above the focus is called the **epicenter** (**FIG. 16.1**). In most cases, the epicenter, being closest to the focus, is the site of greatest ground motion and damage.

- There are two kinds of body waves, distinguished by how particles move as the wave passes through rocks: P-waves (primary waves) and S-waves (secondary waves). P-waves (**FIG. 16.2a**) are a seismic wave in which particles in rock vibrate back and

FIGURE 16.1 The focus and epicenter of an earthquake.

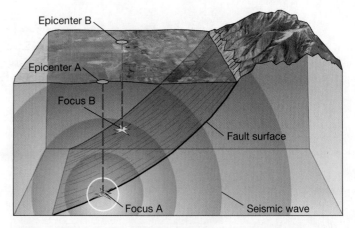

The focus is the point on the fault where slip begins. Seismic energy starts radiating from it. The epicenter is the point on the Earth's surface directly above the focus. Earthquake A just happened; earthquake B happened a while ago.

FIGURE 16.2 Different types of earthquake waves.

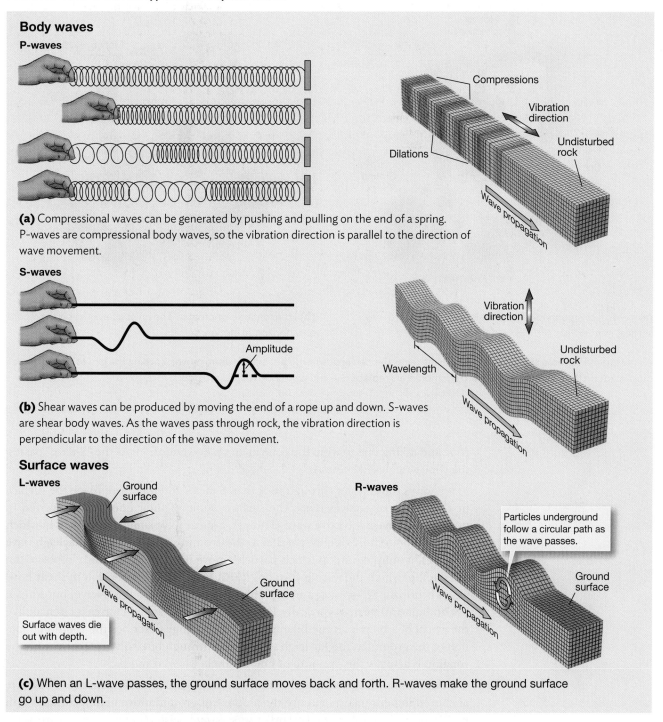

Body waves
P-waves

(a) Compressional waves can be generated by pushing and pulling on the end of a spring. P-waves are compressional body waves, so the vibration direction is parallel to the direction of wave movement.

Compressions

Vibration direction

Undisturbed rock

Dilations

Wave propagation

S-waves

Amplitude

(b) Shear waves can be produced by moving the end of a rope up and down. S-waves are shear body waves. As the waves pass through rock, the vibration direction is perpendicular to the direction of the wave movement.

Vibration direction

Undisturbed rock

Wavelength

Wave propagation

Surface waves
L-waves

Ground surface

Ground surface

Wave propagation

Surface waves die out with depth.

R-waves

Particles underground follow a circular path as the wave passes.

Ground surface

Wave propagation

(c) When an L-wave passes, the ground surface moves back and forth. R-waves make the ground surface go up and down.

forth (red arrow) *in the direction that the wave is traveling* (green arrow). This kind of wave is called a *longitudinal wave.* You can demonstrate this by stretching a Slinky on a table and pushing on one end while keeping the other end in place. As the "P-wave" passes through the Slinky, particles move as shown in Figure 16.2a. Instead of remaining equal distances apart, the coils bunch together in some places and move farther apart in others.

■ S-waves (**FIG. 16.2b**) are seismic waves in which the particle movement is **perpendicular** *to the direction in which the wave is traveling.* This kind of wave is called a *transverse wave.* You can demonstrate this by having two people hold the ends of a

FIGURE 16.3 Vertical and horizontal seismometers.

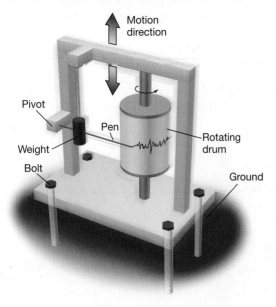

(a) Vertical seismometer: The pivot lets the pen record only vertical motion.

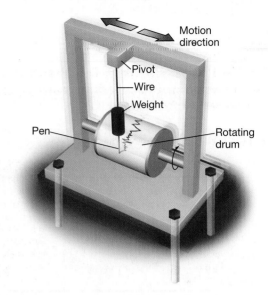

(b) Horizontal seismometer: The pivot allows the pen to record only horizontal motion.

As the ground vibrates, the inertia of the weight keeps the pen in position. The rotating drum moves, causing the pen to trace ground movement on the seismogram. Modern seismometers replace pen and pencil with digital recorders.

rope and asking one to whip the rope in an up-down motion. As the "S-wave" passes, the rope wriggles like a snake.

■ P-waves travel faster through rock than S-waves and therefore reach the surface first. When P- or S-waves reach the surface, some of their energy is converted to two **surface waves:** the **Love wave** (L-wave), a shear wave similar to S-waves in which particles vibrate horizontally, parallel to the ground surface, and the **Rayleigh wave** (R-wave), a unique wave in which particles move in a circular pattern opposite the direction in which the wave is traveling (**FIG. 16.2c**). Particle motion in shear Love waves is horizontal—you can model this with a rope as for the S-wave, but whip it horizontally rather than vertically. Particle motion in Rayleigh waves is circular, like the wheel of a bicycle as the bike moves. This analogy describes the rotational motion of the ground as the Rayleigh wave passes through but isn't perfect because the rotation is actually the opposite of how a bicycle wheel rotates.

■ Seismic waves are detected with instruments called **seismometers**. These are anchored in bedrock to measure the amount of ground movement associated with each type of wave. Seismic stations use separate seismometers to measure vertical and horizontal motion (**FIG. 16.3**). As the ground vibrates, the inertia of the weight keeps the pen in position. The rotating drum moves, causing the pen to trace ground movement on the seismogram. Modern seismometers replace pen and pencil with digital recorders. The printed or digital record of ground motion is a **seismogram**.

■ The strength of an earthquake is measured either by the Richter magnitude scale, based on the amount of ground motion and energy released, or by the Mercalli Intensity Scale, based on the amount of damage to structures.

Exercise 16.1 will help you to understand how seismic waves affect buildings and cause damage.

Name: _____ Section: _____

Course: _____ Date: _____

When P- and S-waves reach the surface, the ground vibrates and anything built on it is shaken in directions that depend on which wave is involved and the distance from the epicenter. The figure below shows the arrival of P-, S-, Love, and Rayleigh waves at a skyscraper.

Ground shaking caused by seismic waves

| P-wave at epicenter | P-wave far from epicenter | S-wave at epicenter | S-wave far from epicenter | Love wave | Rayleigh wave |

(a) Draw arrows to indicate how each of the buildings will move in response to the different types of waves.

(b) Why do the P- and S-waves at an earthquake epicenter make the ground shake differently from the ground in an area far from the epicenter?

_____

_____

_____

16.3 Locating Earthquakes

Locating earthquakes helps us to understand what causes them and to predict if an area will experience more in the future. Most earthquakes occur in linear belts caused by faulting at the three kinds of plate boundaries (**FIG. 16.4**) and are used to define those boundaries. But intraplate earthquakes also occur, and their causes are less well understood. How can we locate earthquake epicenters, especially when they are in remote areas? Seismologists triangulate epicenter locations by using sophisticated mathematical analysis of the arrival times of the different seismic waves from many seismic recording stations.

In this book, we use a much simpler method to locate an epicenter from only three seismic stations and determine precisely when the earthquake occurred. The basis for this method is the fact that the four types of seismic waves travel at different velocities. The same reasoning permits us to locate epicenters and determine the precise time of the faulting that causes individual earthquakes. Let's look at location first, then timing.

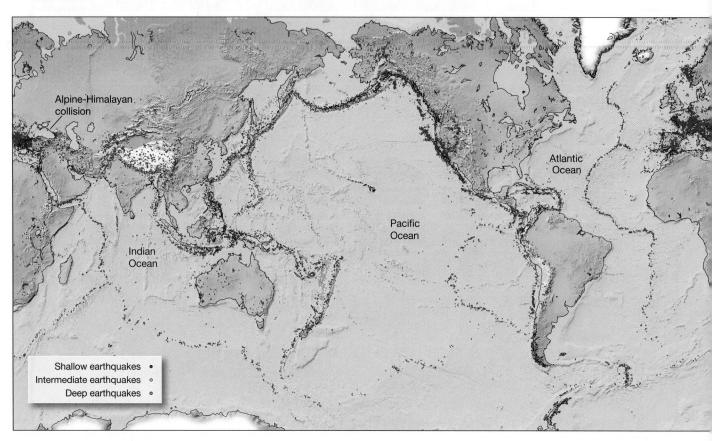

EXERCISE 16.2   Locating Earthquake Epicenters

Name: _____   Section: _____

Course: _____   Date: _____

This exercise leads you through the reasoning used to calculate the distance from a seismic station to an earthquake epicenter. But instead of two different seismic waves, let's see first how this works with two cars that start along a road at exactly the same time (see the figure on the next page). Both use cruise control set at 1 mile per minute (60 miles per hour), but the controls are not exactly the same. Car 1 covers each mile in exactly 60 seconds and Car 2 in 61 seconds. Because Car 1 arrives a second before Car 2 for each mile they travel, the delay between the arrival times of Car 1 and Car 2 increases at each mile marker.

To see how this works, complete the illustration on the next page:

(a) In each of the boxes to the right of Car 1 and Car 2, note the travel time (in seconds) needed to arrive at the different mileposts—the first one has been completed.

(b) In the boxes at the bottom of the figure, indicate the delay time between arrival of Car 1 and Car 2 at each milepost.

(continued)

Name: _____   Section: _____
Course: _____   Date: _____

The basis for locating earthquake epicenters based on the arrival times of different seismic waves.

Of course, geologists don't know at first how far seismic waves traveled or when they were generated by an earthquake. But as with the cars in this exercise, geologists do know (1) the velocities of the different seismic waves; (2) that waves started at the same time (when the fault moved); and (3) the precise time when the different waves were recorded by seismometers. To figure out how far away an epicenter was and when the earthquake happened, we work backwards to derive the information.

You can see how this works by continuing to use the two imaginary cars instead of seismic waves and working backwards like a seismologist. Imagine you are looking out the window of a room waiting for Car 1 and Car 2 to drive by a streetlight outside your window. You don't know where they came from or what time they left, but you do know (1) their velocities (you have been told that Car 1 will arrive traveling at 1 mile in 60 seconds and Car 2 at 1 mile in 61 seconds); and (2) that they started at exactly the same time. As they drive by, you (3) record the precise delay between the times they passed the streetlight outside your window.

(c) Using that information, **how far** did the two cars travel if Car 2 passed the streetlight
   i.   25 seconds after Car 1? _____ miles
   ii. 45 seconds after Car 1? _____ miles

(d) Now determine the **precise time** that the cars started their trip (hour:minute:second) if Car 1 passed the streetlight at 12:25:00 and was
   i.   25 seconds before Car 2: _____:_____:_____
   ii. 45 seconds before Car 2: _____:_____:_____
   (**Hint:** Determine how far the cars traveled and then use the known velocity of Car 1 to estimate how much travel time it needed to go that distance.)

Now put the two pieces of reasoning together.

(e) If Car 1 arrives at precisely 1:45:22, followed by Car 2 at 1:46:37, how far did they travel? _____ miles

(f) At what precise time did they leave? _____:_____:_____

This simple exercise illustrates the reasoning used to determine the distance from a seismometer (the streetlight in this exercise) to an earthquake epicenter (the starting line of each car). But, in this exercise, you have been given the velocity of the cars. How do seismologists determine these measurements?

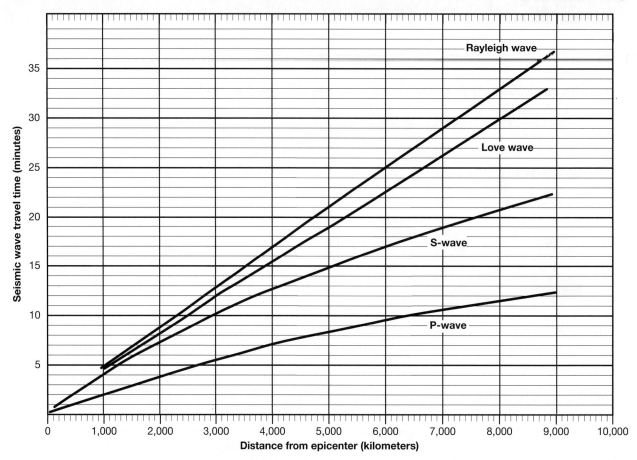

FIGURE 16.5 Travel-time curves show the relationship between distance traveled and relative seismic wave velocity.

Seismologists have measured velocities of P-, S-, Love, and Rayleigh waves in different rock types and know that P-waves average approximately 6.3 kilometers per *second* at the surface and S-waves about 3.6 km/s. A graph of these speeds called a **travel-time diagram** is used to calculate the distance to an epicenter (**FIG. 16.5**). The black curves in Figure 16.5 show how much time it takes (vertical axis) the four types of seismic waves to travel the distances shown on the horizontal axis. To find the time it takes a P-wave to reach a point 4,000 km from an epicenter, find the intersection of the 4,000-km vertical line with the P-wave travel-time curve. Draw a horizontal line from the intersection to the time axis on the left and read the time required—7 minutes in this example. Exercise 16.3 provides practice in reading travel-time curves and then Exercise 16.4 takes you through the steps of how to locate an earthquake's epicenter.

EXERCISE 16.3	Reading a Travel-Time Curve

Name: _____   Section: _____
Course: _____   Date: _____

Refer to the travel-time curve in Figure 16.5.

(a) How long does it take a P-wave to travel 5,000 km? _____ minutes

(b) How long does it take an S-wave to travel 5,000 km? _____ minutes

(c) How long does it take a Love wave to travel 5,000 km? _____ minutes

(d) How long does it take a Rayleigh wave to travel 5,000 km? _____ minutes

Name: _____   Section: _____
Course: _____   Date: _____

Park the cars, and let's tackle an earthquake. Use the following series of steps to get all the information you need to identify the location of the earthquake and pinpoint when it occurred:

1. Identify the four different seismic waves on seismograms from three stations.

2. Determine the arrival times and measure the delays between different waves.

3. Use these data and the travel-time diagram to estimate each station's distance from the epicenter.

4. Use triangulation to locate the epicenter.

5. Determine the time of faulting with the travel-time diagram.

Step 1: Reading a Seismogram

Recognizing Types of Waves

You will need to know what P-, S-, Love, and Rayleigh waves look like on a seismogram in order to identify them correctly. Each seismic wave causes the ground to shake differently (Fig. 16.2), and this produces a different appearance on a seismogram. The differences are in wave amplitude (height) and frequency (the time between adjacent wave peaks). These differences are summarized in the figure below and will help you interpret the seismograms in the next part of this exercise.

This close-up of a seismogram shows the signals generated by different kinds of seismic waves.

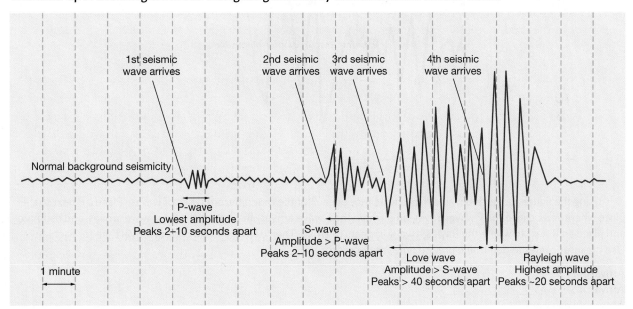

Now that you know what different types of waves look like on a seismogram, let's see how to measure their arrival times. The figure on the next page shows how waves look on a typical seismogram printout. The dashed vertical lines are time markers and are 1 minute apart. The waves reflect time moving forward from left to right on each row, and as a row ends, the time starts again on the left side of the next row below it (just like lines in a book). Seismic waves rarely arrive precisely on a minute marker, so you have to estimate the number of seconds before or after each minute. [*Note:* To avoid confusion involving time zones and changes to and from Daylight Savings Time, seismogram times are recorded in Greenwich Mean Time (GMT)].

(continued)

Name: _____   Section: _____

Course: _____   Date: _____

Recognizing seismic wave arrivals and measuring time on a seismogram.

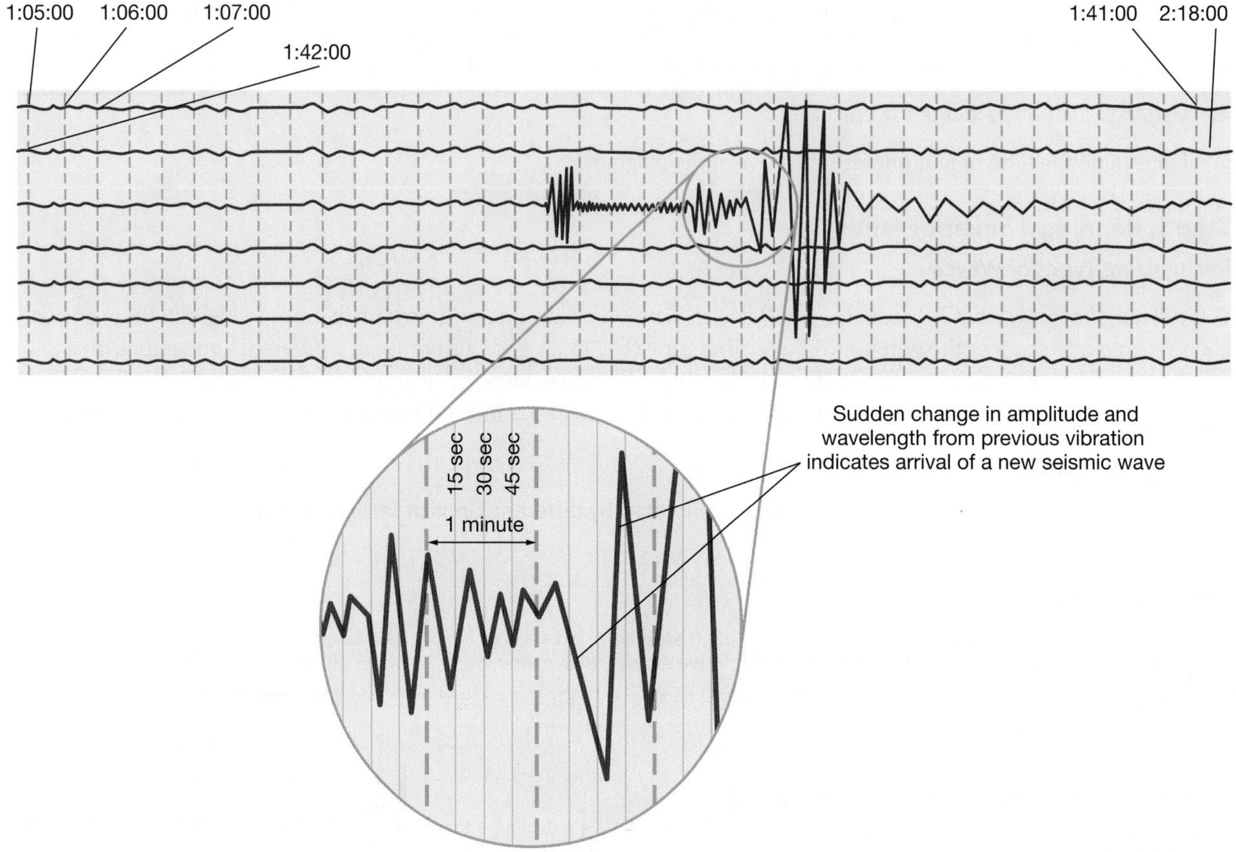

Sudden change in amplitude and wavelength from previous vibration indicates arrival of a new seismic wave

Using the figure above, determine the arrival times of each of the four different types of waves. Place a P next to the place where you believe the P wave arrives, an S where the S wave arrives, an L where the Love wave arrives, and estimate and place an R next to where the Rayleigh wave appears to arrive. Then determine the correct arrival time for each wave. Use the inset circles to determine precise times as necessary.

Arrival of P-wave _____ : _____ : _____

Arrival of S-wave _____ : _____ : _____

Arrival of Love wave _____ : _____ : _____

Estimated arrival of Rayleigh wave _____ : _____ : _____

Looking at the size of the amplitudes of the measurements on the seismograph, which wave types might cause the most damage in an earthquake? Explain. _____

_____

_____

(continued)

Name: _____   Section: _____
Course: _____   Date: _____

Step 2: Measuring the Delay between Arrival of Different Waves

The following figure shows seismograms from three stations. Identify each of the waves and record their arrival times in the table provided on the next page. Then calculate the times between arrivals by subtracting the P-wave arrival time from that of the S-wave, the S-wave from the Love wave, and the Love from the Rayleigh.

Seismograms for Exercise 16.4.

(a) Seattle, Washington

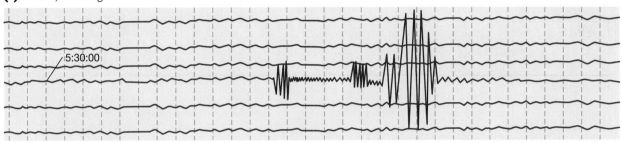

(b) Boston, Massachusetts

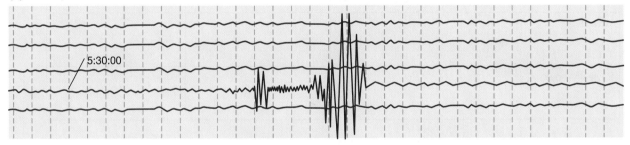

(c) Los Angeles, California

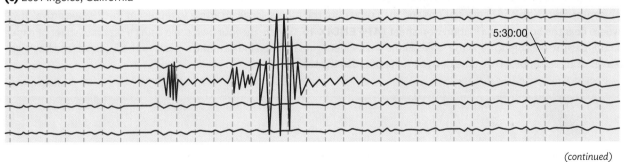

(continued)

Name: _____ Section: _____

Course: _____ Date: _____

Arrival times and delays between seismic waves.			
Seismic wave arrival times	Seattle	Boston	Los Angeles
P-Wave			
S-Wave			
Love Wave			
Rayleigh Wave			
Delays between seismic waves			
S–P			
Love–S			
Rayleigh–Love			

Step 3: Estimating Distance from the Epicenter to Each Station Using the Travel-Time Diagram

The next figure in this exercise shows how to use seismic wave delay data to locate the distance from each station to the epicenter. Start with one of the stations and then repeat for the others. Draw an arrow to represent the P-wave arrival anywhere near the bottom on a station worksheet (**APPENDIX 16.1**) or create your own scale using a separate sheet of paper. In this case, make sure to copy the scales in the worksheets *exactly*. Then indicate the arrivals of the S-, Love, and Rayleigh waves using the appropriate time delays you recorded in the table.

(continued)

Name: _____   Section: _____
Course: _____   Date: _____

Using a travel-time diagram to determine distance to an earthquake epicenter and time the earthquake occurred. Slide the station worksheet across the travel-time diagram until all four arrows coincide with the appropriate travel-time curves (possible positioned worksheet is shown in the figure below).

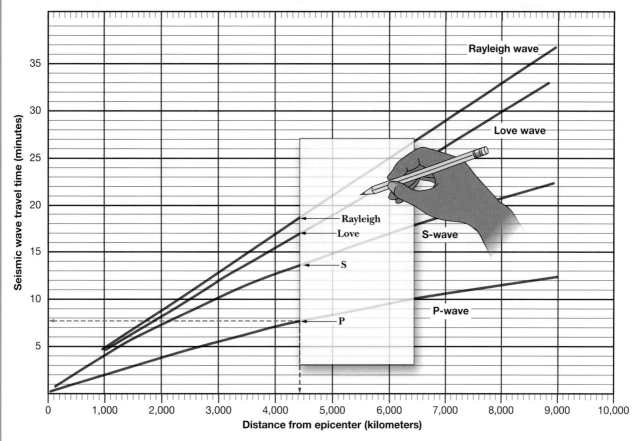

Draw a vertical line from the edge of the graph for that station to the horizontal distance scale in the figure above and read the distance from the epicenter. Repeat for the other two stations and record the data.

Distance to epicenter from Seattle: _____ km   Boston: _____ km   Los Angeles: _____ km

Step 4: Determining the Location of an Earthquake with Triangulation

It's one thing to know *how far* an epicenter is from a seismometer station, but quite a different thing to know exactly *where* it is. On the map of North America on the following page, seismologists from Nova Scotia's Cape Breton Island calculated that an earthquake occurred 4,000 km from their station. The epicenter must lie somewhere on a circle with a radius of 4,000 km centered on their station—but that could be in the Atlantic Ocean, Hudson Bay, Mexico's Yucatán Peninsula, or the front ranges of the Rocky Mountains.

(continued)

Name: _____ Section: _____

Course: _____ Date: _____

Locating an earthquake.

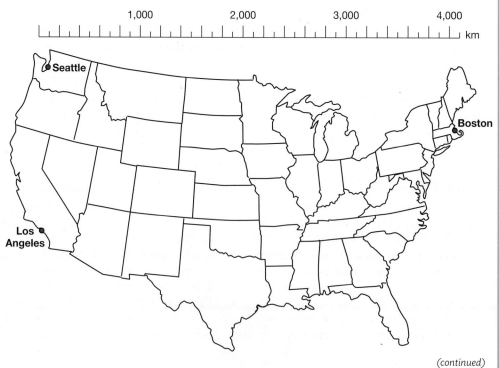

A second seismometer station in Caracas, Venezuela, estimates a distance of 6,000 km to the epicenter—somewhere on a circle with a radius of 6,000 km centered on Caracas. The two circles cross in two places, one in the Rockies, the other somewhere in the eastern Atlantic Ocean (outside the figure). The epicenter must be at one of these intersections, but both locations are equally possible. A third station is needed to settle the question, in this case pinpointing the epicenter in the front ranges of the Rocky Mountains. *Data from at least three seismic stations is needed to locate any epicenter, and the process is called* **triangulation**. The more stations used, the more accurate the location.

To locate the approximate epicenter (at last!), use an architect's compass or piece of string scaled to the appropriate distance for the figure. Draw an arc representing the distance for the first station and repeat the process for the other two stations as well. Review your work. Because of the tools you are using, your three arcs might not perfectly intersect, but they should be near each other. Draw a small circle with a 300 km diameter showing the area where you would expect to find the epicenter.

Map and scale for locating the earthquake epicenter.

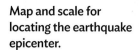

(continued)

Name: _____ Section: _____
Course: _____ Date: _____

Step 5: Determining the Time When an Earthquake Occurred

When, exactly, did the earthquake occur at the epicenter? Align each data graph from Appendix 16.1 with the travel-time curves in Figure 16.5. For each, draw a horizontal line from the **P-wave intercept** on your graph to the time scale on the left side of the travel-time diagram, as shown by the green dashed arrow in the travel-time curve on page 419. Read directly how many minutes the P-wave took to travel to the station, using a millimeter ruler to estimate the number of seconds—in this case, 9 minutes 35 seconds.

Subtract that amount of time from the P-wave arrival time in the table of this exercise to get the time when the earthquake occurred. Repeat for the other two stations—the answer should be the same for each.

Time of the earthquake based on: Seattle _____ : _____ : _____   Boston _____ : _____ : _____ Los Angeles _____ : _____ : _____

16.4 Measuring the Strength of an Earthquake

The last piece of information we need is to determine the strength of the earthquake. There are two ways to describe the strength of an earthquake. The *Mercalli Intensity Scale* is based on the amount of damage sustained by buildings. But because damage to buildings can depend on factors unrelated to the energy released by an earthquake—such as the quality and nature of construction and type of ground beneath the buildings—this method is not very useful in our study of the Earth. In 1935, Charles Richter designed the first widely accepted method for estimating energy released in an earthquake—the Richter *magnitude scale*, which is based on the amount of energy released during faulting and calculated from the amount of actual bedrock motion. Each level of magnitude indicates an earthquake with ground motion *10 times greater* than the next lower level. Thus, a magnitude 4 earthquake has 10 times more ground motion than that of a magnitude 3 and one-tenth that of a magnitude 5.

We now know that Richter's method is accurate only for local, shallow-focus earthquakes. Modern estimates of earthquake strength use different methods involving body waves, surface waves, and a *moment magnitude* scale to calculate accurately the magnitude of shallow- and deep-focus, local and distant, and large and small earthquakes. They also depend on complex analyses of the rock type that broke in the fault, how much offset took place at the fault, and other factors that can't be determined from seismic records alone. In Exercise 16.5, we will use a simple graphical method to determine the magnitude of an earthquake.

Name: _____ Section: _____
Course: _____ Date: _____

In this simplified exercise, you will estimate m_b, the magnitude based on the amplitude of a body wave (P-wave). Because ground motion decreases the farther a seismic station is from an earthquake, distance from the epicenter must also be taken into account. This is done graphically (see next page). To determine m_b of the earthquake you measured in Exercise 16.4, mark the left-hand scale of the chart at the appropriate S-P delay for one of your stations to account for distance from the epicenter. Measure the maximum P-wave amplitude and mark it on the right-hand scale. Now draw a line connecting these two points. The value for m_b is where the line intersects the center magnitude scale.

(continued)

Name: _____   Section: _____
Course: _____   Date: _____

(a) What is m_b for an earthquake with exactly the same **P-wave amplitude** as the example in the figure below but with an S-P delay of 30 seconds? _____ 10 seconds? _____

(b) What is the relationship between wave amplitude, magnitude, and distance from the epicenter?

_____

_____

_____

(c) What is m_b for an earthquake with exactly the same **S-P delay time** as the example in the figure below but with a P-wave amplitude of 2 mm? _____ 100 mm? _____

(d) What is the magnitude of the earthquake based on the following data from three stations close to the epicenter determined in Exercise 16.4? Station A: P-S lag = 20 s, P-wave amplitude = 100 mm; Station B: P-S lag = 40 s; P-wave amplitude = 12 mm; Station C: P-S lag = 50 s, P-wave amplitude = 7 mm. _____

Note: The seismograms in Exercise 16.4 are artificial. If they were from a real earthquake, the values for m_b calculated from each should be nearly identical.

Determining the body wave magnitude of an earthquake.

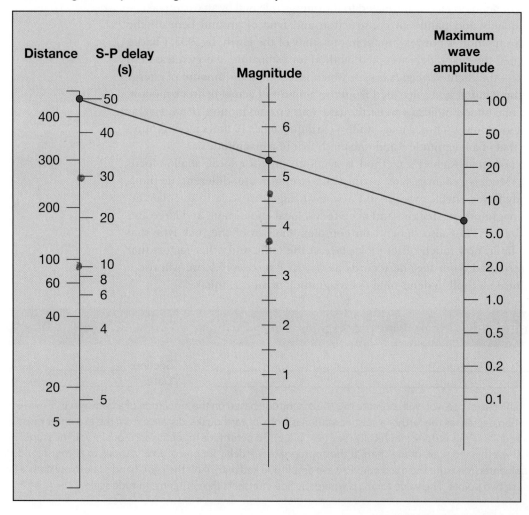

Name: _____   Section: _____

Course: _____   Date: _____

With the modern worldwide network of seismic stations, any earthquake epicenter can be located quickly and accurately. But what about earthquakes that took place *before* seismometers were invented? How can we define Earth's zones of seismic activity if we can't include earthquakes that occurred as (geologically) recently as 100 or 200 years ago? If there are records of damage associated with those events, geologists can estimate epicenter locations by making *isoseismal maps* based on the modified Mercalli Intensity Scale that show the geographic distribution of damage. First, historic reports of damage are analyzed for each location and given an approximate intensity value. These data are plotted on a map, and the map is contoured to show the variability of damage. Ideally, the epicenter is located within the area of greatest damage.

    The figure below shows Mercalli intensity values for an 1872 earthquake that shook much of the Pacific Northwest. Contour the map to show areas of equal damage (isoseismal areas) and indicate with an x the location for the epicenter of this earthquake.

Modified Mercalli intensity values for the December 1872 Pacific Northwest earthquake.

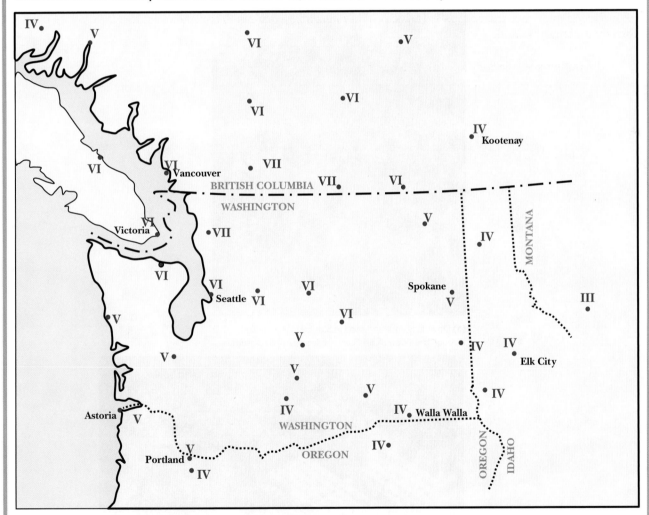

16.5 Predicting Earthquake Hazards: Liquefaction

One of the hazards associated with earthquakes is **liquefaction**, a process by which seismic vibration causes friction between sand grains in waterlogged sediment to be reduced so the sediment loses its ability to support overlying weight and flows like a liquid. At a small scale, this can be annoying (**FIG. 16.6a**); but on a larger scale, it can be catastrophic (**FIG. 16.6b**).

Liquefaction requires several key conditions: unconsolidated sediment, pores saturated or nearly saturated with water, and ground vibration. The sediment may be natural, such as rapidly deposited deltaic or other coastal sediments, or sandy fill used to create new land for shoreline development.

Landfill is increasingly used in crowded cities for new housing and business districts, but these could be severely damaged by liquefaction during an earthquake. Emergency planners must prepare for such an event, and they begin by identifying areas in which conditions for liquefaction are likely. The next exercise examines one such case.

FIGURE 16.6 Results of liquefaction.

(a) Upwelling liquefied sand in San Francisco broke through the sidewalk and curbs, causing minor disruption.

(b) Apartment houses in Niigata, Japan, rotated and sank when an earthquake caused liquefaction of the ground beneath them.

Name: _____   Section: _____
Course: _____   Date: _____

The figure below shows data from a study of liquefaction potential for San Francisco County. It separates areas where bedrock is exposed at the surface from those underlain by unconsolidated sediment and shows the depth to the water table in the sediment. Examine the map carefully to locate places where liquefaction has occurred in the past and for clues to why it happened in those locations.

(a) Briefly describe how the water table relates to past episodes and locations of liquefaction in San Francisco. Why does it seem important to study the water table when considering the possibility of liquefaction?

_____

_____

Liquefaction history in San Francisco related to depth to bedrock and water table height.

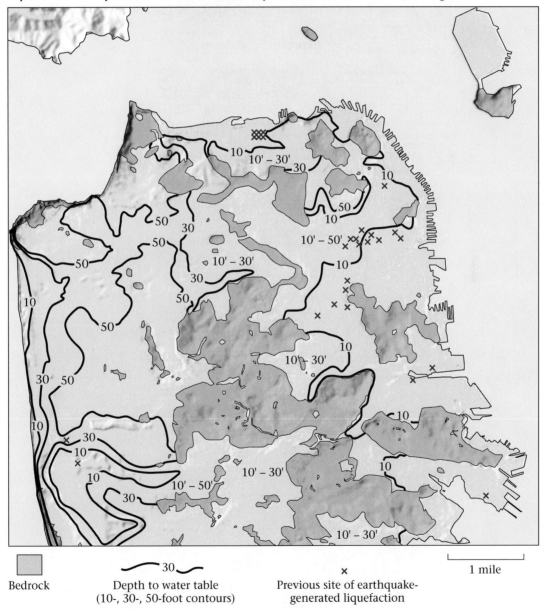

Bedrock

Depth to water table
(10-, 30-, 50-foot contours)

× Previous site of earthquake-
generated liquefaction

1 mile

(continued)

Name: _____   Section: _____

Course: _____   Date: _____

(b) Compare the sites of previous liquefaction events with the water table information in the previous map. What range of water table depths is associated with those events? _____ Shade areas of the previous page with a colored pencil to show where liquefaction is likely to take place in the future.

Topographic map of the San Francisco area, 1899.

(continued)

Name: _____   Section: _____
Course: _____   Date: _____

(c) San Francisco grew dramatically between 1899 and 1999. To make room for the growing population and major new facilities, land was reclaimed from the sea, and these landfills are potential liquefaction sites. After comparing the map on the previous page with the satellite image that follows, color in areas on the photograph where there is a potential for liquefaction.

Satellite view of the San Francisco area, 1999.

(continued)

Name: _____ Section: _____
Course: _____ Date: _____

? **What do you think** If San Francisco is hit by another major earth-quake, many roads will be blocked, and much of the relief effort will have to come by ship and plane. As adviser to the San Francisco disaster preparedness group, what might happen that could make supplies by ship and plane difficult to receive in San Francisco? On a separate piece of paper, write a memo to the City Council describing the steps that you would propose the city take in order to deal with these risks.

16.6 Tsunami!

The Japanese word *tsunami* is now familiar to the entire world because of the devastation caused by these waves along Indian Ocean shorelines in 2004 and more recently in Japan in 2011. A **tsunami** is a seismic shock wave transmitted through ocean water when the floor of the ocean is offset *vertically* during an earthquake. This displacement causes a series of shock waves to travel across the ocean at about 500 miles per hour—but if you were sitting in a small boat in the middle of the ocean, you wouldn't even know it passed beneath you. Unlike a typical wind-generated wave, a tsunami's wave height may be a foot or less; but its wavelength is thousands of feet.

These waves pose no danger in mid-ocean but can result in the worst coastal disasters when, like ordinary waves, they begin to "break" as they near land (see Figure 14.7). The word *tsunami* means "harbor wave," and it is in harbors and estuaries that the damage is magnified (**FIG. 16.7**). Tsunamis striking a relatively straight coastline distribute their energy along the entire shore. When the wave enters an embayment like a harbor or river mouth, the water is funneled into a narrow space and the wave can build to heights greater than 10 m (more than 33 feet) and cause unimaginable damage (see Fig. 14.24).

FIGURE 16.7 Concentration of tsunami energy along coastlines. Tsunami waves are concentrated into embayments, resulting in walls of water much higher than along straight coastal segments.

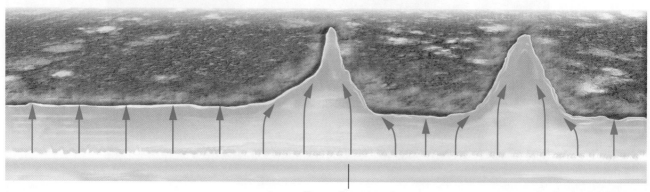

Tsunami wave crest

After watching dramatic images of tsunamis coming ashore near Sendai, Japan, some of our students asked how "just water" could cause so much damage. The amount of energy carried by a tsunami is truly unimaginable, but Exercise 16.8 offers a comparison that may help.

EXERCISE 16.8   Why Is a Tsunami So Powerful?

Name: _____   **Section:** _____
Course: _____   **Date:** _____

A cube of water 1 foot on a side (1 cubic foot, or 1 ft^3) weighs approximately 62 pounds. Imagine being hit by a 62-pound weight—it would certainly hurt.

(a) Now imagine a low wall of water 10 feet wide, 1 foot deep, and 1 foot high: it would weigh _____ lb.

(b) A wall of water 30 feet high (like the Sendai tsunami), 1 foot deep, and 10 feet wide would weigh _____ lb.

(c) A wall of water 30 feet high, 5,280 feet wide (a mile), and 1 foot deep would weigh _____ lb.

(d) To approximate a tsunami better: a wall of water 30 feet high, 5,280 feet wide, and "only" 2,640 feet deep would weigh _____ lb.

(e) One of the most powerful man-made objects is a modern railroad locomotive. A large locomotive weighs about 200 **tons** = 400,000 pounds. The impact of the conceptual tsunami in (d) above is equivalent to being hit by _____ locomotives.

This simple calculation should help you understand why "just water" can cause so much damage. And remember that a tsunami moves much faster than a locomotive.

Seismic Analysis Worksheets

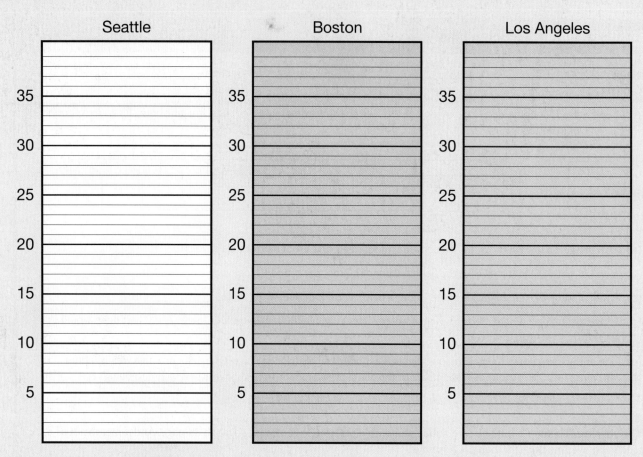

Seattle

Boston

Los Angeles

Cut out and use separately for the appropriate seismograph station on the travel-time curve on p. 414.

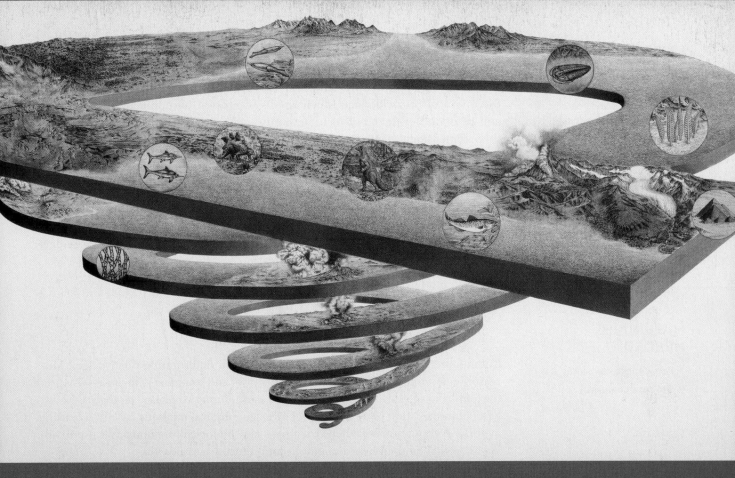

17

Interpreting Geologic History: What Happened, and When Did It Happen?

The Earth has an over 4.5 billion-year history and has undergone many changes, as depicted in this spiral timeline.

- Determine the relative ages of rocks and geologic processes and use these methods to interpret complex geological histories
- Learn how numerical (absolute) ages of rocks are calculated and apply them to dating geologic materials and events
- Understand how geologists piece together Earth history from widely separated areas

MATERIALS
NEEDED

- Pen, pencils, calculator
- Drafter's compass

17.1 Introduction

You've learned to identify minerals, use mineralogy and texture to interpret the origin of rocks, deduce which agents of erosion have affected a given area, and recognize evidence of tectonic events. With these skills you can construct a three-dimensional picture of Earth, using topography and surface map patterns to infer underground relationships.

This chapter adds the fourth dimension—time: the ages of rocks and processes. Geologists ask two different questions about age: "Is a rock or process older or younger than another?" (their **relative** ages) and "Exactly how many years old are they?" (their **numerical** ages). We look first at how relative ages are determined, then at methods for calculating numerical age, and finally combine them to decipher geologic histories of varying complexity.

17.2 Physical Criteria for Determining Relative Age

Common sense is the most important resource for determining relative ages. Most reasoning used in relative age dating is intuitive, and the basic principles were used for hundreds of years before we could measure numerical ages. Geologists use two types of information to determine relative age: **physical methods** based on features in rocks and relationships between them, and **biological methods** that use fossils. We focus first on the physical methods and return to fossils later.

17.2.1 Principles of Original Horizontality and Superposition

The **principle of original horizontality** states that *most* sedimentary rocks are deposited in horizontal beds (there are exceptions, such as inclined sedimentation in alluvial fans, dunes, and deltas). Deposition as horizontal beds makes it easy to determine relative ages in a sequence of sedimentary beds *if the rocks are still in their original horizontal position.* In such cases, the oldest bed will be at the bottom of the sequence overlain by younger layers, with the youngest at the top.

This bit of common sense was first applied to sedimentary rocks by Nils Stensen (also known as Nicolaus Steno) almost 400 years ago and is called the **principle of superposition**. It cannot be used if rocks have been tilted or folded because, in some cases, the rocks may have been completely overturned so that the oldest is on top. In Exercise 17.1, you will look at an outcrop of rock and decide, like Steno, where the oldest and youngest layers are.

EXERCISE 17.1	Relative Ages of Horizontal Rocks

Name: _____ Section: _____

Course: _____ Date: _____

(a) The photograph on the next page shows sedimentary rocks in Painted Desert National Park. Assuming original horizontality, label the oldest and the youngest rocks. Where will the oldest rocks be in any sequence of undeformed, horizontal sedimentary rocks? Explain.

_____

_____

(continued)

Horizontal strata in Painted Desert National Park, Arizona.

(b) Would your answers be different if the rocks were volcanic ash deposits or lava flows? Explain.

_____

_____

17.2.2 Principle of Cross-Cutting Relationships

Superposition cannot be used if sedimentary rocks or lava flows have been tilted or folded from their original horizontal attitudes, nor for most metamorphic rocks and igneous rocks that weren't originally horizontal layers. The principle of cross-cutting relationships helps to determine the relative ages of such rocks: if one rock cuts across another, it must be younger than the rock that it cuts (**FIG. 17.1**).

FIGURE 17.1 Cross-bedded sandstones, Zion National Park, Utah. The person in the photograph is pointing to a contact that cuts across several inclined beds.

We use the term "cross-cutting" broadly, not just for features that physically cut others, but also for processes that have *affected* materials. For example, although the folding shown in photo (c) of Exercise 17.2 doesn't cut across the sedimentary layers, it certainly affected them and therefore had to have occurred after they were deposited. Contact metamorphism is another process that affects rocks without cutting across them. Thus, a lava flow bakes the rock or sediment that it flows upon.

EXERCISE 17.2   Relative Ages in Cross-Cutting Situations

Name: _____   Section: _____
Course: _____   Date: _____

Refer to Figure 17.1 to answer questions (a) and (b).

(a) Which is younger? The inclined beds in Figure 17.1 or the layer that cuts them?

_____

(b) Why can't the principle of superposition be used to determine the relative age of the rock layers in Figure 17.1?

_____

_____

This **principle of cross-cutting relationships** is useful for many types of geologic materials and processes, such as those shown in the photographs below. Refer to these photographs in order to answer the remaining questions of this Exercise.

Applying the principle of cross-cutting relationships.

(a) Dikes intruding granite of the Sierra Nevada batholith in Yosemite National Park, California.

(b) Vertical fault offsetting volcanic ash deposits in Kingman, Arizona.

(c) Folded sedimentary rocks on the island of Crete.

(continued)

Name: _____   Section: _____

Course: _____   Date: _____

(c) An intrusive igneous rock that cuts across other rock units as in photo (a) must be _____ than the units it cuts across. Using this principle, label the order of intrusion of the granite and dikes in the photos above.

(d) A fault, like in photo (b), must be _____ than the rocks it offsets. While you're at it, use the principle of original horizontality to label the oldest and youngest (horizontal) volcanic ash layers that the fault cuts.

(e) The process that folded the sedimentary rocks in photo (c) must be _____ than the rocks. Why can't you use the principle of original horizontality to determine the relative ages of the sedimentary layers in this photograph?

_____

_____

(f) Using your knowledge of the cross-cutting nature of contact metamorphism, how could you tell whether a basalt layer between two shale beds was a lava flow or a sill?

_____

_____

_____

_____

17.2.3 Principle of Inclusions

Pieces of one rock type are sometimes enclosed (included in) another rock. This is most common in clastic sedimentary rocks where fragments of older rocks are incorporated in conglomerates, breccias, and sandstones. It also happens where igneous rocks intrude an area and enclose pieces of the host rock. Inclusions of different rock types in an igneous rock are called **xenoliths** (from the Greek *xenos*, meaning stranger, and *lith*, meaning rock). The principle of inclusions states that such inclusions must have been there before the intrusion or before the sedimentary rock formed, and are therefore older than the rock that included them (**FIG. 17.2**).

FIGURE 17.2 Inclusions in igneous and sedimentary rocks.

(a) Gabbro inclusions in the Baring Granite in Maine.

(b) Fragments of rhyolite tuff in conglomerate from Maine.

(a) Which is older, the granite in Figure 17.2a or the xenoliths? Explain.

_____

_____

_____

_____

(b) Clasts in sedimentary rock must be _____ than the sedimentary rock in which they are included. Explain.

_____

_____

_____

_____

17.2.4 Sedimentary Structures

Some sedimentary structures indicate where the top or bottom of a bed was at the time it was deposited, so that it is possible to tell whether a bed is older or younger than another even if the rocks are vertical or completely overturned. These "top-and-bottom" features include cross-bedding, mud cracks, graded bedding, ripple marks, and impressions such as animal footprints. **FIGURE 17.3** illustrates some of the features that help determine the relative ages of sedimentary rocks.

Cross-bedding as we saw in Figure 17.1 is produced when grains are deposited by an air or water current. Inclined (cross-) beds truncated by overlying layers were used as an example of cross-cutting relationships, and in many cases the inclined layers cut older inclined layers as well.

FIGURE 17.3 Sedimentary "top-and-bottom" indicators of relative age.

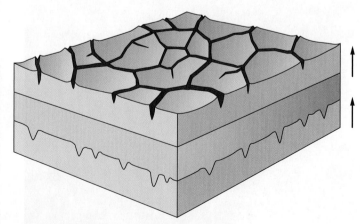

(a) Mud cracks: Mud cracks are widest at the top and narrow downward. The diagram at right is therefore right-side up, indicating that the bottom-most bed is older than the bed in the middle.

(b) Graded beds: The coarsest grains settle first and lie at the bottom of the bed. They are followed by progressively smaller grains, producing a size gradation. Arrows in the diagram at right show how a geologist would interpret the upright nature of the graded beds.

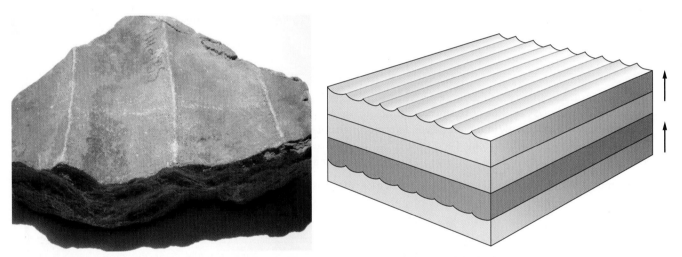

(c) Symmetrical ripple marks: The sharp points of the ripple marks point toward the top of the bed.

(d) Impressions: Features such as dinosaur footprints (left) and raindrop impressions (right) formed when something (here, a dinosaur and raindrops) sank into soft sediment.

Name: _____ Section: _____

Course: _____ Date: _____

(a) Sketch diagrams showing how the three sedimentary structures indicated below would appear in beds that had been turned upside down.

(i) Mud cracks   **(ii)** Graded beds   **(iii)** Symmetrical ripple marks

(b) Are the sedimentary features in Figure 17.3d right-side up or upside down as shown? Explain your reasoning.

_____

_____

_____

The figure below shows the value of sedimentary structures in relative age dating. A cliff face exposing horizontal rock (left side) might be incorrectly interpreted as undeformed horizontal strata without the top-and-bottom information that proves the layers have been folded and some must have been overturned. The right side of the figure shows what the geologist might have seen had the adjacent rocks not been eroded away or covered by glacial deposits.

(c) Number the layers, with 1 representing the oldest rock. Which is the youngest?

_____

_____

Reversals of top-and-bottom features reveal folding.

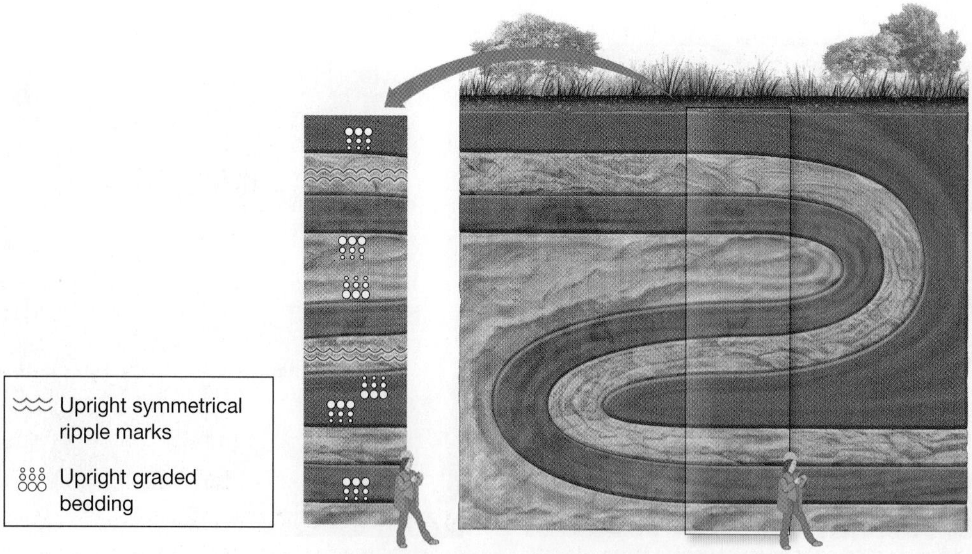

〰 Upright symmetrical
ripple marks

⣿ Upright graded
bedding

17.2.5 Unconformities: Evidence for a Gap in the Geologic Record

When rocks are deposited continuously in a basin without interruption by tectonic activity, uplift, or erosion, the result is a stack of parallel beds. Beds in a continuous sequence are said to be **conformable** because each conforms to, or has the same shape and orientation of, the others, as in the photograph of the Painted Desert that appears in Exercise 17.1.

Tilting, folding, and uplift leading to erosion interrupt this simple history and break the continuity of deposition. In these cases, younger layers are not parallel to the older folded or tilted beds, and erosion may remove large parts of the rock record, leaving a gap in an area's history. We may recognize that deposition was interrupted but can't always tell how long the interruption lasted or what happened during it. A contact indicating a gap in the geologic record is called an **unconformity** because the layer above it is not conformable with those below.

There are three kinds of unconformity (**FIG. 17.4**). A **disconformity** separates parallel beds when some older rocks below the disconformity were removed by erosion. A **nonconformity** is an erosion surface separating older igneous rocks from sedimentary rocks deposited after the pluton was eroded. When rocks above an unconformity cut across folded or tilted rocks below it, the contact is called an **angular unconformity**.

FIGURE 17.5 shows an angular unconformity in the Grand Canyon that separates two conformable sequences of sedimentary rock. Rocks above the unconformity are a horizontal sequence of sandstone, siltstone, and shale that is essentially undeformed and within which the principle of superposition can be applied. The sequence of sedimentary rock below the unconformity was originally horizontal but was tilted and eroded before the oldest overlying bed was deposited.

FIGURE 17.4 The three kinds of unconformities and their formation.

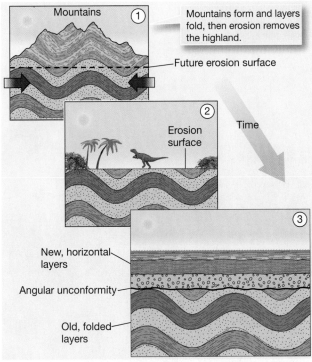

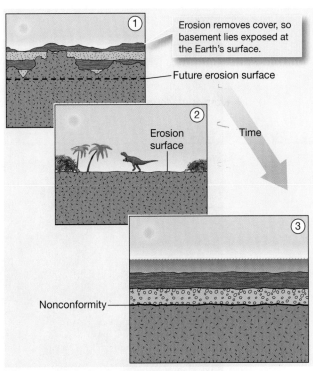

(a) An angular unconformity: (1) layers undergo folding; (2) erosion produces a flat surface; (3) sea level rises and new layers of sediment accumulate.

(b) A nonconformity: (1) a pluton intrudes; (2) erosion cuts down into the crystalline rock; (3) new sedimentary layers accumulate above the erosion surface.

FIGURE 17.4 The three kinds of unconformities and their formation. (continued)

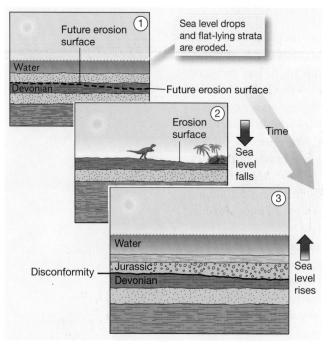

(d) This roadcut in Utah shows a sand-filled channel cut down into floodplain mud. The mud was exposed between floods, and a soil formed on it. When later buried, all the sediment turned into rock; the channel floor is an unconformity, and the ancient soil is a "paleosol." Note that the channel cut across the paleosol. The paleosol also represents an unconformity, a time during which deposition did not occur.

(c) A disconformity: (1) layers of sediment accumulate; (2) sea level drops and an erosion surface forms; (3) sea level rises and new sedimentary layers accumulate.

FIGURE 17.5 Angular unconformity in the Grand Canyon (highlighted by the red dashed line). The lower beds are tilted gently to the right and are separated by an angular unconformity from the horizontal beds that lie above them.

The only way to know how long it took for the tilting and erosion is to determine the ages of the youngest tilted rocks and of the oldest horizontal rock above the unconformity.

Name: _____   Section: _____
Course: _____   Date: _____

In this exercise, you will apply the principles of relative age dating and your knowledge of geologic structures (Chapter 15) to interpret geologic histories of various degrees of complexity. Rock units in the four structural cross sections that follow are labeled with letters that have no meaning with respect to relative age (i.e., A can be older or younger than B). In all diagrams, sedimentary and metamorphic rocks are labeled with *uppercase* letters, igneous rocks with *lowercase* letters. Note that the rocks and some contacts are labeled but the *processes affecting those rocks* are not.

In the timeline on the right side of each cross section, arrange the materials in order from oldest (at the bottom—remember superposition) to youngest. Where an event has occurred (e.g., folding, tilting, erosion), draw an arrow to indicate the place in the sequence and label the event.

In many cases, you will have no doubt about the sequence of events, but in some there may be more than one possibility. In the space provided after each diagram, briefly explain why you ordered your choices the way you did.

Geologic cross section 1.

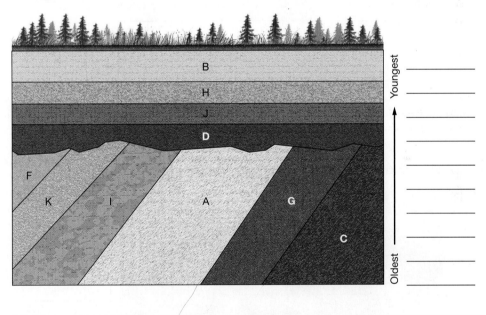

_____

_____

_____

_____

(continued)

Name: _____   Section: _____

Course: _____   Date: _____

Geologic cross section 2.

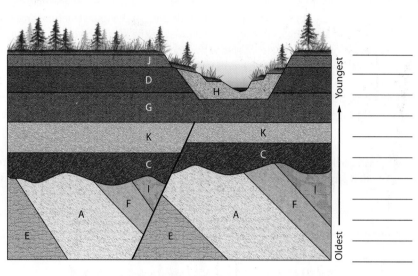

Geologic cross section 3.

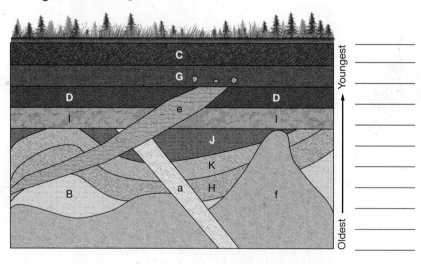

(continued)

Name: _____   Section: _____
Course: _____   Date: _____

Geologic cross section 4.

_____
_____
_____
_____

17.3 Biological Methods for Relative Age Dating and Correlation

In 1793, British canal builder and amateur naturalist William Smith noticed different fossils in the layers he was excavating. He discovered wherever he worked that some fossils were always found in rocks that lay above others, and he suggested that fossils could be used to tell the relative ages of the rocks.

17.3.1 Principle of Faunal and Floral Succession

Geologists confirmed Smith's hypothesis by showing that fossil animals (fauna) and plants (flora) throughout the world record an increasing complexity of life-forms from old rocks to young ones. This is the **principle of faunal and floral succession**. Not all fossils can be used to date rocks because some, like blue-green algae, have existed over most of geologic time and are not specific to a narrow time span.

Index fossils are remains of plants or animals that were distributed widely throughout the world but existed for only a short span of geologic time before becoming extinct. When we find an index fossil, we know that the rock in which it is found dates from that unique segment of time. This is like knowing that a Ford Edsel could only have been made in the three years between 1957 and 1960 or the Model A between 1903 and 1931. *Tyrannosaurus rex*, for example, lived only in the span of geologic time known as the Cretaceous Period; it should not have been in a Jurassic (an earlier time period) park.

17.3.2 The Geologic Time Scale

Determining the sequence of rock units by physical methods and using index fossils to place those units in their correct position in time resulted in the geologic time scale (**FIG. 17.6**). The chart divides geologic time into progressively smaller segments called eons, eras, periods, and epochs (not shown). Era names reveal the complexity of their life forms: Paleozoic, meaning *ancient* life; Mesozoic, meaning *middle* life; and Cenozoic, meaning *recent* life. The end of an era is marked by a major change in life-forms, such as an extinction in which most life-forms disappear and others fill their ecologic niches. For example, nearly 90% of fossil genera became extinct at the end of the Paleozoic Era, making room for the dinosaurs, and the extinction of the dinosaurs at the end of the Mesozoic Era made room for us mammals. Several period names come from areas where rocks of that particular age were best documented: Devonian from Devonshire in England; Permian from the Perm Basin in Russia; and the Mississippian and Pennsylvanian from U.S. states.

FIGURE 17.6 The geologic time scale.

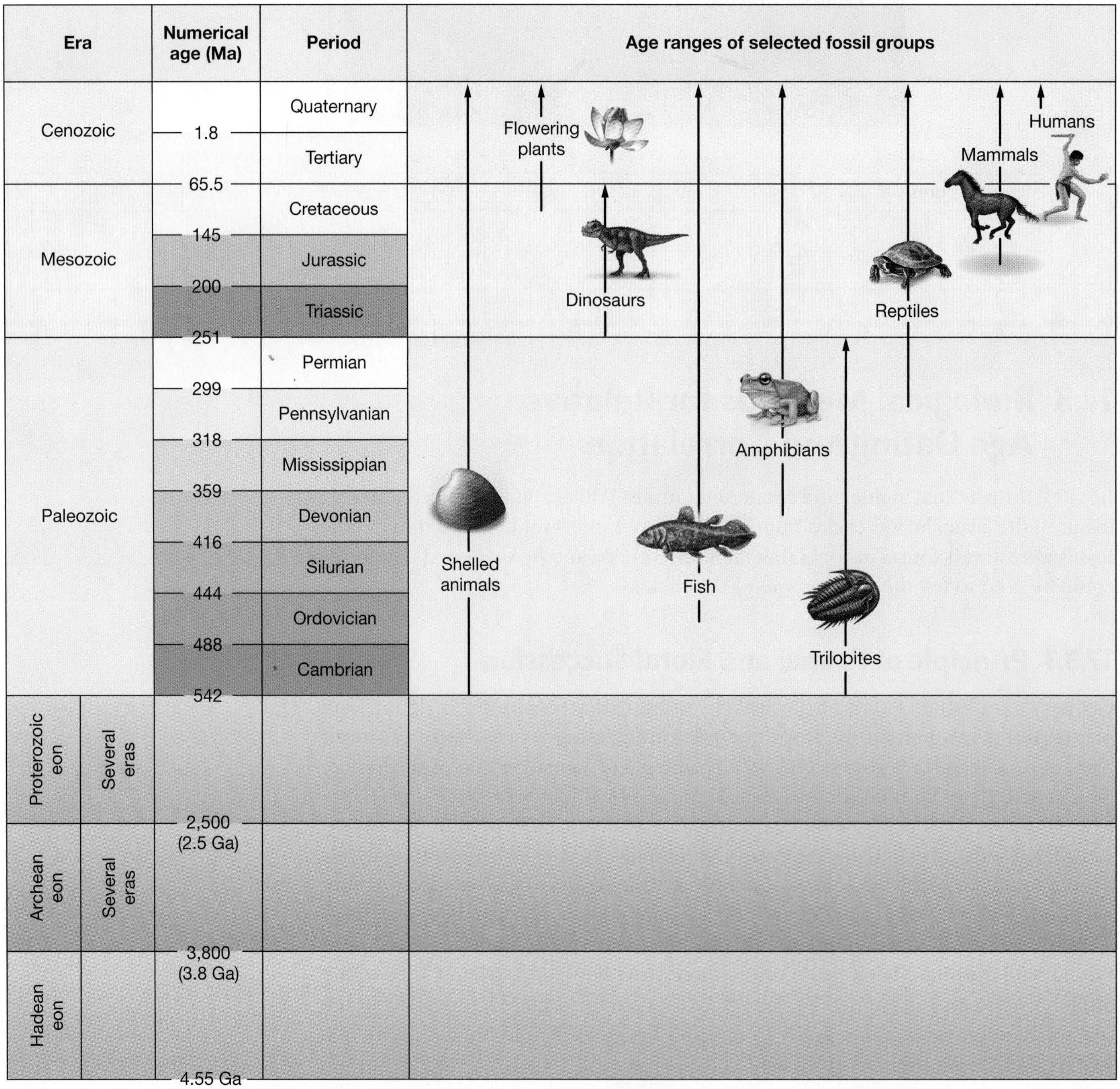

The geologic time scale was originally based entirely on *relative age* dating. We knew that Ordovician rocks and fossils were older than Silurian rocks and fossils, but had no way to tell *how much* older. The ability to calculate the numerical ages came nearly 100 years after the original time scale. Section 17.4 explores methods of numerical age dating.

17.3.3 Fossil Age Ranges

Some index fossil ages are very specific. The trilobites *Elrathia* and *Redlichia* lived only during the Middle and Early Cambrian, respectively. Others lived over a longer span, like *Cybele* (Ordovician and Silurian) and the brachiopod *Leptaena* (Middle Ordovician to Mississippian). But even index fossils with broad ranges can yield specific information if they occur with other index fossils whose overlap in time limits the possible age of the rock. This can be seen even in the broad fossil groups shown in Figure 17.6. Trilobite, fish, and reptile fossils each span several periods of geologic time, but if specimens of all three are found together, the rock that contains them could only have been formed during the Pennsylvanian or Permian.

EXERCISE 17.6	Dating Rocks by Overlapping Fossil Range

Name: _____   Section: _____
Course: _____   Date: _____

The following figures show (a) selected Paleozoic brachiopod species and (b) graphs their ranges within the geologic record.

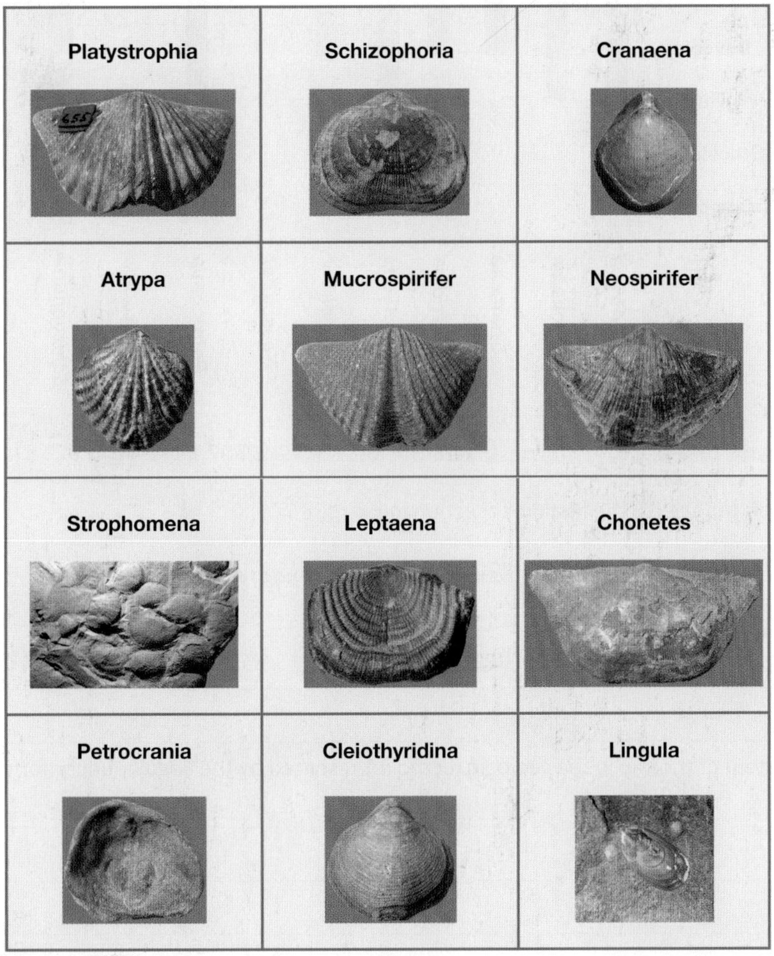

(a) Selected brachiopod species.

(continued)

(a) Based on the overlaps in their ranges shown in the graph, what brachiopod fossil assemblage would indicate:

   (i) a Permian age?

_____

_____

   (ii) a Silurian age?

_____

_____

   (iii) an Ordovician age?

_____

_____

_____

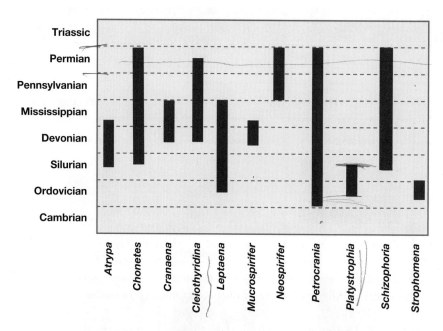

(b) Age ranges of brachiopod species.

(b) Now apply overlapping ranges to the first cross section in Exercise 17.5.

   (i) If *Neospirifer* is found in Unit D, *Platystrophia* in F, and *Strophomena* in A, suggest an age for C. Explain your reasoning.

_____

_____

_____

   (ii) What is the extent of the gap in the geologic record represented by the angular unconformity below Layer D?

_____

_____

_____

(continued)

Name: _____   Section: _____

Course: _____   Date: _____

(c) In the second cross section of Exercise 17.5, *Strophomena* is found in E, *Platystrophia* and *Petrocrania* in F, and *Petrocrania*, *Chonetes*, and *Neospirifer* in C.

   (i) When were Units E, A, F, and I tilted? Explain.

   _____

   _____

   (ii) What is the extent of the gap in the geologic record represented by the contact between C and the tilted rocks beneath it? Explain.

   _____

   _____

   (iii) When in geologic time did the fault cutting Units K, C, I, F, A, and E occur? Explain.

   _____

   _____

   _____

17.4 Numerical Age Dating

The geologic time scale (Fig. 17.6) was used for relative age dating for about 100 years before numerical ages could be added. Numerical age dating is based on the fact that the nuclei of some atoms of elements found in minerals (**parent elements**) decay to form atoms of new elements (**daughter elements**) at a fixed rate *regardless of conditions*. This decay is called radioactivity and the atoms that decay are radioactive isotopes.

The decay "clock" begins when a mineral containing the parent element crystallizes during igneous and metamorphic processes. With time, the amount of the parent decreases and the amount of the daughter increases (**FIG. 17.7**). The amount of time the process takes depends on the *decay rates* of the different isotopes. The **half-life** is the amount of time it takes for half of the parent atoms in a mineral sample to decay to an equal number of daughter atoms (the center hourglass).

FIGURE 17.7 Parent:daughter ratios during radioactive decay.

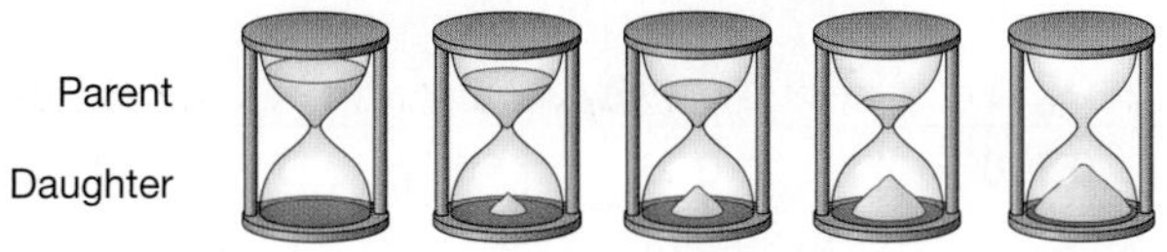

Parent
Daughter

Percent parent	100	75	50	25	0
Percent daughter	0	25	50	75	100
Parent:daughter ratio	—	3:1	1:1	1:3	—

TABLE 17.1 Geologically important radioactive decay schemes.

Parent isotope	Daughter decay product	Half-life (years)	Useful dating range (years)	Dateable materials
^{147}Samarium	^{143}Neodymium	106 billion	>10,000,000	Garnets, micas
^{87}Rubidium	^{87}Strontium	48.8 billion	>10,000,000	Potassium-bearing minerals (mica, feldspar, hornblende)
^{238}Uranium	^{206}Lead	4.5 billion	>10,000,000	Uranium-bearing minerals (zircon, apatite, uraninite)
^{235}Uranium	^{207}Lead	713 million	>10,000,000	Uranium-bearing minerals (zircon, apatite, uraninite)
^{40}Potassium	^{40}Argon	1.3 billion	>10,000	Potassium-bearing minerals (mica, feldspar, hornblende)
^{14}Carbon	^{14}Nitrogen	5,730	100 to 70,000	Organic materials

TABLE 17.1 lists the isotopes used commonly in numerical age dating, their half-lives, and the minerals used in dating. In general, an isotope can date ages as old as ten of its half-lives; any older than that, and there wouldn't be enough parent atoms to measure accurately. Isotopes with short half-lives are therefore best used for relatively recent ages. Conversely, isotopes with very long half-lives, like ^{147}Samarium, decay so slowly that they can be used only to date very old rocks.

To calculate the numerical age of a rock, geologists crush it to separate minerals containing the desired isotope. A mass spectrometer determines the parent:daughter ratio and then a logarithmic equation is solved to calculate the age. Your calculation is easier: once you know the parent:daughter ratio, you can use **TABLE 17.2** to calculate age through simple multiplication.

TABLE 17.2 Calculating the numerical age of a rock from decay scheme half-lives.

Percent of parent atoms remaining	Parent: daughter ratio	Number of half-lives elapsed	Multiply half-life by ___ to determine age	Percent of parent atoms remaining	Parent: daughter ratio	Number of half-lives elapsed	Multiply half-life by ___ to determine age
100	—	0	0	35.4	0.547	1½	1.500
98.9	89.90	1/64	0.016	25	0.333	2	2.000
97.9	46.62	1/32	0.031	12.5	0.143	3	3.000
95.8	22.81	1/16	0.062	6.2	0.066	4	4.000
91.7	11.05	1/8	0.125				
84.1	5.289	1/4	0.250				
70.7	2.413	1/2	0.500	0.05		11	Don't bother! There are too few parent atoms to measure accurately enough.
50	1.000	1	1.000	0.025		12	

Name: _____   **Section:** _____
Course: _____   **Date:** _____

(a) First, get practice calculating ages using Tables 17.1 and 17.2. How old is a rock if it contains:

- (i) a ^{235}uranium:^{207}lead ratio of 46.62? _____ years
- (ii) a ^{87}rubidium:^{87}strontium ratio of 89.9? _____ years
- (iii) 6.2% of its original ^{14}carbon? _____ years
- (iv) 97.9% of its original ^{40}potassium? _____ years

Now add more detail to the cross sections in Exercise 17.5.

(b) In the third cross section of Exercise 17.5, Dike E has zircon with a ^{235}uranium:^{207}lead ratio of 11.05; Dike A has zircon with a ^{238}uranium:^{206}lead ratio of 22.81; and Pluton F has hornblende with 84.1% of its parent ^{40}potassium.

- (i) How old is E? _____ A? _____ F? _____
- (ii) How old (in years and using period names) are Layers B, H, K, and J?

_____

_____

- (iii) When were these layers folded?

_____

_____

- (iv) When did the unconformity separating Layer I from the underlying rocks form?

_____

_____

- (v) How old are Layers I and D?

_____

_____

_____

(c) In the fourth cross section of Exercise 17.5, Layer H has a thin volcanic ash bed at its base and zircon with a ^{235}uranium:^{207}lead ratio of 46.62; Layer D has zircon with a ^{235}uranium:^{207}lead ratio of 2.413; and Dike F has hornblende with 50% of its parent ^{40}potassium.

- (i) How old is H? _____ D? _____ F? _____
- (ii) When (in years and using period names) were Layers C, G, and J folded?

_____

_____

- (iii) How large a gap in the geologic record is represented by the unconformity below Layer B?

_____

_____

- (iv) How old is Layer I? _____
- (v) If Layer I contains the brachiopods *Neospirifer*, *Chonetes*, and *Schizophoria*, how does this help narrow its possible age?

_____

_____

- (vi) In that case, how large a gap in the geologic record is represented by the unconformity below Layer H?

_____

_____

17.5 Correlation: Fitting Pieces of the Puzzle Together

Imagine how difficult it would be for an alien geologist visiting Earth 200 million years from now to reconstruct today's geography. Plate tectonics could have moved some continents, split some and sutured others together; opened new oceans and shrunk or closed others. In what is now North America, there would be rock and fossil evidence of mangrove swamps; forests; grassy plains; large lakes and rivers; shoreline features; alpine glaciers and deserts; active volcanoes and other mountains. Numerical age dating would show rocks that today range from more than 3 billion to as little as 1 year old.

This is the challenge facing geologists today as we try to read the record of Earth history. There is no single place where all 4.6 billion years of Earth history are revealed—not even in the Grand Canyon. And at any one point in time, we can find evidence of all of the environments listed above—plus the deep marine and oceanic island settings. To work out a history for the entire planet, we first decipher the records of local areas and combine them into regional and larger-scale stories. This requires just about all of the skills you've learned during this course, from interpreting ancient environments and processes from the rocks they produce to deciphering the sequence of events and how long ago they occurred. Exercise 17.8 introduces you to the kind of reasoning used in combining information from two different areas into a regional picture.

EXERCISE 17.8 Lithologic Correlation

Name: _____   Section: _____
Course: _____   Date: _____

Five years ago, geologists determined the relative ages of rocks in two parts of the midcontinent, but the sequences were not the same (see the figure on the facing page). Unfortunately, similar rocks appear at several places in each section, making it difficult to know exactly which limestone, for example, in the western section correlates with a particular limestone in the east. It is better to compare *sequences* of units based on similar sequences of depositional environments rather than similarities in single rock units.

(a) Suggest correlations between layers of the two sequences by connecting the tops and bottoms of layers in the western section to those you believe are their matching layers in the eastern section. As in the real world, there may be more than one hypothesis, so explain your choice(s) and how you might further test them.

_____

_____

_____

(b) Do you think the environments indicated by the sedimentary rocks are the same in both the east and the west? Was deposition continuous (conformable) in both areas? Explain why or why not.

_____

_____

_____

_____

(continued)

Name: _____ **Section:** _____
Course: _____ **Date:** _____

Correlation of lithologic sequences.

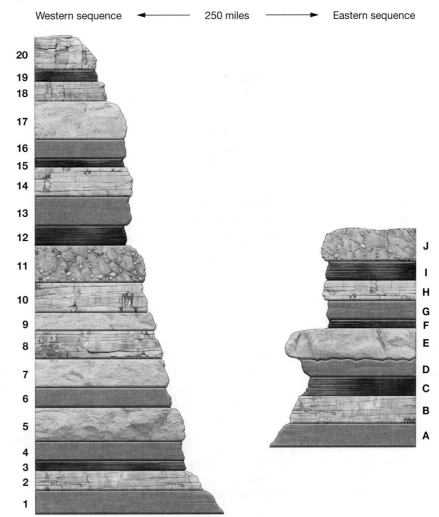

Western sequence ◄——— 250 miles ———► Eastern sequence

During field mapping last year, trilobites were found in limestone units in both areas (see the figure on the next page). Paleontologists reported that these were identical index fossils from a narrow span of Middle Cambrian time.

(c) Using this additional information, draw lines indicating matching layers on the figure.

(d) Why does the presence of index fossils make correlation more accurate than correlation based on lithologic and sequence similarities alone?

_____

_____

_____

(continued)

EXERCISE 17.8 Lithologic Correlation (continued)

Name: ______________________________ Section: ____________
Course: ______________________________ Date: ____________

Correlation assisted by index fossils.

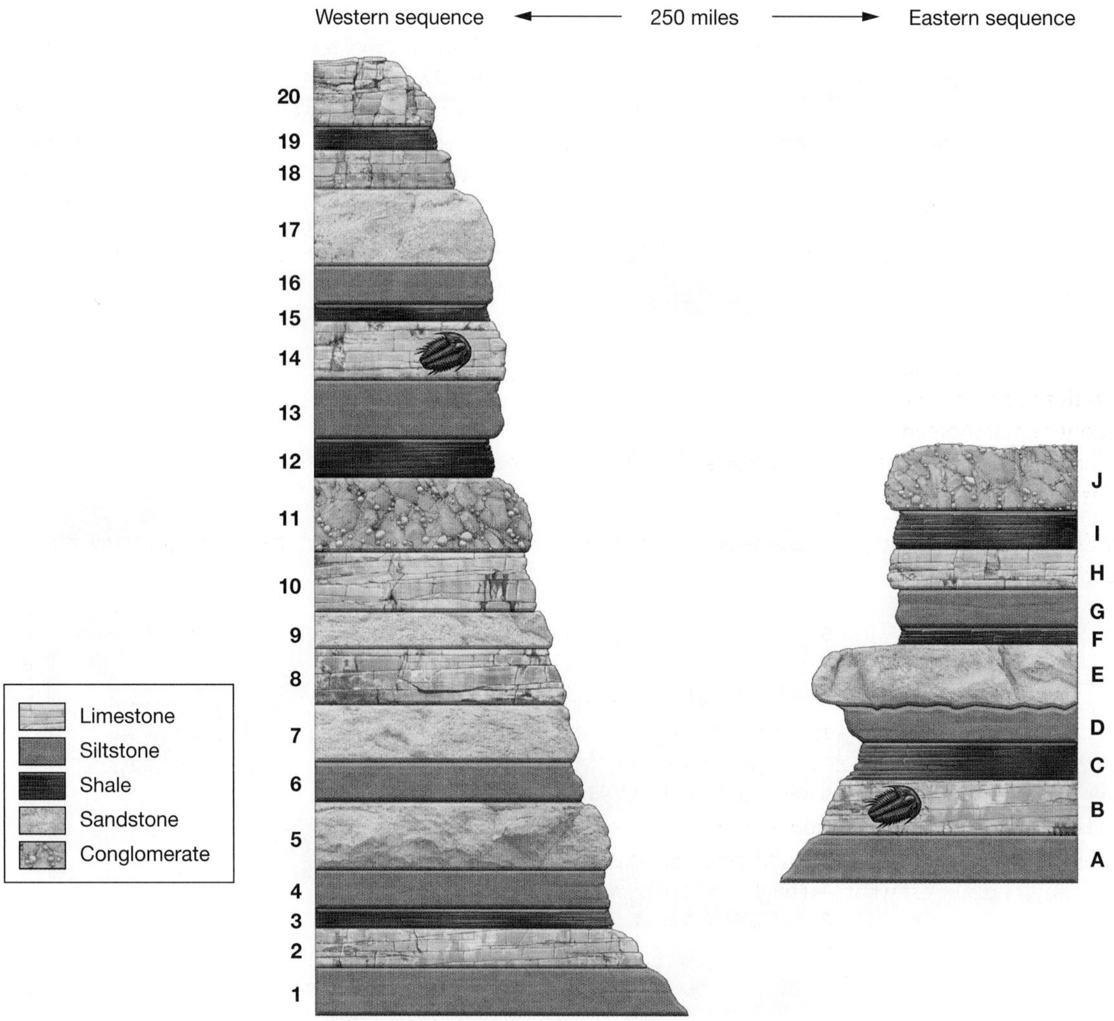

Western sequence
250 miles
Eastern sequence
20
19
18
17
16
15
14
13
12
11
10
9
8
7
6
5
4
3
2
1
J
I
H
G
F
E
D
C
B
A
Limestone
Siltstone
Shale
Sandstone
Conglomerate

Name: _____ Section: _____
Course: _____ Date: _____

After more than 200 years of study, geologists have gathered an enormous amount of data about the geologic history of North America. This exercise presents some of that data and asks **what do you think** the geography of the continent looked like toward the end of the Cretaceous Period, just before an event that led to extinction of most life-forms—including, most famously, the dinosaurs?

FIGURE 17.8 on the next page summarizes evidence from the distribution of sedimentary rocks that form in specific environments, like those you examined in Chapter 6; fossil and numerical age dating of sedimentary and igneous rocks; and interpretation of the folding, faulting, and metamorphic history of Cretaceous and older rocks. We saw earlier that index fossils are valuable for pinpointing the relative age of the rocks in which they are found. Here, we also use **facies fossils**, fossils of organisms that could only live in a narrowly restricted environment and therefore reveal the surface geography at the time they lived.

(a) On Figure 17.8, draw boundaries between the following North American geographic settings in the late Cretaceous Period:

- Continental interior
- Old, eroded mountain ranges
- Active tectonic zone
- Continental shoreline
- Shallow sea (continental shelf and parts of the continent flooded by shallow seas)
- Deep ocean

Evidence for these settings is described here. The facies fossils used as symbols on the map are, wherever possible, Cretaceous organisms.

? What Do You Think We all have impressions of what the Earth may have looked like during the cretaceous from watching movies and television. But using the evidence from Figure 17.8, describe on a separate sheet of paper what you think the United States and Canada looked like at that time, by region—north, south, central, and coastal areas. Then find your state or province and describe what you think it was like at that time, and suggest what you might want to explore to learn more.

FIGURE 17.8 Map of North America today showing evidence for Cretaceous paleogeography.

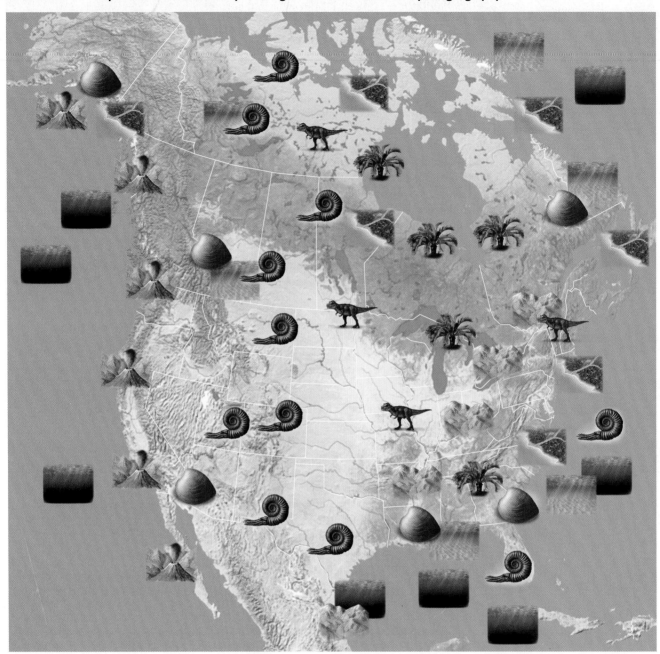

Continental Land Environments

Continental Interior	Active Tectonic Zone	Old Mountains	Shorelines
Terrestrial plants and animals	Volcanic arc rocks	Coarse clastic sedimentary rocks	Beach sandstones and coastal organisms

Marine Environments

Shallow Ocean	Deep Ocean
Shallow marine organisms and sedimentary rocks	Deep marine organisms and sedimentary rocks

CREDITS

Text Permission

Figure 11.18: Joanne Nova: Figure: "The 800 year lag – graphed," from Joannenova.com.au. Reprinted by permission of Joanne Nova.

Photos

Chapter 1

Page 1: Gary Crabbe/Aurora Photos; **p. 2:** Corey Templeton; **p. 7:** NASA; **p. 14 (left):** G.R. Roberts, NSIL; **p. 14 (right):** Belgian Federal Science Policy Office; **p. 18 (left):** Dr. Stephen Hughes, Coastal Hydraulic Laboratories; **p. 18 (right):** US Army Corps of Engineers/Steven Hughes; **p. 21 (both):** Courtesy of Stephen Marshak; **p. 22 (left):** Courtesy of Allan Ludman; **p. 22 (right):** Courtesy of Stephen Marshak.

Chapter 2

Page 25: Buurserstraat386/Dreamstime.com; **p. 29:** NGDC/NOAA; **p. 36:** Unavco.org; **p. 38 (both):** Marine Geoscience Data System/Natural Science Foundation; **p. 40:** NGDC/NOAA; **p. 50:** © Marie Tharp 1977/2003. Reproduced by permission of Marie Tharp Maps, LLC, 8 Edward Street, Sparkill, New York 10976.

Chapter 3

Page 55: Javier Trueba/MSF/Photo Researchers, Inc.; **p. 60 (top-left):** Ross Frid/Visuals Unlimited, Inc.; **p. 60 (top-right):** Mark Schneider/Visuals Unlimited, Inc.; **p. 60 (bottom-left):** Ken Larsen/Smithsonian, National Museum of Natural History; **p. 60 (bottom-right):** Mark Schneider/Visuals Unlimited, Inc.; **p. 61 (both):** Courtesy of Allan Ludman; **p. 63 (top 4):** Courtesy of Paul Brandes; **p. 63 (bottom-left):** Richard P. Jacobs/JLM Visuals; **p. 63 (bottom-right):** Scientifica/Visuals Unlimited, Inc.; **p. 64 (top-left):** Courtesy of Stephen Marshak; **p. 64 (top-center):** Mark Schneider/Visuals Unlimited, Inc.; **p. 64 (top-right):** Courtesy of Allan Ludman; **p. 64 (bottom-left):** 1995-1998 by Amethyst Galleries, Inc. http://mineral.galleries.com; **p. 64 (bottom-center):** Mark Schneider/Visuals Unlimited, Inc.; **p. 64 (bottom-right):** 1993 Jeff Scovil; **p. 65 (left):** Scientifica/Visuals Unlimited, Inc.; **p. 65 (right):** Biophoto Associates/Photo Researchers Inc.; **p. 66 (1):** Scientifica/Visuals Unlimited, Inc.; **p. 66 (2-3):** Courtesy of Paul Brandes; **p. 66 (4):** Courtesy of Allan Ludman; **p. 66 (5):** Courtesy of Paul Brandes.

Chapter 4

Page 85 (left): Rafael Laguillo/Dreamstime.com; **p. 85 (center):** Dirk Wiersma/Science Source; **p. 85 (right):** Les Palenik/Dreamstime.com; **p. 89 (top-left):** Courtesy of Stephen Marshak; **p. 89 (top-right):** Visualphotos.com; **p. 89 (center-left):** Courtesy of Stephen Marshak; **p. 89 (center-right):** Courtesy of Stephen Marshak; **p. 89 (bottom):** Mark Schneider/Visuals Unlimited, Inc.; **p. 90 (top-left):** Marli Bryant Miller, University of Oregon; **p. 90 (top-right):** Michael C. Rygel; **p. 90 (bottom-left & right):** Marli Bryant Miller, University of Oregon; **p. 91 (top-left & right):** Courtesy of Stephen Marshak; **p. 91 (bottom-left):** Rafael Laguillo/Dreamstime.com; **p. 91 (bottom-right):** Courtesy of Allan Ludman; **p. 91 (bottom):** Courtesy of Allan Ludman; **p. 104:** © Ron Chapple/Corbis.

Chapter 5

Page 105: Patricia Marroquin/Dreamstime.com; **p. 109 (left):** Courtesy of Allan Ludman; **p. 109 (center):** Albert Copley/Visuals Unlimited, Inc.; **p. 109 (right):** Courtesy of Allan Ludman; **p. 111 (top-left):** Courtesy of Allan Ludman; **p. 111 (top-right):** Courtesy of Allan Ludman; **p. 111 (bottom-left):** Courtesy of Stephen Marshak; **p. 111 (bottom-right):** Marli Miller/Visuals Unlimited, Inc.; **p. 112:** Scientifica/Visuals Unlimited, Inc.; **p. 113 (left):** Courtesy of Stephen Marshak; **p. 113 (right):** Marli Bryant Miller, University of Oregon; **p. 114 (left):** Courtesy of Allan Ludman; **p. 114 (top):** Courtesy of Douglas W. Rankin; **p. 114 (bottom):** Courtesy of Allan Ludman; **p. 115 (left to right):** Courtesy of Allan Ludman; Albert Copley/Visuals Unlimited, Inc.; Courtesy of Allan Ludman; Courtesy of Allan Ludman; Courtesy of Stephen Marshak; Courtesy of Allan Ludman; Courtesy of Allan Ludman; **p. 121 (left):** Marli Miller/Visuals Unlimited; **p. 121 (right):** Courtesy of Allan Ludman; **p. 123:** Courtesy of Allan Ludman; Courtesy of Allan Ludman; **p. 127:** Alamy; **p. 128 (top-left):** Wikimedia Commons; **p. 128 (top-right):** Dr. Richard Roscoe/Visuals Unlimited, Inc.; **p. 128 (bottom-left):** Dr. Richard Roscoe; **p. 128 (bottom-right):** Shutterstock; **p. 129:** USGS.

Chapter 6

Page 135 Dreamstime.com; **p. 137 (top):** R.Weller/Cochise College; **p. 137 (bottom):** Joel Arem/Science Source; **p. 143(left):** Marli Miller/Visuals Unlimited, Inc.; **p. 143 (right):** © Kavring/Dreamstime.com; **p. 150 (left):** Courtesy of Allan Ludman; **p. 150 (center):** Courtesy of Allan Ludman; **p. 150 (right):** Michael P. Gadomski/Science Source; **p. 151 (both):** Courtesy of Allan Ludman; **p. 154 (top):** Steven Kazlowski/Science Faction/Getty Images; **p. 154 (bottom):** Glowimages/Corbis; **p. 155 (top):** Courtesy of Allan Ludman; **p. 155 (bottom-left & right):** Courtesy of Allan Ludman; **p. 156 (top-right):** Stephen Marshak; **p. 156 (bottom-left):** Dr. Marli Miller/Visuals Unlimited, Inc.; **p. 156 (bottom-right):** Dr. Marli Miller/Visuals Unlimited, Inc.; **p. 157 (left):** Courtesy of Stephen Marshak; **p. 157 (right):** Omikron/Photo Researchers, Inc.; **p. 158 (top-left):** Kandi Traxel/Dreamstime.com; **p. 158 (top-right):** brachiopod: American Museum of Natural History; **p. 158 (bottom-left):** Dr. John D. Cunningham/Visuals Unlimited, Inc.; **p. 158 (center-right):** Imv/Dreamstime.com; **p. 158 (bottom-right):** Bloopiers/Dreamstime.com; **p. 159 (top-left):** American Museum of Natural History; **p. 159 (right):** Xiao Bin Lin/Dreamstime.com; **p. 159 (bottom-left):** Sam Pierson/Photo Researchers, Inc; **p. 160 (clockwise from top):** Emma Marshak; Stephen Marshak; Marli Miller/Visuals Unlimited; Stephen Marshak; Stephen Marshak; Stephen Marshak; Yahn Arthus-Bertrand/Corbis;John S. Shelton; **p. 161 (left):** Allan Ludman; **p. 161 (right):** Stephen Marshak; **p. 162:** (Science Museum in Logroño). Photo by jynus, October 2005.

Chapter 7

Page 165: R.Weller/Cochise College; **p. 167 (top row):** Ashley B. Staples, courtesy Appalachian State University; Scientifica/Visuals Unlimited, Inc.; **p. 167 (top-center row):** Courtesy of Allan Ludman; **p. 167 (bottom-center row:** Courtesy of Allan Ludman; Biophoto Associates/Science Source; **p. 167 (bottom row):** L.S. Stephanowicz/Visuals Unlimited; Biophoto Associates/Science Source; **p. 168 (top-left):** Biophoto Associates;

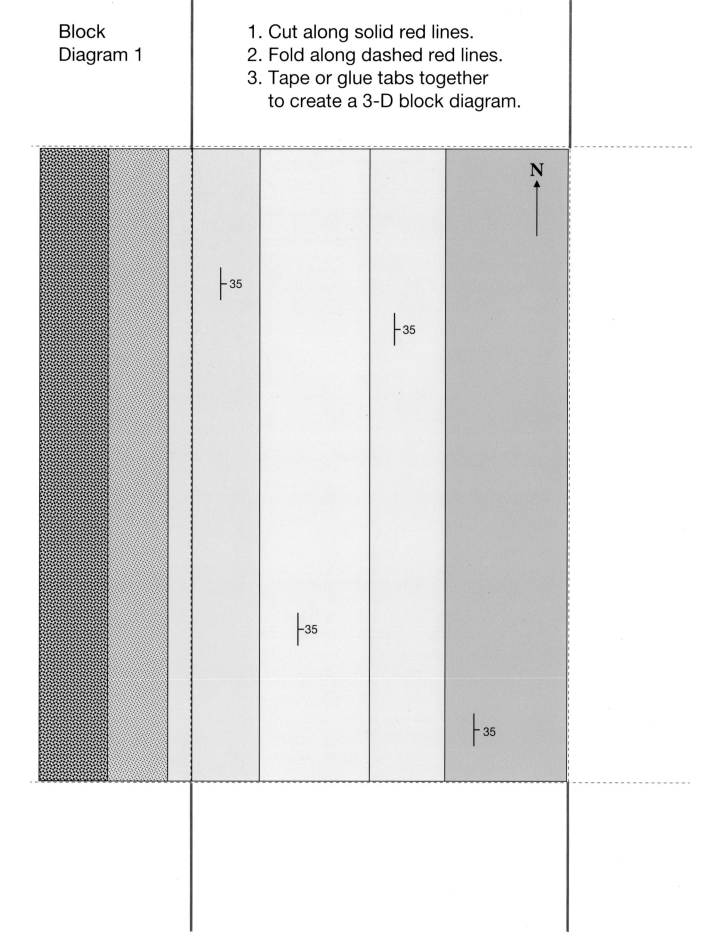

Block
Diagram 1

1. Cut along solid red lines.
2. Fold along dashed red lines.
3. Tape or glue tabs together
   to create a 3-D block diagram.

N

35

35

35

35

Block
Diagram 2

1. Cut along solid red lines.
2. Fold along dashed red lines.
3. Tape or glue tabs together
   to create a 3-D block diagram.

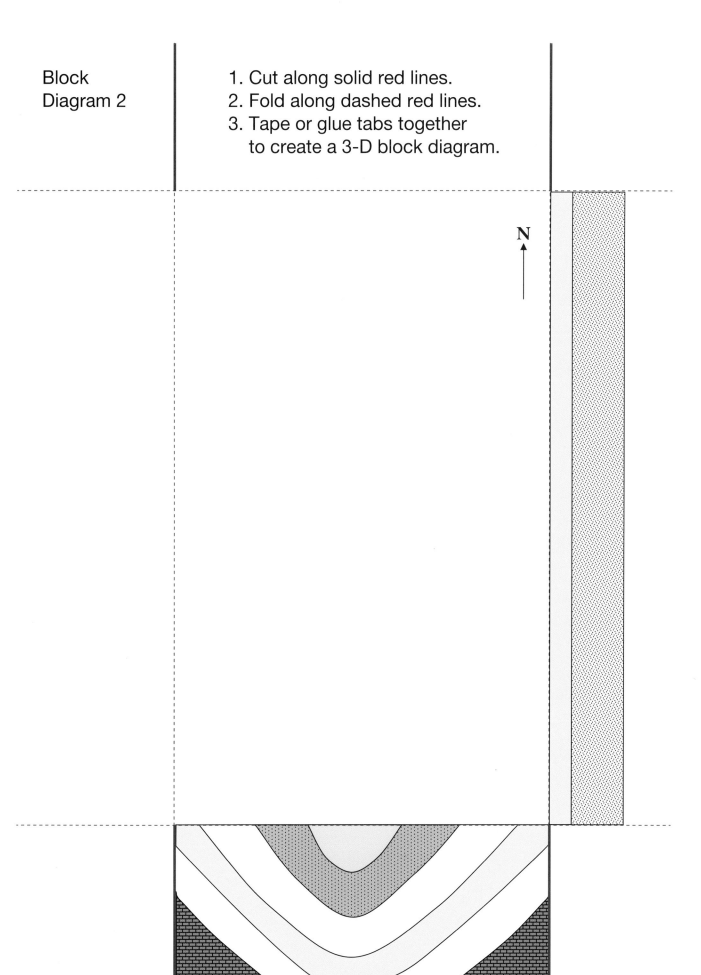

N

Block
Diagram 3

1. Cut along solid red lines.
2. Fold along dashed red lines.
3. Tape or glue tabs together
   to create a 3-D block diagram.

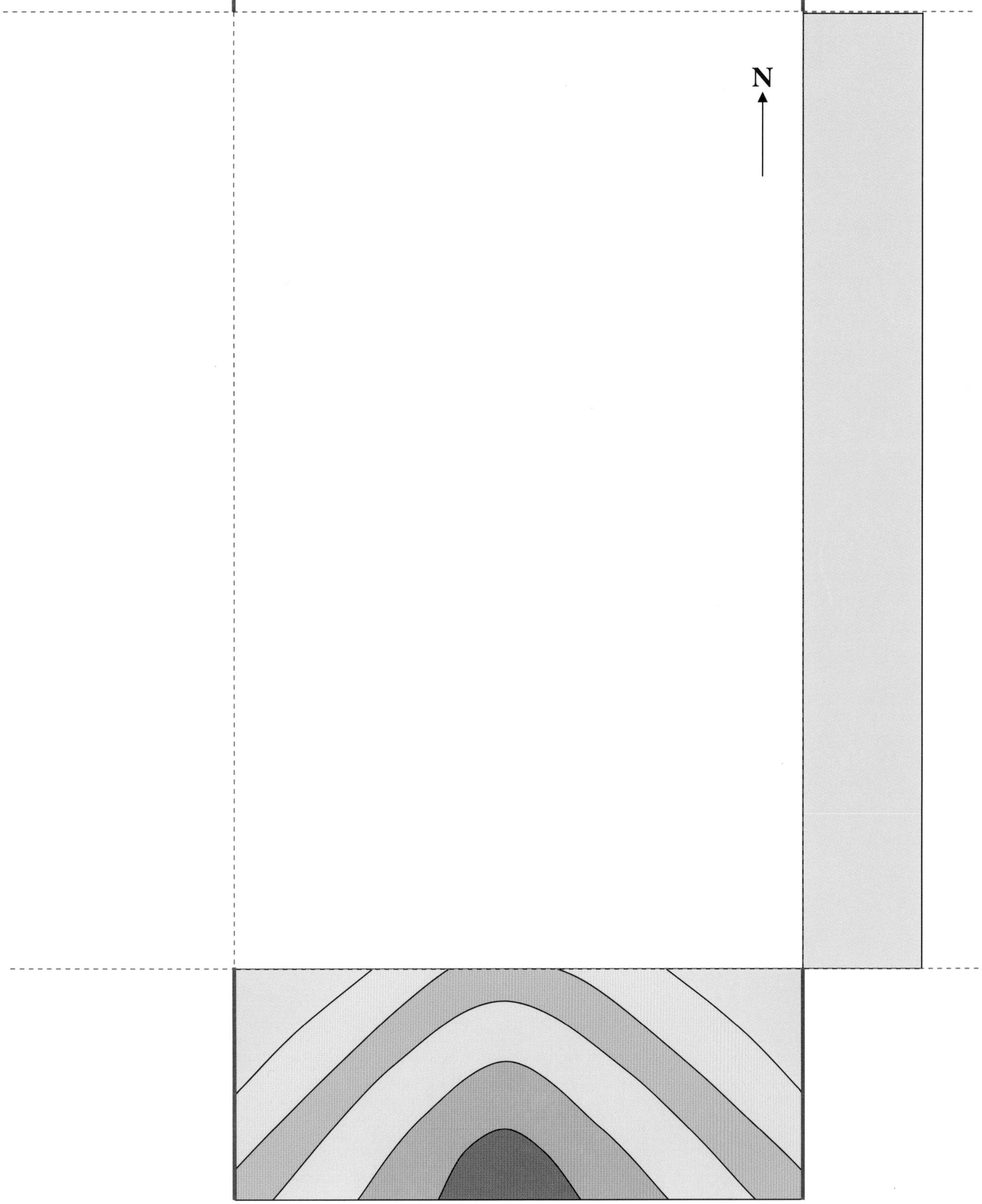

N

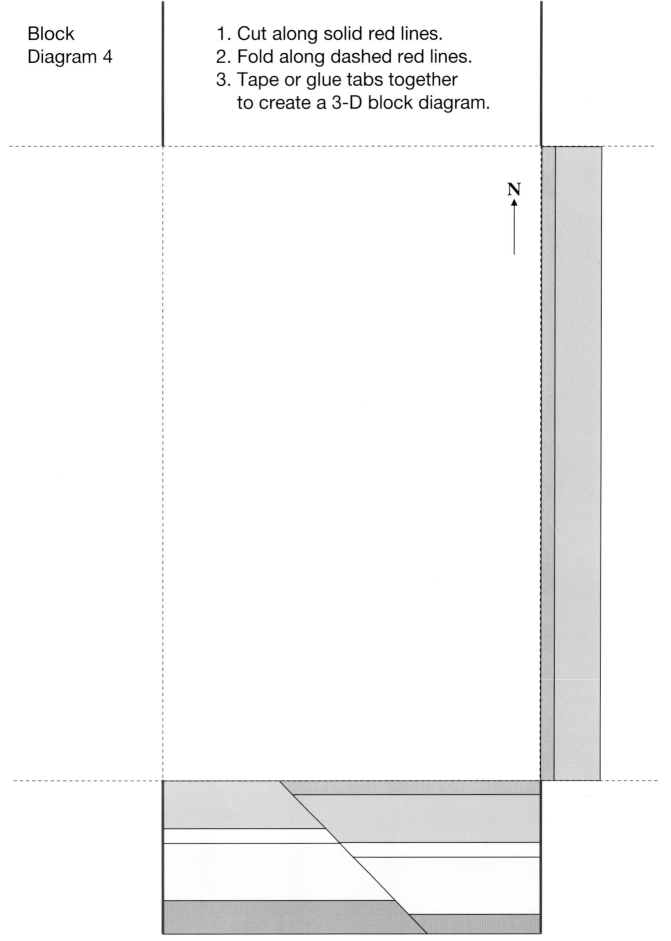

Block
Diagram 4

1. Cut along solid red lines.
2. Fold along dashed red lines.
3. Tape or glue tabs together
   to create a 3-D block diagram.

N